Annals of Mathematics Studies
Number 184

The Gross–Zagier Formula on Shimura Curves

Xinyi Yuan, Shou-Wu Zhang, and Wei Zhang

PRINCETON UNIVERSITY PRESS

PRINCETON AND OXFORD

2013

Published by Princeton University Press, 41 William Street,
Princeton, New Jersey 08540

In the United Kingdom: Princeton University Press, 6 Oxford Street,
Woodstock, Oxfordshire OX20 1TW

press.princeton.edu

Library of Congress Cataloging-in-Publication Data

Yuan, Xinyi, 1981-
    The Gross–Zagier formula on Shimura curves / Xinyi Yuan, Shou-Wu Zhang,
    and Wei Zhang.
        p. cm. – (Annals of mathematics studies ; no. 184)
    Includes bibliographical references and index.
    ISBN 978-0-691-15591-3 (hardcover : alk. paper) –
    ISBN 978-0-691-15592-0 (pbk. : alk. paper)
    1. Shimura varieties. 2. Arithmetical algebraic geometry. 3. Automorphic
    forms. 4. Quaternions. I. Zhang, Shou-Wu. II. Zhang, Wei, 1981- III. Title.
    QA242.5.Y83 2013
    516.3'52–dc23

                                                                    2012010981

British Library Cataloging-in-Publication Data is available

This book has been composed in LATEX.

The publisher would like to acknowledge the author of this volume for providing
the camera-ready copy from which this book was printed.

Printed on acid-free paper ∞

Printed in the United States of America

10 9 8 7 6 5 4 3 2 1

Contents

Preface

The aim of this book is to prove a complete Gross–Zagier formula on quaternionic Shimura curves over totally real fields. The original formula proved by Benedict Gross and Don Zagier in 1983 relates the Néron–Tate heights of Heegner points on $X_0(N)$ to the central derivatives of some Rankin–Selberg L-functions under the Heegner condition, which is an assumption of mild ramification. Since then, some generalizations were given in various works by Shou-Wu Zhang. The proofs of Gross–Zagier and Shou-Wu Zhang depend on some newform theories. There are essential difficulties to removing all ramification assumptions by these methods. This book is a completion of those generalizations in which all ramification restrictions are removed.

The Gross–Zagier formula in this book is an analogue of the central value formula of Jean-Loup Waldspurger, and has been speculated by Benedict Gross in terms of representation theory in a lecture at MSRI in 2002. In fact, the Waldspurger formula concerns periods of automorphic forms on quaternion algebras over number fields, while the Gross–Zagier formula may be viewed as a formula of periods of "automorphic forms" on *incoherent quaternion algebras*. These incoherent automorphic forms are functions on Shimura curves with values in some abelian varieties.

Besides many ideas of Gross–Zagier and Shou-Wu Zhang, one main new ingredient of this book is to construct the analytic kernel and the geometric kernel systematically using Weil representations and the generating series of Hecke correspondences of Stephen S. Kudla constructed in 1997, though we do not use his program on the *arithmetic Siegel–Weil formula*. The construction is inspired by the Waldspurger formula mentioned above. To simplify many computations in both automorphic forms and arithmetic geometry, we take advantage of representation theory and make use of the concepts of *degenerate Schwartz functions*, *coherence of pseudo-theta series*, and *modularity of generating functions*.

Acknowledgments. The authors are extremely grateful to Benedict Gross for his MSRI lecture, and for his constant help and support. This MSRI lecture is the main motivation of this book, and many new ideas in this book are derived from discussions with him. The book would have been impossible without the generating series of Hecke operators constructed by Stephen S. Kudla in 1997. We thank him for explaining his work to us during a joint FRG project supported by the NSF. We would also like to acknowledge our debt to the work of Jean-

Loup Waldspurger. We have directly adapted his strategy of proving special value formula to our incoherent situation.

The authors are also indebted to many important discussions with and crucial comments of Pierre Deligne, Dorian Goldfeld, Ming-Lun Hsieh, Hervé Jacquet, Dihua Jiang, Jian-Shu Li, Yifeng Liu, Peter Sarnak, Richard Taylor, Ye Tian, Andrew Wiles, and Tonghai Yang.

The authors would like to thank the Morningside Center of Mathematics at the Chinese Academy of Sciences for its hospitality and constant support. Xinyi Yuan was supported by a research fellowship from the Clay Mathematics Institute. Shou-Wu Zhang is very grateful to the Institute for Advanced Studies at Tsinghua University and the Institute for Advanced Studies at the Hong Kong University of Science and Technology for their hospitality and support during the major revision of this book. He was also partially supported by a Guggenheim Fellowship and NSF grants DMS-0354436, DMS-0700322, DMS-0970100, and DMS-1065839. Wei Zhang is partially supported by NSF Grant DMS-1001631.

The Gross–Zagier Formula
on Shimura Curves

Chapter One

Introduction and Statement of Main Results

In this chapter, we will state the main result (Theorem 1.2) of this book and describe the main idea of our proof. Let us start with the original work of Gross and Zagier.

1.1 GROSS–ZAGIER FORMULA ON MODULAR CURVES

1.1.1 The original formula

Let N be a positive integer and $f \in S_2(\Gamma_0(N))$ a newform of weight 2. Let $K \subset \mathbb{C}$ be an imaginary quadratic field and χ a character of $\mathrm{Pic}(O_K)$. Form the L-series $L(f, \chi, s)$ as the Rankin–Selberg convolution of the L-series $L(f, s)$ and the L-series $L(\chi, s)$. This L-series $L(f, \chi, s)$ has a holomorphic continuation to the whole complex plane and satisfies a functional equation relating s to $2 - s$.

Assume that K has an odd fundamental discriminant D, and satisfies the following *Heegner condition*:

> *every prime factor of N is split in K.*

Then the sign of the functional equation of the L-series $L(f, \chi, s)$ is -1 and hence $L(f, \chi, 1) = 0$.

Let $X_0(N)$ be the modular curve over $\mathbb{Q}$, whose $\mathbb{C}$-points parametrize isogenies $E_1 \to E_2$ of elliptic curves over $\mathbb{C}$ with kernel isomorphic to $\mathbb{Z}/N\mathbb{Z}$. By the Heegner condition, there exists an ideal $\mathcal{N}$ of O_K such that $O_K/\mathcal{N} \simeq \mathbb{Z}/N\mathbb{Z}$. For every nonzero ideal $\mathcal{I}$ of O_K, let $P_{\mathcal{I}}$ denote the point on $X_0(N)(\mathbb{C})$ representing the isogeny $\mathbb{C}/\mathcal{I} \longrightarrow \mathbb{C}/\mathcal{I}\mathcal{N}^{-1}$. Then $P_{\mathcal{I}}$ is defined over the Hilbert class field H and depends only on the class of $\mathcal{I}$ in $\mathrm{Pic}(O_K)$. Form a point in the Jacobian $J_0(N)$ of $X_0(N)$ using the cusp ∞ by

$$P_\chi = \sum_{[\mathcal{I}] \in \mathrm{Pic}(O_K)} [P_{\mathcal{I}} - \infty] \otimes \chi([\mathcal{I}]) \in J_0(N)(H) \otimes_{\mathbb{Z}} \mathbb{C}.$$

Denote by $P_\chi(f)$ the f-isotypical component of P_χ in $J_0(N)(H) \otimes_{\mathbb{Z}} \mathbb{C}$ under the action of the Hecke operators. The seminal Gross–Zagier formula proved in [GZ] is as follows:

THEOREM 1.1 (Gross–Zagier). *Denote by h the class number of K, and u half of the number of units of O_K. Then*

$$\langle P_\chi(f), P_\chi(f) \rangle_{\mathrm{NT}}^H = \frac{h u^2 |D|^{1/2}}{8\pi^2 (f, f)} \cdot L'(f, \chi, 1).$$

Here (f, f) is the Peterson inner product of f, and $\langle P_\chi(f), P_\chi(f)\rangle_{NT}^H$ denotes the Néron–Tate height over H.

The construction of Heegner points $P_\chi(f)$ involves the idempotent in the Hecke algebra for $\Gamma_0(N)$, thus involves some denominators. For application to the BSD conjecture, one should construct such a point on abelian varieties without denominators. This has been achieved when f corresponds to an elliptic curve over $\mathbb{Q}$ which we are going to explain as follows.

1.1.2   Application to elliptic curves

Let E be an elliptic curve defined over $\mathbb{Q}$ with conductor N. By the landmark modularity theorem of Wiles [Wi] completed by Taylor–Wiles [TW], and the generalization of Breuil–Conrad–Diamond–Taylor [BCDT], E is modular in the sense that $L(E, s) = L(f, s)$ for some newform f of weight 2 for $\Gamma_0(N)$. By the isogeny theorem of Faltings [Fa1], there is a finite $\mathbb{Q}$-morphism

$$\phi : \; X_0(N) \longrightarrow E$$

which takes the cusp ∞ on $X_0(N)$ to the identity 0 on E. Denote

$$P_\chi(\phi) = \sum_{[\mathcal{I}] \in \mathrm{Pic}(O_K)} \phi(P_\mathcal{I}) \otimes \chi([\mathcal{I}]) \in E(H) \otimes_{\mathbb{Z}} \mathbb{C}.$$

Then the above formula gives

$$\langle P_\chi(\phi), P_\chi(\phi)\rangle_{NT}^H = \deg \phi \cdot \frac{hu^2 |D|^{1/2}}{8\pi^2 (f, f)} \cdot L'(f, \chi, 1).$$

The generalization of the Gross–Zagier formula in this book will be written similar to the above form. We will replace $X_0(N)$ by a Shimura curve X over a totally real field F, and replace ϕ by a pair of parametrizations $f_1 : X \to A$ and $f_2 : X \to A^\vee$ for a dual pair of abelian varieties $(A, A^\vee)$ over F. Then we consider a totally imaginary quadratic extension E of F. Let $P \in X$ be a fixed point under the Hecke action of $E^\times$. For a finite character χ on $\mathrm{Gal}(E^{\mathrm{ab}}/E)$, define $P_\chi(f_1) \in A$ and $P_{\chi^{-1}}(f_2) \in A^\vee$ as the twisted integral of $f_1(P)$ and $f_2(P)$ by χ or χ^{-1}. The formula will relate the Néron–Tate height pairing $\langle P_\chi(f_1), P_{\chi^{-1}}(f_2)\rangle$ with the derivative of $L(A, \chi, s)$ at $s = 1$.

1.2   SHIMURA CURVES AND ABELIAN VARIETIES

1.2.1   Incoherent quaternion algebras and Shimura curves

Let F be a number field with adele ring $\mathbb{A} = \mathbb{A}_F$ and let $\mathbb{A}_f$ be the ring of finite adeles. Let Σ be a finite set of places of F. Up to isomorphism, let $\mathbb{B}$ be the unique $\mathbb{A}$-algebra, free of rank 4 as an $\mathbb{A}$-module, whose localization $\mathbb{B}_v := \mathbb{B} \otimes_{\mathbb{A}} F_v$ is isomorphic to $M_2(F_v)$ if $v \notin \Sigma$ and to the unique division quaternion algebra over F_v if $v \in \Sigma$. We call $\mathbb{B}$ *the quaternion algebra over* $\mathbb{A}$ *with ramification set* $\Sigma(\mathbb{B}) := \Sigma$.

If $\#\Sigma$ is even then $\mathbb{B} = B \otimes_F \mathbb{A}$ for a quaternion algebra B over F unique up to an F-isomorphism. In this case, we call $\mathbb{B}$ a *coherent* quaternion algebra. If $\#\Sigma$ is odd, then $\mathbb{B}$ is not the base change of any quaternion algebra over F. In this case, we call $\mathbb{B}$ an *incoherent* quaternion algebra. This terminology is inspired by Kudla's notion of *incoherent collections of quadratic spaces* (cf. [Ku2]).

Now assume that F is a totally real number field and that $\mathbb{B}$ is an incoherent quaternion algebra over $\mathbb{A}$, totally definite at infinity in the sense that $\mathbb{B}_\tau$ is the Hamiltonian algebra for every archimedean place τ of F.

For each open compact subgroup U of $\mathbb{B}_f^\times := (\mathbb{B} \otimes_\mathbb{A} \mathbb{A}_f)^\times$, we have a (compactified) Shimura curve X_U over F. For any embedding $\tau : F \hookrightarrow \mathbb{C}$, the complex points of X_U at τ forms a Riemann surface as follows:

$$X_{U,\tau}(\mathbb{C}) \simeq B(\tau)^\times \backslash \mathcal{H}^\pm \times \mathbb{B}_f^\times / U \cup \{\text{cusps}\}.$$

Here $B(\tau)$ is the unique quaternion algebra over F with ramification set $\Sigma \backslash \{\tau\}$, $\mathbb{B}_f$ is identified with $B(\tau)_{\mathbb{A}_f}$ as an $\mathbb{A}_f$-algebra, and $B(\tau)^\times$ acts on $\mathcal{H}^\pm$ through an isomorphism $B(\tau)_\tau \simeq M_2(\mathbb{R})$. The set $\{\text{cusps}\}$ is non-empty if and only if $F = \mathbb{Q}$ and $\Sigma = \{\infty\}$.

For any two open compact subgroups $U_1 \subset U_2$ of $\mathbb{B}_f^\times$, one has a natural surjective morphism

$$\pi_{U_1, U_2} : X_{U_1} \to X_{U_2}.$$

Let X be the projective limit of the system $\{X_U\}_U$. It is a regular scheme over F, locally noetherian but not of finite type. In terms of the notation above, it has a uniformization

$$X_\tau(\mathbb{C}) \simeq B(\tau)^\times \backslash \mathcal{H}^\pm \times \mathbb{B}_f^\times / D \cup \{\text{cusps}\}.$$

Here D denotes the closure of $F^\times$ in $\mathbb{A}_f^\times$. If $F = \mathbb{Q}$, then $D = F^\times$. In general, D is much larger than $F^\times$.

The Shimura curve X is endowed with an action T_x of $x \in \mathbb{B}^\times$ given by "*right multiplication by* x_f." The action T_x is trivial if and only if $x_f \in D$. Each X_U is just the quotient of X by the action of U. In terms of the system $\{X_U\}_U$, the action gives an isomorphism $\mathrm{T}_x : X_{xUx^{-1}} \to X_U$ for each U.

The induced action of $\mathbb{B}_f^\times$ on the set $\pi_0(X_{U,\overline{F}})$ of geometrically connected components of X_U factors through the norm map $q : \mathbb{B}_f^\times \to \mathbb{A}_f^\times$ and makes $\pi_0(X_{U,\overline{F}})$ a principal homogeneous space over $F_+^\times \backslash \mathbb{A}_f^\times / q(U)$. There is a similar description for X.

1.2.2 Hodge classes

The curve X_U has a *Hodge class* $L_U \in \mathrm{Pic}(X_U)_\mathbb{Q}$. It is the line bundle whose global sections are holomorphic modular forms of weight two. The system $L =$

$\{L_U\}_U$ is a direct system in the sense that it is compatible under the pull-back via the projection $\pi_{U_1,U_2} : X_{U_1} \to X_{U_2}$. See §3.1.3 for a precise definition.

Here are some basic explicit descriptions. If X_U is a modular curve, which happens exactly when $F = \mathbb{Q}$ and $\Sigma = \{\infty\}$, then L_U is linearly equivalent to some linear combination of cusps on X_U. If $F \neq \mathbb{Q}$ or $\Sigma \neq \{\infty\}$, then X_U has no cusps and L_U is isomorphic to the canonical bundle of X_U over F for sufficiently small U.

For each component $\alpha \in \pi_0(X_{U,\overline{F}})$, denote by $L_{U,\alpha} = L_U|_{X_{U,\alpha}}$ the restriction to the connected component $X_{U,\alpha}$ of $X_{U,\overline{F}}$ corresponding to α. It is also viewed as a divisor class on X_U via push-forward under $X_{U,\alpha} \to X_U$. Denote by
$$\xi_{U,\alpha} = \frac{1}{\deg L_{U,\alpha}} L_{U,\alpha}$$ the normalized Hodge class on $X_{U,\alpha}$, and by $\xi_U = \sum_\alpha \xi_{U,\alpha}$ the normalized Hodge class on X_U.

We remark that $\deg L_{U,\alpha}$ is independent of α since all geometrically connected components are Galois conjugate to each other. It follows that $\deg L_{U,\alpha} = \deg L_U / |F_+^\times \backslash \mathbb{A}_f^\times / q(U)|$. The degree of L_U can be further expressed as the volume of X_U.

For any open compact subgroup U of $\mathbb{B}_f^\times$, define
$$\mathrm{vol}(X_U) := \int_{X_{U,\tau}(\mathbb{C})} \frac{dxdy}{2\pi y^2}.$$

Here the measure $\dfrac{dxdy}{2\pi y^2}$ on $\mathcal{H}$ descends naturally to a measure on $X_{U,\tau}(\mathbb{C})$ via the complex uniformization for any $\tau : F \hookrightarrow \mathbb{C}$. Then Lemma 3.1 asserts that $\deg L_U = \mathrm{vol}(X_U)$. In particular, the volume is always a positive rational number.

For any $U_1 \subset U_2$, the projection $\pi_{U_1,U_2} : X_{U_1} \to X_{U_2}$ has degree
$$\deg(\pi_{U_1,U_2}) = \mathrm{vol}(X_{U_1})/\mathrm{vol}(X_{U_2}).$$

It follows from the definition. Because of this, we will often use $\mathrm{vol}(X_U)$ as a normalizing factor.

1.2.3   Abelian varieties parametrized by Shimura curves

Let A be a simple abelian variety defined over F. We say that A *is parametrized by X* if there is a non-constant morphism $X_U \to A$ over F for some U. By the Eichler–Shimura theory, if A is parametrized by X, then A *is of strict $GL(2)$-type* in the sense that
$$M = \mathrm{End}^0(A) := \mathrm{End}_F(A) \otimes_{\mathbb{Z}} \mathbb{Q}$$

is a field and $\mathrm{Lie}(A)$ is a free module of rank one over $M \otimes_{\mathbb{Q}} F$ by the induced action. See §3.2 for more details.

Define
$$\pi_A = \mathrm{Hom}_\xi^0(X, A) := \varinjlim_U \mathrm{Hom}_{\xi_U}^0(X_U, A),$$

where $\mathrm{Hom}^0_{\xi_U}(X_U, A)$ denotes the morphisms in $\mathrm{Hom}_F(X_U, A) \otimes_{\mathbb{Z}} \mathbb{Q}$ using ξ_U as a base point. More precisely, if ξ_U is represented by a divisor $\sum_i a_i x_i$ on $X_{U,\overline{F}}$, then $f \in \mathrm{Hom}_F(X_U, A) \otimes_{\mathbb{Z}} \mathbb{Q}$ is in π_A if and only if $\sum_i a_i f(x_i) = 0$ in $A(\overline{F})_{\mathbb{Q}}$.

Since any morphism $X_U \to A$ factors through the Jacobian variety J_U of X_U, we also have

$$\pi_A = \mathrm{Hom}^0(J, A) := \varinjlim_U \mathrm{Hom}^0(J_U, A).$$

Here $\mathrm{Hom}^0(J_U, A) = \mathrm{Hom}_F(J_U, A)_{\otimes_{\mathbb{Z}} \mathbb{Q}}$. The direct limit of $\mathrm{Hom}(J_U, A)$ defines an integral structure on π_A but we will not use this.

The space π_A admits a natural $\mathbb{B}^\times$-module structure. It is an *automorphic representation of $\mathbb{B}^\times$ over $\mathbb{Q}$*. We will see the natural identity $\mathrm{End}_{\mathbb{B}^\times}(\pi_A) = M$ and that π_A has a decomposition $\pi = \otimes_M \pi_v$ where π_v is an absolutely irreducible representation of $\mathbb{B}_v^\times$ over M. Using the Jacquet–Langlands correspondence, one can define L-series

$$L(s, \pi) = \prod_v L_v(s, \pi_v) \in M \otimes_{\mathbb{Q}} \mathbb{C}$$

as an entire function of $s \in \mathbb{C}$. Let

$$L(s, A, M) = \prod L_v(s, A, M) \in M \otimes_{\mathbb{Q}} \mathbb{C}$$

be the L-series defined using ℓ-adic representations with coefficients in $M \otimes_{\mathbb{Q}} \mathbb{Q}_\ell$, completed at archimedean places using the Γ-function. Then $L(s, A, M)$ converges absolutely in $M \otimes \mathbb{C}$ for $\mathrm{Re}(s) > 3/2$. The Eichler–Shimura theory asserts that, for almost all finite places v of F, the local L-function of A is given by

$$L_v(s, A, M) = L(s - \frac{1}{2}, \pi_v).$$

Conversely, by the Eichler–Shimura theory and the isogeny theorem of Faltings [Fa1], if A is of strict GL(2)-type, and if for some automorphic representation π of $\mathbb{B}^\times$ over $\mathbb{Q}$, $L_v(s, A, M)$ is equal to $L(s - 1/2, \pi_v)$ for almost all finite places v, then A is parametrized by the Shimura curve X.

If A is parametrized by X, then the dual abelian variety $A^\vee$ is also parametrized by X. Denote by $M^\vee = \mathrm{End}^0(A^\vee)$. There is a canonical isomorphism $M \to M^\vee$ sending a homomorphism $m : A \to A$ to its dual $m^\vee : A^\vee \to A^\vee$.

There is a perfect $\mathbb{B}^\times$-invariant pairing

$$\pi_A \times \pi_{A^\vee} \longrightarrow M$$

given by

$$(f_1, f_2) = \mathrm{vol}(X_U)^{-1}(f_{1,U} \circ f_{2,U}^\vee), \quad f_{1,U} \in \mathrm{Hom}(J_U, A), \ f_{2,U} \in \mathrm{Hom}(J_U, A^\vee)$$

where $f_{2,U}^\vee : A \to J_U$ is the dual of $f_{2,U}$ composed with the canonical isomorphism $J_U^\vee \simeq J_U$. It follows that $\pi_{A^\vee}$ is dual to π_A as representations of $\mathbb{B}^\times$ over M.

In the case that A is an elliptic curve, we have $M = \mathbb{Q}$ and π_A is self-dual. For any morphism $f \in \pi_A$ represented by a direct system $\{f_U\}_U$, we have

$$(f, f) = \mathrm{vol}(X_U)^{-1} \deg f_U.$$

Here $\deg f_U$ denotes the degree of the finite morphism $f_U : X_U \to A$.

1.2.4   Height pairing

The usual theory of Néron–Tate height gives a $\mathbb{Q}$-bilinear non-degenerate pairing

$$\langle \cdot, \cdot \rangle_{\mathrm{NT}} : A(\overline{F})_\mathbb{Q} \times A^\vee(\overline{F})_\mathbb{Q} \longrightarrow \mathbb{R}.$$

We refer to §7.1.1 for a quick review.

Recall that the field $M = \mathrm{End}^0(A)$ acts on $A(\overline{F})_\mathbb{Q}$ by definition, and acts on $A^\vee(\overline{F})_\mathbb{Q}$ through the duality. By the adjoint property of the height pairing in Proposition 7.3, the pairing $\langle \cdot, \cdot \rangle_{\mathrm{NT}}$ descends to a $\mathbb{Q}$-linear map

$$\langle \cdot, \cdot \rangle_{\mathrm{NT}} : A(\overline{F})_\mathbb{Q} \otimes_M A^\vee(\overline{F})_\mathbb{Q} \longrightarrow \mathbb{R}.$$

For any fixed $(x, y) \in A(\overline{F})_\mathbb{Q} \otimes_M A^\vee(\overline{F})_\mathbb{Q}$, the correspondence

$$a \in M \longmapsto \langle ax, y \rangle_{\mathrm{NT}} \in \mathbb{R}$$

define an element in $\mathrm{Hom}(M, \mathbb{R})$. One has an isomorphism

$$\mathrm{Hom}(M, \mathbb{R}) \cong M \otimes_\mathbb{Q} \mathbb{R}$$

using the trace map. Let $\langle x, y \rangle_M$ denote the corresponding element in $M \otimes_\mathbb{Q} \mathbb{R}$. Then we have just defined an M-bilinear pairing

$$\langle \cdot, \cdot \rangle_M : A(\overline{F})_\mathbb{Q} \otimes_M A^\vee(\overline{F})_\mathbb{Q} \longrightarrow M \otimes_\mathbb{Q} \mathbb{R}$$

such that

$$\langle \cdot, \cdot \rangle_{\mathrm{NT}} = \mathrm{tr}_{M \otimes \mathbb{R}/\mathbb{R}} \langle \cdot, \cdot \rangle_M.$$

We call this new pairing *an M-linear Néron–Tate height pairing*.

1.3   CM POINTS AND GROSS–ZAGIER FORMULA

1.3.1   CM points

Let E/F be a totally imaginary quadratic extension, with a fixed embedding $E_\mathbb{A} \hookrightarrow \mathbb{B}$ over $\mathbb{A}$. Then $E_\mathbb{A}^\times$ acts on X by the right multiplication via $E_\mathbb{A}^\times \hookrightarrow \mathbb{B}^\times$. Let $X^{E^\times}$ be the subscheme of X of fixed points of X under $E^\times$. Up to translation by $\mathbb{B}^\times$, the subscheme $X^{E^\times}$ does not depend on the choice of the embedding $E_\mathbb{A} \hookrightarrow \mathbb{B}$. The scheme $X^{E^\times}$ is defined over F. The theory of complex multiplication asserts that every point in $X^{E^\times}(\overline{F})$ is defined over E^{ab} and that the Galois action is given by the Hecke action under the reciprocity law.

Fix a point $P \in X^{E^\times}(E^{\mathrm{ab}})$ throughout this book. It induces a point $P_U \in X_U(E^{\mathrm{ab}})$ for every U. We can normalize the complex uniformization

$$X_{U,\tau}(\mathbb{C}) = B(\tau)^\times \backslash \mathcal{H}^\pm \times \mathbb{B}_f^\times / U \cup \{\mathrm{cusps}\}$$

so that the point P_U is exactly represented by the double coset of $[z_0, 1]_U$. Here $z_0 \in \mathcal{H}$ is the unique fixed point of $E^\times$ in $\mathcal{H}$ via the action induced by the embedding $E \hookrightarrow B(\tau)$. A similar description can be made on X.

Let A be an abelian variety over F parametrized by X with $M = \mathrm{End}^0(A)$ and let $\chi : \mathrm{Gal}(E^{\mathrm{ab}}/E) \to L^\times$ be a character of finite order, where L is a finite field extension of M. For any $f \in \pi_A$, the image $f(P)$ is a well-defined point in $A(E^{\mathrm{ab}})_\mathbb{Q}$. Consider the integration

$$P_\chi(f) = \int_{\mathrm{Gal}(\overline{E}/E)} f(P^\tau) \otimes_M \chi(\tau) d\tau \ \in A(E^{\mathrm{ab}})_\mathbb{Q} \otimes_M L,$$

where we use the Haar measure on $\mathrm{Gal}(\overline{E}/E)$ of total volume 1. It is essentially a finite sum, and it is easy to see that

$$P_\chi(f) \in A(\chi) := (A(E^{\mathrm{ab}})_\mathbb{Q} \otimes_M L_\chi)^{\mathrm{Gal}(E^{\mathrm{ab}}/E)}.$$

Here L_χ denotes the M-vector space L with the action of $\mathrm{Gal}(E^{\mathrm{ab}}/E)$ given by the multiplication by the character χ. It is also clear that $P_\chi(f) \neq 0$ only if the central character ω_A of π_A is compatible with χ in the sense that

$$\omega_A \cdot \chi|_{\mathbb{A}_F^\times} = 1.$$

Let $L(s, A_E, \chi) \in L \otimes_\mathbb{Q} \mathbb{C}$ be the L-series which is obtained by ℓ-adic representation twisted by χ. Define

$$L(s, \pi_A, \chi) = L(s, \pi_{A,E} \otimes \chi).$$

Here $\pi_{A,E}$ denotes the base change of π_A to E, and χ is considered as a character of $E^\times \backslash \mathbb{A}_E^\times$ via the reciprocity law

$$E^\times \backslash \mathbb{A}_E^\times \longrightarrow \mathrm{Gal}(E^{\mathrm{ab}}/E)$$

which maps uniformizers to geometric Frobenii. As an identity in $L \otimes_\mathbb{Q} \mathbb{C}$, we have

$$L(s, A_E, \chi) = L\left(s - \frac{1}{2}, \pi_A, \chi\right).$$

As a refinement of the Birch and Swinnerton-Dyer conjecture, it is conjectured that the leading term of $L(s, A_E, \chi)$ is invertible in $L \otimes_\mathbb{Q} \mathbb{C}$ and that

$$\mathrm{ord}_{s=1} L(s, A_E, \chi) = \dim_L A(\chi).$$

1.3.2   Gross–Zagier formula

Assume that $\omega_A \cdot \chi|_{\mathbb{A}_F^\times} = 1$. Define a linear space over L by

$$\mathcal{P}(\pi_A, \chi) := \mathrm{Hom}_{E_{\mathbb{A}}^\times}(\pi_A \otimes \chi, L).$$

Then the correspondence $f \mapsto P_\chi(f)$ defines an element

$$P_\chi \in \mathcal{P}(\pi_A, \chi) \otimes_L A(\chi).$$

Thus $P_\chi(f) \neq 0$ for some f only if $\mathcal{P}(\pi_A, \chi) \neq 0$.

By the following Theorem 1.3 of Saito–Tunnel, $\mathcal{P}(\pi_A, \chi)$ is at most one-dimensional, and it is one-dimensional if and only if the ramification set $\Sigma(\mathbb{B})$ of $\mathbb{B}$ is equal to

$$\Sigma(A, \chi) = \left\{ \text{ places } v \text{ of } F : \ \epsilon(\frac{1}{2}, \pi_{A,v}, \chi_v) \neq \chi_v(-1)\eta_v(-1) \right\}.$$

In that case, $\epsilon(1/2, \pi_A, \chi) = -1$ and thus $L(1/2, \pi_A, \chi) = 0$.

The next problem is to find a nonzero element of $\mathcal{P}(\pi_A, \chi)$ if it is one-dimensional. Denote $\pi = \pi_A$ for simplicity. The contragredient $\widetilde{\pi} = \pi_{A^\vee}$ by the duality map

$$\pi_A \times \pi_{A^\vee} \longrightarrow M.$$

It is more convenient to work with $\mathcal{P}(\pi, \chi) \otimes_L \mathcal{P}(\widetilde{\pi}, \chi^{-1})$. Here $\mathcal{P}(\widetilde{\pi}, \chi^{-1})$ has the same dimension as $\mathcal{P}(\pi, \chi)$ by Theorem 1.3. When $\mathcal{P}(\pi, \chi) \otimes \mathcal{P}(\widetilde{\pi}, \chi^{-1})$ is nonzero, we would like to define a canonical generator denoted by α. Decompose $\pi = \otimes \pi_v$ and $\chi = \otimes \chi_v$. Then we have a decomposition $\mathcal{P}(\pi, \chi) = \otimes \mathcal{P}(\pi_v, \chi_v)$, where the space $\mathcal{P}(\pi_v, \chi_v)$ is defined analogously. Then α will have a decomposition $\alpha = \otimes \alpha_v$ for some $\alpha_v \in \mathcal{P}(\pi_v, \chi_v) \otimes \mathcal{P}(\widetilde{\pi}_v, \chi_v^{-1})$ to be defined.

Fix Haar measures dt_v on $E_v^\times / F_v^\times$ such that the product measure over all v gives the Tamagawa measure on $E_{\mathbb{A}}^\times / \mathbb{A}^\times$. We can further assume that the maximal compact subgroup $O_{E_v}^\times / O_{F_v}^\times$ has a volume in $\mathbb{Q}$ for all non-archimedean place v. Then α_v is defined formally by

$$\alpha_v(f_1 \otimes f_2) = \frac{L(1, \eta_v)L(1, \pi_v, \mathrm{ad})}{\zeta_{F_v}(2)L(\frac{1}{2}, \pi_v, \chi_v)} \int_{E_v^\times / F_v^\times} (\pi_v(t)f_1, f_2)_v \, \chi_v(t)dt,$$

$$f_1 \in \pi_v, \ f_2 \in \widetilde{\pi}_v.$$

More precisely, we may take an embedding $\iota : L \hookrightarrow \mathbb{C}$ and define the above integral with value in $\mathbb{C}$. It turns out that, for all places v, the value of $\alpha_v(f_1 \otimes f_2)$ lies in L, and does not depend on the choice of the embedding ι.

It is worth mentioning that in the definition of α_v, the local equivariant pairing $(\cdot, \cdot)_v : \pi_v \times \widetilde{\pi}_v \to M$ satisfies the compatibility condition that

$$(\cdot, \cdot) = \otimes_v(\cdot, \cdot)_v.$$

Here the global pairing $(\cdot, \cdot) : \pi_A \times \widetilde{\pi}_{A^\vee} \to M$ is the duality map introduced above.

The following is the main theorem of this book.

THEOREM 1.2. *Assume* $\omega_A \cdot \chi|_{\mathbb{A}_F^\times} = 1$. *For any* $f_1 \in \pi_A$ *and* $f_2 \in \pi_{A^\vee}$,

$$\langle P_\chi(f_1), \ P_{\chi^{-1}}(f_2) \rangle_L = \frac{\zeta_F(2) L'(1/2, \pi_A, \chi)}{4 L(1, \eta)^2 L(1, \pi_A, \mathrm{ad})} \alpha(f_1, f_2)$$

as an identity in $L \otimes_{\mathbb{Q}} \mathbb{C}$. *Here* $\langle \cdot, \cdot \rangle_L : A(\chi) \times A^\vee(\chi^{-1}) \to L \otimes_{\mathbb{Q}} \mathbb{R}$ *is the L-linear Néron–Tate height pairing induced by the M-linear Néron–Tate height pairing* $\langle \cdot, \cdot \rangle_M$ *between* $A(\overline{F})$ *and* $A^\vee(\overline{F})$.

The theorem compares two elements

$$\alpha, \quad \langle P_\chi(\cdot), \ P_{\chi^{-1}}(\cdot) \rangle_L$$

of the *L*-linear space

$$\mathcal{P}(\pi_A, \chi) \otimes_L \mathcal{P}(\pi_{A^\vee}, \chi^{-1}).$$

The space is at most one-dimensional.

If $\Sigma(\mathbb{B}) \neq \Sigma(A, \chi)$, the linear space is zero and thus both sides of the formula are zero. The theorem is vacuous in this case.

In the essential case $\Sigma(\mathbb{B}) = \Sigma(A, \chi)$, the linear space is one-dimensional and α is a generator. It follows that $\langle P_\chi(\cdot), \ P_{\chi^{-1}}(\cdot) \rangle_L$ must be a constant multiple of α. Then the theorem can be viewed as an explicit expression of the multiple in terms of special values and special derivatives of *L*-functions. Note that $\epsilon(1/2, \pi_A, \chi) = -1$ in this case.

In the original Gross–Zagier formula, the Heegner condition implies that $\Sigma(f, \chi) = \{\infty\}$, so that the Heegner points are constructed from the modular curve. Here $\Sigma(f, \chi)$ is similarly defined in terms of local root numbers.

1.4 WALDSPURGER FORMULA

1.4.1 Linear forms over local fields

Let F be a local field and B a quaternion algebra over F. Then B is isomorphic to either $M_2(F)$ or the unique division quaternion algebra over F. The Hasse invariant $\epsilon(B) = 1$ if $B \simeq M_2(F)$, and $\epsilon(B) = -1$ if B is the division algebra.

Let E be either $F \oplus F$ or a quadratic field extension over F, with a fixed embedding $E \hookrightarrow B$ of algebras over F. Let $\eta : F^\times \to \mathbb{C}^\times$ be the quadratic character associated to the extension E/F.

Let π be an irreducible admissible representation of $B^\times$ with central character ω_π, and let $\chi : E^\times \to \mathbb{C}^\times$ be a character of $E^\times$ such that

$$\omega_\pi \cdot \chi|_{F^\times} = 1.$$

Define

$$\mathcal{P}(\pi, \chi) := \mathrm{Hom}_{E^\times}(\pi \otimes \chi, \mathbb{C}).$$

The following result asserts that the dimension of this space is determined by the local root number

$$\epsilon(\frac{1}{2}, \pi, \chi) = \epsilon(\frac{1}{2}, \sigma, \chi) = \epsilon(\frac{1}{2}, \sigma_E \otimes \chi).$$

Here σ is the Jacquet–Langlands correspondence of π on $\mathrm{GL}_2(F)$, and σ_E is the base change to E.

THEOREM 1.3 (Tunnell [Tu], Saito [Sa]). *The space* $\mathcal{P}(\pi, \chi)$ *is at most one-dimensional, and it is one-dimensional if and only if*

$$\epsilon(\frac{1}{2}, \pi, \chi) = \chi(-1)\eta(-1)\epsilon(B).$$

The next problem is to find a nonzero element of $\mathcal{P}(\pi, \chi)$ if it is one-dimensional. It is more convenient to work with $\mathcal{P}(\pi, \chi) \otimes \mathcal{P}(\widetilde{\pi}, \chi^{-1})$. Here $\widetilde{\pi}$ denotes the contragredient of π, and $\mathcal{P}(\widetilde{\pi}, \chi^{-1})$ has the same dimension as $\mathcal{P}(\pi, \chi)$ by Theorem 1.3.

Fix a Haar measure dt on $E^\times/F^\times$. Define a bilinear form

$$\alpha : \pi \otimes \widetilde{\pi} \longrightarrow \mathbb{C}$$

by

$$\alpha(f_1 \otimes f_2) = \frac{L(1,\eta)L(1,\pi,\mathrm{ad})}{\zeta_F(2)L(\frac{1}{2},\pi,\chi)} \int_{E^\times/F^\times} (\pi(t)f_1, f_2) \, \chi(t)dt, \quad f_1 \in \pi, \ f_2 \in \widetilde{\pi}.$$

Here $(\cdot, \cdot) : \pi \times \widetilde{\pi} \to \mathbb{C}$ is a fixed $B^\times$-invariant pairing. The integral converges absolutely if both π and χ are unitary.

The normalizing factor before the integration is nonzero, and makes

$$\alpha(f_1 \otimes f_2) = 1$$

in the following spherical case:

- $B = M_2(F)$,

- E is an unramified extension of F,

- π and χ are both unramified,

- dt is normalized such that $\mathrm{vol}(O_E^\times/O_F^\times) = 1$,

- $f_1 \in \pi^{\mathrm{GL}_2(O_F)}$ and $f_2 \in \widetilde{\pi}^{\mathrm{GL}_2(O_F)}$ satisfy $(f_1, f_2) = 1$.

It is easy to see that α defines an element of $\mathcal{P}(\pi, \chi) \otimes \mathcal{P}(\widetilde{\pi}, \chi^{-1})$. It is actually a generator of the space. In other words, $\alpha \neq 0$ if and only if $\mathcal{P}(\pi, \chi) \neq 0$. See [Wa, §III-2, Lemme 10].

1.4.2 Waldspurger formula

We start with the following notations and assumptions:

- F is a number field with adele ring $\mathbb{A} = \mathbb{A}_F$.

- B is a quaternion algebra over F.

- E is a quadratic field extension of F, with a fixed embedding $E \hookrightarrow B$ over F.

- π is an irreducible cuspidal automorphic representation of $B_{\mathbb{A}}^\times$ with central character $\omega_\pi : F^\times \backslash \mathbb{A}^\times \to \mathbb{C}^\times$.

- $\chi : E^\times \backslash E_{\mathbb{A}}^\times \to \mathbb{C}^\times$ is a character with $\omega_\pi \cdot \chi|_{\mathbb{A}^\times} = 1$.

- $\eta : F^\times \backslash \mathbb{A}^\times \to \mathbb{C}^\times$ is the quadratic character associated to the extension E/F.

Define a period integral $P_\chi : \pi \to \mathbb{C}$ by

$$P_\chi(f) = \int_{E^\times \backslash E_{\mathbb{A}}^\times / \mathbb{A}^\times} f(t)\chi(t)dt, \quad f \in \pi.$$

The integral uses the Haar measure with total volume 1 on $E^\times \backslash E_{\mathbb{A}}^\times / \mathbb{A}^\times$. Define $P_{\chi^{-1}} : \tilde{\pi} \to \mathbb{C}$ in the same way.

THEOREM 1.4 (Waldspurger [Wa]). *For any $f_1 \in \pi$ and $f_2 \in \tilde{\pi}$, we have*

$$P_\chi(f_1) \cdot P_{\chi^{-1}}(f_2) = \frac{\zeta_F(2)L(\frac{1}{2}, \pi, \chi)}{8L(1, \eta)^2 \ L(1, \pi, \mathrm{ad})} \alpha(f_1 \otimes f_2).$$

Here the L-function is

$$L(s, \pi, \chi) = L(s, \sigma, \chi) = L(s, \sigma_E \otimes \chi),$$

where σ denotes the Jacquet-Langlands lifting of π, and σ_E denotes the base change of π to E.

The global bilinear form

$$\alpha : \pi \times \tilde{\pi} \longrightarrow \mathbb{C}$$

is defined to be the tensor product of the local bilinear forms $\alpha : \pi_v \otimes \tilde{\pi}_v \longrightarrow \mathbb{C}$ introduced above. The definition depends on the choice of a local invariant pairing $(\cdot, \cdot)_v : \pi_v \times \tilde{\pi}_v \to \mathbb{C}$ at every place v of F. Normalize the local pairings by the compatibility

$$(\cdot, \cdot)_{\mathrm{Pet}} = \otimes_v (\cdot, \cdot)_v.$$

Here the Petersson pairing $(\cdot, \cdot)_{\mathrm{Pet}} : \pi \times \tilde{\pi} \to \mathbb{C}$ is defined by

$$(f_1, f_2)_{\mathrm{Pet}} = \int_{B^\times \backslash \mathbb{B}^\times / \mathbb{A}^\times} f_1(h)f_2(h)dh, \quad f_1 \in \pi, \ f_2 \in \tilde{\pi}.$$

The integration uses the Tamagawa measure, which has volume 2 on $B^\times \backslash \mathbb{B}^\times / \mathbb{A}^\times$.

Note that the pair $(\widetilde{\pi}, \chi^{-1})$ is equal to the complex conjugate $(\overline{\pi}, \overline{\chi})$. Take $f_2 = \overline{f}_1$ in the formula. Then the left-hand side becomes

$$P_\chi(f_1) \cdot P_{\chi^{-1}}(f_2) = P_\chi(f_1) \cdot P_{\overline{\chi}}(\overline{f}_1) = |P_\chi(f_1)|^2.$$

This form is widely used for period formulae in the literature.

Remark. The formula was proved by Waldspurger with a slightly different setting (cf. [Wa, Proposition 7]). Note that we use the probability measure in the period integral, while Waldspurger used the Tamagawa measure. Waldspurger assumed that the central character ω_π is trivial, but his proof goes through to the general case.

There is an interpretation of the theorem similar to that of Theorem 1.2. In fact, Theorem 1.4 compares the two bilinear forms

$$\alpha, \ P_\chi \otimes P_{\chi^{-1}} \in \mathcal{P}(\pi, \chi) \otimes \mathcal{P}(\widetilde{\pi}, \chi^{-1}).$$

The linear space on the right-hand side is nonzero only if $\Sigma(B) = \Sigma(\pi, \chi)$. In that case, the space is one-dimensional and generated by α. The formula can be viewed as an explicit expression of the multiple in terms of special values of L-functions.

One may start with the data (F, E, σ, χ) satisfying $\epsilon(1/2, \sigma, \chi) = 1$, and find the pair (B, π) by the condition $\Sigma(B) = \Sigma(\pi, \chi)$. By the construction of $\mathbb{B}$, the Jacquet–Langlands lifting π of σ to $B_\mathbb{A}^\times$ always exists, and E can be embedded into B. By this way, the formula is non-trivial.

There is a similar point of view for Theorem 1.2. We omit it here.

1.5   PLAN OF THE PROOF

Our proof of Theorem 1.2 is still based on the idea of Gross–Zagier [GZ], namely, to compare an analytic kernel function representing the central derivative of the L-function with a geometric kernel function formed by a generating series of height pairings of CM points. Many ideas of [Zh1, Zh2] are also used in this book.

In the following, we give an outline of our proof along the order of this book. It also gives the structure of this book.

1.5.1   Proof of Waldspurger formula

In Chapter 2, we review some basic results on Weil representations, theta liftings and Eisenstein series. In particular, a proof of the Waldspurger formula is contained in the chapter, which has inspired our proof of the Gross–Zagier formula in the book.

Assume the notations in Theorem 1.4. Recall that B is a quaternion algebra over F. Write $\mathbb{B} = B_\mathbb{A}$ for simplicity. The Weil representation here is an action

r of $\mathrm{GL}_2(\mathbb{A}) \times \mathbb{B}^\times \times \mathbb{B}^\times$ on the space $\mathcal{S}(\mathbb{B} \times \mathbb{A}^\times)$ of Schwartz functions. For any $\Phi \in \mathcal{S}(\mathbb{B} \times \mathbb{A}^\times)$, the theta series is defined by

$$\theta(g, (h_1, h_2), \Phi) = \sum_{u \in F^\times} r(g, (h_1, h_2)) \Phi(x, u).$$

By integrating against σ, it gives the theta lifting

$$\theta : \mathcal{S}(\mathbb{B} \times \mathbb{A}^\times) \otimes \sigma \longrightarrow \pi \otimes \widetilde{\pi}.$$

It is also called the Shimizu lifting.

We start with the period integrals $P_\chi(f_1)$ and $P_{\chi^{-1}}(f_2)$. Assume that under the Shimizu lifting

$$f_1 \otimes f_2 = \theta(\Phi \otimes \varphi), \quad \Phi \in \mathcal{S}(\mathbb{B} \times \mathbb{A}^\times), \ \varphi \in \sigma.$$

Then it is easy to have

$$P_\chi(f_1) \, P_{\chi^{-1}}(f_2) = \left(\theta(\cdot, (\chi, \chi^{-1}), \Phi), \ \varphi \right)_{\mathrm{Pet}}.$$

Here

$$\theta(g, (\chi, \chi^{-1}), \Phi) = \int_{E^\times \backslash E_\mathbb{A}^\times} \int_{E^\times \backslash E_\mathbb{A}^\times / \mathbb{A}^\times} \theta(g, (tt', t'), \Phi) \chi(t) dt' dt.$$

On the other hand, the L-function has an integral representation

$$(I(s, \cdot, \chi, \Phi), \ \varphi)_{\mathrm{Pet}} = (*) \, L\left(\frac{s+1}{2}, \sigma, \chi\right).$$

Here the analytic kernel is

$$I(s, g, \chi, \Phi) = \int_{E^\times \backslash E_\mathbb{A}^\times} I(s, g, r(t) \Phi) \chi(t) dt,$$

where $I(s, g, \Phi)$ is a mixed theta-Eisenstein series defined by

$$I(s, g, \Phi) = \sum_{u \in F^\times} \sum_{x_1 \in E} \sum_{\gamma \in P^1(F) \backslash P^1(\mathbb{A})} \delta(\gamma g)^s r(\gamma g) \Phi(x).$$

The Waldspurger formula follows from the equalities

$$I(0, g, \chi, \Phi) = 2 \, \theta(g, (\chi, \chi^{-1}), \Phi), \tag{1.5.1}$$

$$(I(0, \cdot, \chi, \Phi), \ \varphi)_{\mathrm{Pet}} = (*) \, L\left(\frac{1}{2}, \sigma, \chi\right) \alpha(f_1 \otimes f_2). \tag{1.5.2}$$

Here $(*)$ denotes a nonzero explicit constant. Both of the equalities are implied by the Siegel–Weil formula with some minor extra work. The first equality is given by the Siegel–Weil formula on the quadratic space (E, q), and the second equality is given by the Siegel–Weil formula on the quadratic space (B_0, q). Here B_0 is the space of elements of B with traces equal to zero.

1.5.2   Projectors

Assume that $F, E, \mathbb{B}$ and X are as in Theorem 1.2 from now on. Fix an embedding $\tau : F \hookrightarrow \mathbb{C}$. The cohomology

$$H^{1,0}(X_\tau) = \varinjlim_U H^{1,0}(X_{U,\tau})$$

has a decomposition

$$H^{1,0}(X_\tau) = \bigoplus_{\pi \in \mathcal{A}(\mathbb{B}^\times)} \pi.$$

Here $\mathcal{A}(\mathbb{B}^\times)$ denotes the set of irreducible admissible representations π of $\mathbb{B}^\times$ such that the Jacquet–Langlands correspondence σ of π on $\mathrm{GL}_2(\mathbb{A})$ is a cuspidal automorphic representation of $\mathrm{GL}_2(\mathbb{A})$, discrete of parallel weight two at infinity.

Any $f_1 \otimes f_2 \in \pi \otimes \widetilde{\pi}$ acts on $H^{1,0}(X_\tau)$ by $f \longmapsto (f, f_2)\, f_1$. This action can be represented by a Hecke operator, and thus it is algebraic. Hence, it lifts to a map

$$\mathrm{T} : \pi \otimes \widetilde{\pi} \to \mathrm{Hom}^0(J, J^\vee)_{\mathbb{C}}.$$

Here

$$\mathrm{Hom}^0(J, J^\vee) = \varinjlim_U \mathrm{Hom}^0(J_U, J_U).$$

We call $\mathrm{T}(f_1 \otimes f_2)$ a projector.

1.5.3   Main result in terms of projectors

Resume the above notations. Assume that $\chi : E^\times \backslash E_{\mathbb{A}}^\times \longrightarrow \mathbb{C}^\times$ is a character of finite order with the compatibility that $\chi|_{\mathbb{A}^\times} \cdot \omega_\pi = 1$. Theorem 3.15 asserts that, for any $f_1 \otimes f_2 \in \pi \otimes \widetilde{\pi}$,

$$\langle \mathrm{T}(f_1 \otimes f_2) P_\chi, P_{\chi^{-1}} \rangle_{\mathrm{NT}} = \frac{\zeta_F(2) L'(1/2, \pi, \chi)}{4 L(1, \eta)^2 L(1, \pi, \mathrm{ad})} \alpha(f_1 \otimes f_2). \qquad (1.5.3)$$

Here P is the CM point on X given by E, and P_χ is the χ-eigencomponent on X. The height pairing is on X.

It is easy to show that the theorem is equivalent to Theorem 1.2. The key ingredient is another interpretation of the projector. Let $f_1 \in \pi_A$ and $f_2 \in \pi_{A^\vee}$ be as in Theorem 1.2. We introduce

$$\mathrm{T}_{\mathrm{alg}}(f_1, f_2) = f_2^\vee \circ f_1 \in \mathrm{Hom}^0(J, J^\vee).$$

One may compare it with the duality maps

$$(f_1, f_2) = f_1 \circ f_2^\vee \in M.$$

Then it is easy to derive, for any embedding $\iota : M \hookrightarrow \mathbb{C}$,

$$\mathrm{T}(f_1^\iota \otimes f_2^\iota) = \mathrm{T}_{\mathrm{alg}}(f_1 \otimes f_2)^\iota.$$

Here $\mathrm{T}_{\mathrm{alg}}(f_1 \otimes f_2)^\iota$ is the ι-eigencomponent of $\mathrm{T}_{\mathrm{alg}}(f_1 \otimes f_2)$ by the action of M. By this identity, the equivalence of the two theorems follows from the projection formula of height pairings.

1.5.4   Generating function

Let $\Phi \in \mathcal{S}(\mathbb{B} \times \mathbb{A}^\times)$ be a Schwartz function bi-invariant under an open compact subgroup U of $\mathbb{B}_f^\times$. The generating function on X_U, for any $g \in \mathrm{GL}_2(\mathbb{A})$, is defined by

$$Z(g, \Phi)_U = Z_0(\phi)_U + w_U \sum_{a \in F^\times} \sum_{x \in U \backslash \mathbb{B}_f^\times / U} r(g)\Phi(x, aq(x)^{-1}) \, Z(x)_U.$$

Here $\phi = \overline{\Phi} \in \overline{\mathcal{S}}(\mathbb{B} \times \mathbb{A}^\times)$ is obtained by averaging on $\mathbb{B}_\infty^1$. It only changes the archimedean parts. The constant term $Z_0(\phi)_U$ and the constant w_U are less important here. The key part $Z(x)_U$ is the Hecke correspondence on X_U corresponding to the double coset UxU.

Based on the work of many people, we will show that $Z(g, \Phi)_U$ is an automorphic form of $g \in \mathrm{GL}_2(\mathbb{A})$ with coefficients in $\mathrm{Pic}(X_U \times X_U)_{\mathbb{C}}$. If $X_U = X_0(N)$ is the modular curve, by properly choosing Φ, the series $Z(g, \Phi)_U$ recovers the classical series

$$\sum_{n \geq 0} T_n e^{2\pi i n z}, \quad z \in \mathcal{H}.$$

This classical form is used in the proof of Gross–Zagier and S. Zhang.

The generating function is the counterpart of the theta series in the incoherent case. In fact, we compute its intersection or arithmetic intersection with other cycles and it is easy to express the result in the form of a usual theta series.

1.5.5   Geometric kernel

Multiplying $Z(g, \Phi)_U$ by a suitable constant, we obtain a system $\widetilde{Z}(g, \Phi) = \{\widetilde{Z}(g, \Phi)_U\}_U$ compatible with pull-back maps. Then $\widetilde{Z}(g, \Phi)$ is an element of

$$\mathrm{Pic}(X \times X)_{\mathbb{C}} = \varinjlim_U \mathrm{Pic}(X_U \times X_U)_{\mathbb{C}}.$$

For any $h_1, h_2 \in \mathbb{B}^\times$, define the height series

$$\widetilde{Z}(g, (h_1, h_2), \Phi) := \langle \widetilde{Z}(g, \Phi) \, [h_1]^\circ, \, [h_2]^\circ \rangle_{\mathrm{NT}}, \quad g \in \mathrm{GL}_2(\mathbb{A}).$$

Here $[h]$ denotes the point $\mathrm{T}_h(P)$ obtained by multiplication by h, and $[h_2]^\circ$ denotes the degree zero cycle $[h] - \xi_{q(h)}$.

The geometric kernel is

$$\widetilde{Z}(g, \chi, \Phi) = \int_{T(F) \backslash T(\mathbb{A})/Z(\mathbb{A})}^{*} \widetilde{Z}(g, (t, 1), \Phi) \, \chi(t) dt.$$

Here the integral is a regularized integral. It takes the place of the kernel function $\theta(g, (\chi, \chi^{-1}), \Phi)$ in the incoherent case.

1.5.6 Kernel identity

The analytic kernel $I(s, g, \chi, \Phi)$ is defined by exactly the same formula as in the coherent case. By the integral representation, the derivative $I'(0, g, \chi, \Phi)$ represents $L'(1/2, \pi, \chi)$.

The kernel identity here, analogous to equation (1.5.1), is

$$\left(I'(0, \cdot, \chi, \Phi),\ \varphi \right)_{\mathrm{Pet}} = 2 \left(\widetilde{Z}(\cdot, \chi, \Phi),\ \varphi \right)_{\mathrm{Pet}}, \quad \forall\, \varphi \in \sigma. \qquad (1.5.4)$$

See Theorem 3.21. We will discuss later the impossibility of the exact equality

$$I'(0, g, \chi, \Phi) = 2\, \widetilde{Z}(g, \chi, \Phi).$$

For any $\Phi \in \mathcal{S}(\mathbb{V} \times \mathbb{A}^\times)$ and $\varphi \in \sigma$, define an "arithmetic theta lifting" by

$$\widetilde{Z}(\Phi \otimes \varphi) = \left(\widetilde{Z}(g, \Phi),\ \varphi \right)_{\mathrm{Pet}}.$$

It lies in $\mathrm{Pic}(X \times X)_{\mathbb{C}}$, and thus induces an element of $\mathrm{Hom}^0(J, J^\vee)$ by acting on J_U as push-forward of correspondences.

The main result for arithmetic theta lifting is

$$\widetilde{Z}(\Phi \otimes \varphi) = \frac{L(1, \pi, \mathrm{ad})}{2\zeta_F(2)} \mathrm{T}(\theta(\Phi \otimes \varphi)), \quad \Phi \in \mathcal{S}(\mathbb{V} \times \mathbb{A}^\times),\ \varphi \in \sigma. \qquad (1.5.5)$$

Note that the Shimizu lifting

$$\theta : \mathcal{S}(\mathbb{B} \times \mathbb{A}^\times) \otimes \sigma \longrightarrow \pi \otimes \widetilde{\pi}$$

can be defined locally in the incoherent case.

By this formula, it is easy to prove the equivalence between the formulation in (1.5.3) and the kernel identity in (1.5.4).

1.5.7 Arithmetic theta lifting

The left-hand side of (1.5.5) can be viewed as an "arithmetic theta lifting." The proof of equation (1.5.5) takes up most of Chapter 4, although its coherent counterpart is trivial. We give a rough idea of the proof here.

The first step is to show that the equation is true up to a constant. In fact, it suffices to prove the equation for their induced actions on the cohomology $H^{1,0}(X_\tau)$. By the decomposition

$$H^{1,0}(X_\tau) = \bigoplus_{\pi_1 \in \mathcal{A}(\mathbb{B}^\times)} \pi_1,$$

each side of (1.5.5) induces an element in

$$\bigoplus_{\pi_1, \pi_2 \in \mathcal{A}(\mathbb{B}^\times)} \pi_1 \otimes \widetilde{\pi}_2.$$

As functionals on Φ and φ, we obtain two elements in the one-dimensional space

$$\mathrm{Hom}_{\mathrm{GL}_2(\mathbb{A}) \times \mathbb{B}^\times \times \mathbb{B}^\times}(\mathcal{S}(\mathbb{V} \times \mathbb{A}^\times) \otimes \sigma, \ \pi \otimes \widetilde{\pi}).$$

So they must be equal up to a constant.

It remains to check that the constant is exactly given by (1.5.5). For that, fixing a level U, it suffices to prove that traces of two sides, as an operator on $H^{0,1}(X_{U,\tau})$, are equal. It is easy to see that the trace of $\mathrm{T}(f_1 \otimes f_2)$ on $H^{0,1}(X_{U,\tau})$ is exactly (f_1, f_2). The hard part is to figure out the trace of $\widetilde{Z}(\Phi \otimes \varphi)$ on $H^{0,1}(X_{U,\tau})$.

By the Lefschetz fixed point theorem, the trace of $\widetilde{Z}(\Phi \otimes \varphi)$ is reduced to the intersection number $\Delta_U \cdot \widetilde{Z}(\Phi \otimes \varphi)$. Here $\Delta_U : X_U \to X_U \times X_U$ is the diagonal embedding. It is further reduced to compute the pull-back $\Delta_U^* Z(g, \Phi)_U$. Roughly speaking, the pull-back is given by an explicit generating function whose coefficients are CM points, Hodge classes or cusps on X_U. Then the trace will be given by a mixed theta-Eisenstein series different from our analytic kernel. The computation is very complicated if X_U contains cusps. Similar computations on Hilbert modular surfaces were done by Hirzebruch–Zagier.

1.5.8 Degenerate Schwartz functions

In Chapter 5, we introduce two classes of degenerate Schwartz functions. They simplify the computations and arguments of both kernel functions significantly. Of course, we also prove that these assumptions can recover (1.5.4) in the full case.

For any non-archimedean place v, define

$$\mathcal{S}^1(\mathbb{B}_v \times F_v^\times) := \{\Phi_v \in \mathcal{S}(\mathbb{B}_v \times F_v^\times) :$$
$$\Phi_v(x, u) = 0 \text{ if } v(uq(x)) \geq -v(d_v) \text{ or } v(uq(x_2)) \geq -v(d_v)\}.$$

Here d_v is the local different of F at v, and $x = x_1 + x_2$ according to the orthogonal decomposition $\mathbb{B}_v = E_v + E_v j_v$ (cf. §1.6.4). Define

$$\mathcal{S}^2(\mathbb{B}_v \times F_v^\times) := \{\Phi_v \in \mathcal{S}(\mathbb{B}_v \times F_v^\times) :$$
$$r(g)\Phi_v(0, u) = 0, \quad \forall g \in \mathrm{GL}_2(F_v), \ u \in F_v^\times\}.$$

Assume that Φ_v lies in $\mathcal{S}^1(\mathbb{B}_v \times F_v^\times)$ for all ramified non-archimedean places v nonsplit in E, and assume that Φ_v lies in $\mathcal{S}^1(\mathbb{B}_v \times F_v^\times)$ for at least two non-archimedean places v split in E. Here are some major effects of the assumptions:

- Kill the self-intersections of CM points in the height series $Z(g, (t_1, t_2), \Phi)$.

- Kill the logarithmic singularities coming from both the derivatives and the local heights at v.

- Kill the constant term of the generating series $Z(g, \Phi)_U$.

- Kill the arithmetic intersections coming for the Hodge classes in the height series $Z(g, (t_1, t_2), \Phi)$.

The key to recover (1.5.4) in the full case from these assumptions is Theorem 1.3. In fact, like the interpretation of Theorem 1.2, we interpret (1.5.3) as an identity of two vectors in the complex vector space $\mathcal{P}(\pi, \chi) \otimes \mathcal{P}(\tilde{\pi}, \chi^{-1})$. It is at most one-dimensional by Theorem 1.3. Assume that it is one-dimensional. It follows that the ramification set of $\mathbb{B}$ agrees with the set Σ of places determined by the local root numbers. It suffices to prove (1.5.3) for some (f_1, f_2) with $\alpha(f_1 \otimes f_2) \neq 0$. Hence, it suffices to prove (1.5.4) for some (Φ, φ) with $\alpha(\theta(\Phi \otimes \varphi)) \neq 0$.

Go back to the sufficiency of the assumptions. We only need to prove that there exists a Schwartz function Φ satisfying the degeneracy assumptions and the condition that $\alpha(\theta(\Phi \otimes \varphi)) \neq 0$ for some $\varphi \in \sigma$. It is a local problem and solved in Chapter 5.

1.5.9   Components of the kernel functions

To prove (1.5.4), we need to compute the difference

$$\mathcal{P}r I'(0, g, \chi, \Phi) - 2 \, \widetilde{Z}(g, \chi, \Phi).$$

Here $\mathcal{P}r$ denotes the projection to the space of cusp forms, holomorphic of parallel weight two.

It is easy to write

$$\mathcal{P}r I'(0, g, \chi, \Phi) = \sum_{v \text{ nonsplit}} I'(0, g, \chi, \Phi)(v).$$

Roughly speaking, we write $I(s, g, \chi, \Phi)$ as a Fourier series, which is a sum of Whittaker functions. The Whittaker functions are local products, and the derivative falls into the local Whittaker functions at every place. Then $I'(0, g, \chi, \Phi)(v)$ collects all the terms with the derivative taken at v. We also need to apply $\mathcal{P}r$, which only changes $I'(0, g, \chi, \Phi)(v)$ for archimedean v. This is done in Chapter 6.

On the other hand, we also have

$$\widetilde{Z}(g, \chi, \Phi) = \sum_{v \text{ nonsplit}} \widetilde{Z}(g, \chi, \Phi)(v).$$

It essentially follows from the decomposition of a global arithmetic intersection number to the sum of local intersection numbers. Note that there is no self-intersection by the degenerate Schwartz functions. We also prove that the arithmetic intersection with the Hodge class are zero by the degenerate Schwartz functions. This is done in Chapter 7.

For archimedean v and *good* non-archimedean v, we prove

$$I'(0, g, \chi, \Phi)(v) = 2Z(g, \chi, \Phi)(v)$$

by explicit computations in Chapter 6 and Chapter 8. It is routine after the work of Gross–Zagier and S. Zhang.

For the remaining bad primes, the local components are impossible to compute. This was the essential difficulty to remove the assumptions of mild ramifications in the work of Gross–Zagier and S. Zhang. Our solution is the following approximation argument.

1.5.10   Approximation

We say a function $\Psi : \mathrm{GL}_2(\mathbb{A}) \to \mathbb{C}$ is *approximated* by an automorphic form Ψ_0 on $\mathrm{GL}_2(\mathbb{A})$ if there exists a finite set S of places of F such that $\Psi(g) = \Psi_0(g)$ for all $g \in 1_S \mathrm{GL}_2(\mathbb{A}^S)$. Here is a simple fact. If furthermore Ψ is automorphic, then $\Psi = \Psi_0$ identically. It is true since $\mathrm{GL}_2(F)\mathrm{GL}_2(\mathbb{A}^S)$ is dense in $\mathrm{GL}_2(\mathbb{A})$.

Come back to the comparison of the kernel functions. Take advantage of the degenerate Schwartz functions. In Chapter 6, we prove that, for bad v, $I'(0, g, \chi, \Phi)(v)$ can be approximated by a coherent kernel function of the form $I(0, g, \chi, \Phi(v))$, where

$$\Phi(v) = \Phi^v \otimes \Phi'_v \in \mathcal{S}(B(v)_{\mathbb{A}} \times \mathbb{A}^\times)$$

is a Schwartz function based on *the nearby quaternion algebra $B(v)$ over F* obtained by changing the Hasse invariant of $\mathbb{B}$ at v. In Chapter 8, we prove similar approximation results for $\widetilde{Z}(g, \chi, \Phi)(v)$.

It follows that the difference

$$\mathcal{P}r I'(0, g, \chi, \Phi) - 2\, \widetilde{Z}(g, \chi, \Phi)$$

is approximated by a finite sum of functions of the form $I(0, g, \chi, \Phi(v))$. By the above simple consequence of modularity, we conclude that

$$\mathcal{P}r I'(0, g, \chi, \Phi) - 2\, \widetilde{Z}(g, \chi, \Phi) = \sum_v I(0, g, \chi, \Phi(v)). \qquad (1.5.6)$$

The right-hand side is a priori not zero, but it is perpendicular to σ by Theorem 1.3. In fact, the ramification set of each $B(v)$ does not agree with Σ since we have assumed that the ramification set of $\mathbb{B}$ agrees with Σ. It proves (1.5.4).

Note that (1.5.4) is a weak analogue of (1.5.1). However, the exact analogue

$$I'(0, g, \chi, \Phi) = 2\, \widetilde{Z}(g, \chi, \Phi)$$

does not hold in general. In fact, the right-hand sides satisfies the transfer property $\widetilde{Z}(g, \chi, \Phi) = \widetilde{Z}(1, \chi, r(g)\Phi)$, but the left-hand side does not satisfy such a property unless Φ is invariant under the action of $g \in \mathrm{GL}_2(\widehat{O}_F)$.

On the other hand, (1.5.6) can be viewed as an equality of the two kernel functions modulo coherent kernel functions. In this sense, it is an appropriate analogue of (1.5.1).

1.5.11   Pseudo-theta series

In the end, we explain why these local components can be approximated by the coherent kernel functions easily.

We mainly look at $I'(0, g, \chi, \Phi)(v)$. By a local version of the Siegel–Weil formula, it is easy to write $I'(0, g, \Phi)(v)$ as

$$\sum_u \sum_{y \in B(v)} k_{\Phi_v}(g_v, y, u) r(g^v) \Phi^v(y, u),$$

where $k_{\Phi_v}(g_v, y, u)$ is a function on $g_v \in \mathrm{GL}_2(F_v)$ and $(y, u) \in B(v)_v \times F_v^\times$ determined explicitly by Φ_v.

It looks like a theta series except that at v the function $k_{\Phi_v}(g, y, u)$ is not given by Weil representation on Schwartz function on $B(v)_v$, so we call it a *pseudo-theta series*. The key is to show that $k_{\Phi_v}(1, y, u)$ is a Schwartz function of $(y, u) \in B(v)_v \times F_v^\times$ if Φ_v is *degenerate*. Then we form the "authentic" theta series for Schwartz function $k_{\Phi_v} \otimes \Phi^v$ as follows:

$$\sum_u \sum_{y \in B(v)} r(g) k_{\Phi_v}(1, y, u) r(g) \Phi^v(y, u).$$

It approximates the original series since they are the same for $g \in \mathrm{GL}_2(\mathbb{A})$ with $g_v = 1_v$. Hence, $I'(0, g, \chi, \Phi)(v)$ is approximated by the coherent kernel function $I(0, g, \chi, \Phi^v \otimes k_{\Phi_v})$ on the nearby quaternion algebra.

As for the local height $\widetilde{Z}(g, \chi, \Phi)(v)$, we can also write it as a series over $B(v)$. Roughly speaking, the local formal neighborhoods of the integral model of Shimura curve X_U can be uniformized as the quotient of some universal deformation space by the action of $B(v)^\times$. Then the local height pairing on the Shimura curve is a summation of intersections of points in the corresponding orbit indexed by $B(v)^\times$.

1.6   NOTATION AND TERMINOLOGY

We always denote by F the base number field, and by E a quadratic field extension of F. Except in Chapter 1 and Chapter 2, it is assumed that F is totally real and E is totally imaginary.

We normalize the absolute values, additive characters, and measures following Tate's thesis. To be precise, we start with the local case.

1.6.1   Local fields

In the following, k denotes a local field of a number field.

- Normalize the absolute value $|\cdot|$ on k as follows:

    It is the usual one if $k = \mathbb{R}$.

    It is the square of the usual one if $k = \mathbb{C}$.

If k is non-archimedean, it maps the uniformizer to N^{-1}. Here N is the cardinality of the residue field.

- Normalize the additive character $\psi : k \to \mathbb{C}^\times$ as follows:

    If $k = \mathbb{R}$, then $\psi(x) = e^{2\pi i x}$.

    If $k = \mathbb{C}$, then $\psi(x) = e^{4\pi i \operatorname{Re}(x)}$.

    If k is non-archimedean, then it is a finite extension of $\mathbb{Q}_p$ for some prime p. Take $\psi = \psi_{\mathbb{Q}_p} \circ \operatorname{tr}_{k/\mathbb{Q}_p}$. Here the additive character $\psi_{\mathbb{Q}_p}$ of $\mathbb{Q}_p$ is defined by $\psi_{\mathbb{Q}_p}(x) = e^{-2\pi i \iota(x)}$, where $\iota : \mathbb{Q}_p/\mathbb{Z}_p \hookrightarrow \mathbb{Q}/\mathbb{Z}$ is the natural embedding.

- We take the measure dx on k to be the unique Haar measure on k self-dual with respect to ψ in the sense that the Fourier transform

$$\widehat{\Phi}(y) := \int_k \Phi(x)\psi(xy)dx$$

satisfies the inversion formula $\widehat{\widehat{\Phi}}(x) = \Phi(-x)$. The measures are determined explicitly as follows:

    If $k = \mathbb{R}$, then dx is the usual Lebesgue measure.

    If $k = \mathbb{C}$, then dx is twice of the usual Lebesgue measure.

    If k is non-archimedean, then $\operatorname{vol}(O_k) = |d_k|^{\frac{1}{2}}$. Here O_k is the ring of integers and $d_k \in k$ is the different of k over $\mathbb{Q}_p$.

- We take the Haar measure $d^\times x$ on $k^\times$ by

$$d^\times x = \zeta_k(1)|x|^{-1}dx.$$

Recall that $\zeta_k(s) = (1-N^{-s})^{-1}$ if k is non-archimedean whose residue field has N elements, $\zeta_{\mathbb{R}}(s) = \pi^{-s/2}\Gamma(s/2)$, and $\zeta_{\mathbb{C}}(s) = 2(2\pi)^{-s}\Gamma(s)$. With this normalization, if k is non-archimedean, then $\operatorname{vol}(O_k^\times, d^\times x) = \operatorname{vol}(O_k, dx)$.

1.6.2 Haar measures for algebraic groups

Algebraic groups in this book are mainly associated to quadratic extensions and quaternion algebras over the base field. We are going to normalize the Haar measures on them based on the normalization above. Let k be a local field as above.

If (V, q) is a quadratic space over k, the self-dual measure on V with respect to (V, q) (and ψ) is the unique Haar measure dx on V such that the Fourier transform

$$\widehat{\Phi}(y) := \int_V \Phi(x)\psi(\langle x, y\rangle)dx$$

satisfies the inversion formula $\widehat{\widehat{\Phi}}(x) = \Phi(-x)$. Here $\langle x, y\rangle = q(x+y)-q(x)-q(y)$ is the corresponding bilinear pairing.

Let K be a quadratic etale algebra extension over k or a quaternion algebra over k, and let $q : K \to k$ be the reduced norm. One has one of the following:

- $K = k \oplus k$ and $q(x_1, x_2) = x_1 x_2$, in this case we say K is a split quadratic extension;

- K is a quadratic field extension over k and $q = N_{K/k}$;

- K is equal to a (either split ornonsplit) quaternion algebra B over k and q is the reduced norm.

In all cases we can write $q(x) = x\bar{x}$ where $\bar{x}$ is the main involution of x. The pair (K, q) forms a quadratic space over k.

Endow K with the self-dual Haar measure dx with respect to (K, q).

Endow the multiplicative group $K^\times$ with a Haar measure $d^\times x$ as follows. If K is a quadratic etale algebra extension, then

$$d^\times x = \zeta_K(1) \, |q(x)|^{-1} dx.$$

Here $\zeta_K(s) = \zeta_k(s)^2$ if $K = k \oplus k$, and $\zeta_K(s)$ is the zeta function of the local field K if K isnonsplit. If K is a quaternion algebra, then

$$d^\times x = \zeta_k(1) \, |q(x)|^{-2} dx.$$

Endow $K^\times / k^\times$ with the quotient Haar measure.

Endow the subgroup

$$K^1 := \{h \in K^\times : q(h) = 1\}$$

with the Haar measure dh determined by the exact sequence

$$1 \longrightarrow K^1 \longrightarrow K^\times \xrightarrow{\;q\;} q(K^\times) \longrightarrow 1.$$

In other words, it makes the Haar measure on $q(K^\times)$, obtained by the restriction of the Haar measure $d^\times x$ on $k^\times$, equal to the quotient of the Haar measure $d^\times x$ on $K^\times$ by dh on K^1.

Next, we consider some explicit forms of these measures.

If K is a quadratic, then we have a homomorphism

$$K^\times / k^\times \longrightarrow K^1, \quad t \longmapsto t/\bar{t}.$$

It is an isomorphism by Hilbert's Theorem 90. The Haar measures are compatible under this isomorphism.

If K is a quadratic field extension, then the Haar measures dx and $d^\times x$ respectively on K and $K^\times$ are the same as the measures normalized in the last subsection by viewing K as a local field. If $K = k \oplus k$ is split, then the Haar measure on $K = k \oplus k$ (resp. $K^\times = k^\times \times k^\times$) is compatible with the Haar measure dx (resp. $d^\times x$).

Still, consider the case of quadratic extension. If k is non-archimedean, denote by $d \in k$ the different of k over $\mathbb{Q}_p$, and by $D \in O_k$ the discriminant of K in k. Then

$$\text{vol}(O_K, dx) = \text{vol}(O_K^\times, d^\times x) = |D|^{\frac{1}{2}}|d|.$$

Furthermore,

$$\text{vol}(K^1) = \begin{cases} 2 & \text{if } k = \mathbb{R} \text{ and } K = \mathbb{C}, \\ |d|^{\frac{1}{2}} & \text{if } K/k \text{ is nonsplit and unramified}, \\ 2|D|^{\frac{1}{2}}|d|^{\frac{1}{2}} & \text{if } K/k \text{ is ramified}. \end{cases}$$

Now assume that K is a quaternion algebra over k. Note that by the split case $K = M_2(k)$, we have normalized Haar measures on $\text{GL}_2(k), \text{SL}_2(k)$ and $\text{PGL}_2(k)$. If k is non-archimedean, then the normalization gives

$$\text{vol}(\text{GL}_2(O_k)) = \zeta_k(2)^{-1}\text{vol}(O_k)^4, \quad \text{vol}(\text{SL}_2(O_k)) = \zeta_k(2)^{-1}\text{vol}(O_k)^3.$$

If $k = \mathbb{R}$ and K is the Hamiltonian algebra, then the self-dual measure dx on K is four times the usual Lebesgue measure under the natural isometry $K = \mathbb{R}^4$. In this case $\text{vol}(K^1) = 4\pi^2$.

1.6.3 Global case

Throughout this book we fix a global field F and denote by $\mathbb{A} = \mathbb{A}_F$ the ring of adeles of F. We also fix a quadratic field extension E of F, and denote by $\eta : F^\times \backslash \mathbb{A}^\times \to \mathbb{C}^\times$ the quadratic character determined by this extension. Except in Chapter 1 and Chapter 2, F is assumed to be totally real and E is assumed to be totally imaginary.

We always use v to denote a place of F. For each place v of F, we choose $|\cdot|_v$, ψ_v, dx_v, $d^\times x_v$ as above. By tensor products, they induce global $|\cdot|_\mathbb{A}$, ψ, dx, $d^\times x$. The absolute values satisfy the product formula, the sum $\psi = \otimes_v \psi_v$ is actually a character on $\mathbb{A}/F$, and the volume of $\mathbb{A}/F$ is exactly one under the product measure.

For non-archimedean v, we usually denote by O_{F_v} the integer ring of F_v, by p_v the corresponding prime ideal, by N_v the cardinality of its residue field O_{F_v}/p_v, by ϖ_v a uniformizer of F_v, and by $d_v \in F_v$ the local different of F over $\mathbb{Q}$. Denote by $D_v \in F_v$ the discriminant of the quadratic extension E_v in F_v. We use the convention that $D_v = d_v = 1$ if v is archimedean.

Denote by $Z = \text{GL}_1$, viewed as an algebraic group over F. We denote by $T = E^\times$ the algebraic group over F, and

$$E^1 = \{y \in E^\times : q(y) = 1\}$$

also defines an algebraic group over F. At every place v, the Haar measures on $Z(F_v)$, $T(F_v)$ and $E^1(F_v)$ are defined and thus give product Haar measures on $Z(\mathbb{A})$, $T(\mathbb{A})$ and $E^1(\mathbb{A})$. All of them are Tamagawa measures. The total volume $\text{vol}(E^1 \backslash E^1(\mathbb{A})) = 2L(1, \eta)$.

View $V_1 = (E, q = N_{E/F})$ as a two-dimensional vector space over F. Let E^1 act on V_1 by multiplication. It induces an isomorphism $\mathrm{SO}(V_1) \simeq E^1$ of algebraic groups over F. Hilbert Theorem 90 gives an isomorphism $\mathrm{SO}(V_1) \simeq T/Z$ by

$$E^\times/F^\times \longrightarrow E^1, \quad t \longmapsto t/\bar{t}.$$

Let B be a quaternion algebra over F with reduced norm q. Denote

$$B^1 = \{y \in B^\times : q(y) = 1\}.$$

We have defined Haar measures on $B_v^\times$ and B_v^1, and they give product Haar measures on $B_{\mathbb{A}}^\times$ and $B_{\mathbb{A}}^1$. Both of them are Tamagawa measures. It follows that

$$\mathrm{vol}(B^1 \backslash B_{\mathbb{A}}^1) = 1, \quad \mathrm{vol}(B^\times \backslash B_{\mathbb{A}}^\times / Z(\mathbb{A})) = 2.$$

In particular, we have measures on SL_2 and GL_2 by taking B to be the matrix algebra.

More generally, let G be an algebraic group over F. If G is semisimple, endow $G(\mathbb{A})$ with the Tamagawa measure. If $Z = \mathrm{GL}_1$ is contained in the center of G such that G/Z is semisimple, endow $G(\mathbb{A})$ with the Haar measure which induces the Tamagawa measure on $G(\mathbb{A})/Z(\mathbb{A})$ via the quotient by $(\mathbb{A}^\times, d^\times x)$.

1.6.4   Notation on quaternion algebras

Throughout this book, Σ denotes a finite set of places of F. It comes from local root numbers of the Rankin-Selberg L-function. It is assumed to contain all the archimedean places except in Chapter 1 and Chapter 2.

We also denote by $\mathbb{B}$ the unique quaternion algebra over $\mathbb{A}$ such that for every place v of F, the quaternion algebra $\mathbb{B}_v := \mathbb{B} \otimes_{\mathbb{A}} F_v$ over F_v is isomorphic to the matrix algebra if and only if $v \notin \Sigma$. Alternatively, one can define $\mathbb{B}_v$ according to Σ, and $\mathbb{B}$ as a restricted product of $\mathbb{B}_v$.

We say $\mathbb{B}$ is *coherent* if it is a base change of a quaternion algebra over F; otherwise, we say $\mathbb{B}$ is *incoherent*. It follows that $\mathbb{B}$ is coherent if and only if the cardinality of Σ is even.

The reduced norm q makes $\mathbb{B}$ a quadratic space $\mathbb{V} = (\mathbb{B}, q)$ over $\mathbb{A}$. Fix an embedding $E_{\mathbb{A}} \hookrightarrow \mathbb{B}$ if it exists. It gives an orthogonal decomposition

$$\mathbb{B} = E_{\mathbb{A}} + E_{\mathbb{A}}\mathrm{j}, \quad \mathrm{j}^2 \in \mathbb{A}^\times.$$

Then we get two induced subspaces $\mathbb{V}_1 = (E_{\mathbb{A}}, q)$ and $\mathbb{V}_2 = (E_{\mathbb{A}}\mathrm{j}, q)$. Apparently $\mathbb{V}_1$ is the base change of the F-space $V_1 = (E, q)$. We usually write $x = x_1 + x_2$ for the corresponding orthogonal decomposition of $x \in \mathbb{V}$.

Assume that the cardinality of Σ is odd. We will keep this assumption throughout Chapters 3–8 in this book. Then $\mathbb{B}$ is incoherent. But we will get a coherent one by increasing or decreasing Σ by one element. For any place v of F, denote by $B(v)$ the quaternion algebra over F obtained from $\mathbb{B}$ by switching the Hasse invariant at v. We call $B(v)$ the *nearby quaternion algebra* corresponding to v. Throughout this book, we will fix an identification $B(v) \otimes_F \mathbb{A}^v \cong \mathbb{B}^v$. Fix an embedding $E \hookrightarrow B(v)$ if v isnonsplit in E. In this case, such an embedding always exists. Then we also have an orthogonal decomposition $B(v) = V_1 \oplus V_2(v)$.

For any quaternion algebra B over F_v with a fixed embedding $E_v \hookrightarrow B$, we define

$$\lambda : B^\times \longrightarrow F_v, \qquad x \longmapsto \frac{q(x_2)}{q(x)}$$

where $x = x_1 + x_2$ is the orthogonal decomposition induced by $E_v \hookrightarrow B$. This definition applies to all the quaternion algebras above locally and globally.

The measures on $\mathbb{B}_v^\times$ and $\mathbb{B}_v^1$ normalized above define Haar measures on $\mathbb{B}^\times$ and

$$\mathbb{B}^1 := \{x \in \mathbb{B} : q(x) = 1\}.$$

They can be viewed as analogues of Tamagawa measures.

1.6.5 L-functions

All global L-functions in this book are the complete L-functions containing the archimedean parts. These include

$$\zeta_F(s), \quad L(s, \eta), \quad L(s, \pi, \chi), \quad L(s, \pi, \mathrm{ad}).$$

For example, the Dedekind zeta function

$$\zeta_F(s) = \prod_v \zeta_{F_v}(s)$$

is a product over all places v of F. The local L-function $\zeta_{F_v}(s)$ has been introduced in the definition of Haar measures over local fields above.

1.6.6 Subgroups of GL(2)

We introduce the matrix notation:

$$m(a) = \begin{pmatrix} a & \\ & a^{-1} \end{pmatrix}, \quad d(a) = \begin{pmatrix} 1 & \\ & a \end{pmatrix}, \quad d^*(a) = \begin{pmatrix} a & \\ & 1 \end{pmatrix}$$

$$n(b) = \begin{pmatrix} 1 & b \\ & 1 \end{pmatrix}, \quad k_\theta = \begin{pmatrix} \cos\theta & \sin\theta \\ -\sin\theta & \cos\theta \end{pmatrix}, \quad w = \begin{pmatrix} & 1 \\ -1 & \end{pmatrix}.$$

We denote by $P \subset \mathrm{GL}_2$ and $P^1 \subset \mathrm{SL}_2$ the subgroups of upper triangular matrices, and by N the standard unipotent subgroup of them. Denote by $A \subset \mathrm{GL}_2$ the subgroup of diagonal matrices.

We have canonical isomorphisms $N \simeq \mathbb{G}_a$ and $A \simeq \mathbb{G}_m^2$ and thus they are endowed with the Haar measures induced from F_v and $F_v^\times$.

For any local field F_v, the character

$$\delta_v : P(F_v) \longrightarrow \mathbb{R}^\times, \qquad \begin{pmatrix} a & b \\ & d \end{pmatrix} \longmapsto \left| \frac{a}{d} \right|_v^{\frac{1}{2}}$$

extends to a function $\delta_v : \mathrm{GL}_2(F_v) \to \mathbb{R}^\times$ by Iwasawa decomposition.

For the global field F, the product $\delta = \prod_v \delta_v$ gives a function on $\mathrm{GL}_2(\mathbb{A})$.

1.6.7 Averages and regularized integrations

Let G be a topological group. Assume that G has a left Haar measure dg with finite total volume $\mathrm{vol}(G)$. For any function f on G, define

$$\fint_G f(g)dg := \frac{1}{\mathrm{vol}(G)} \int_G f(g)dg.$$

It is independent of the choice of the measure dg. If G is a finite group, then

$$\fint_G f(g)dg = \frac{1}{|G|} \sum_{g \in G} f(g).$$

Let F be a totally real field. For a function f on $F^\times \backslash \mathbb{A}^\times$ which is invariant under the archimedean part $F_\infty^\times$, denote the regularized average of f on $\mathbb{A}^\times$ by

$$\fint_{\mathbb{A}^\times} f(z)dz := \fint_{F^\times \backslash \mathbb{A}^\times / F_\tau^\times} f(z)dz.$$

Here τ is any archimedean place of F. The definition is independent of the choice of τ. The quotient group $F^\times \backslash \mathbb{A}^\times / F_\tau^\times$ is a compact with total volume

$$\mathrm{vol}(F^\times \backslash \mathbb{A}^\times / F_\tau^\times) = \frac{1}{2} \mathrm{Res}_{s=1} \zeta_F(s)$$

if we use the Haar measure normalized in §1.6.1.

If f is further invariant under some open compact subgroup U of $\mathbb{A}_f^\times$, then we have

$$\fint_{\mathbb{A}^\times} f(z)dz = \frac{1}{|F^\times \backslash \mathbb{A}^\times / F_\infty^\times U|} \sum_{z \in F^\times \backslash \mathbb{A}^\times / F_\infty^\times \cdot U} f(z).$$

The right-hand side is just a finite sum and the average does not depend on choice of U.

Let G be a reductive group over F with an embedding of $Z = \mathbb{G}_m$ into the center of G. Assume that the volume of $G(F) \backslash G(\mathbb{A}) / Z(\mathbb{A})$ is finite under some Haar measure dg of $G(\mathbb{A}) / Z(\mathbb{A})$. Let f be an automorphic function on $G(\mathbb{A})$ which is invariant under the action of $Z(F_\infty)$. We define the regularized integration

$$\int_{G(F) \backslash G(\mathbb{A}) / Z(\mathbb{A})}^* f(g)dg := \int_{G(F) \backslash G(\mathbb{A}) / Z(\mathbb{A})} \fint_{Z(\mathbb{A})} f(zg)dzdg. \qquad (1.6.1)$$

We also define the regularized average

$$\fint_{G(F) \backslash G(\mathbb{A}) / Z(\mathbb{A})} f(g)dg$$

$$:= \frac{1}{\mathrm{vol}(G(F) \backslash G(\mathbb{A}) / Z(\mathbb{A}))} \int_{G(F) \backslash G(\mathbb{A}) / Z(\mathbb{A})} \fint_{Z(\mathbb{A})} f(zg)dzdg.$$

In these definitions, f is not required to be invariant under the whole $Z(\mathbb{A})$. We sometimes abbreviate $[G] := G(F)\backslash G(\mathbb{A})/Z(\mathbb{A})$ in the domain of the integrations.

For the application in this book, the definition applies to the algebraic groups GL_2 and $T = E^\times$ over F. In the case $T = E^\times$, if f is further invariant under $T(F_\infty)$, then

$$\fint_{T(F)\backslash T(\mathbb{A})/Z(\mathbb{A})} f(t)dt = \frac{1}{|T(F)\backslash T(\mathbb{A})/T(F_\infty)U|} \sum_{t\in T(F)\backslash T(\mathbb{A})/T(F_\infty)U} f(t).$$

Here U is an open compact subgroup of $T(\mathbb{A}_f)$ which acts trivially on f, and the right-hand side does not depend on the choice of U.

1.6.8 Vector spaces over number fields

Let M be a number field, and V be a vector space over M. The canonical isomorphism

$$M \otimes_{\mathbb{Q}} \mathbb{C} = \bigoplus_{\iota\in\mathrm{Hom}(M,\mathbb{C})} \mathbb{C}$$

induces a canonical decomposition

$$V \otimes_{\mathbb{Q}} \mathbb{C} = \bigoplus_{\iota\in\mathrm{Hom}(M,\mathbb{C})} V^\iota.$$

Here we denote $V^\iota = V \otimes_{(M,\iota)} \mathbb{C}$ for any embedding $\iota : M \hookrightarrow \mathbb{C}$. In particular, there is a canonical embedding $V^\iota \hookrightarrow V \otimes_{\mathbb{Q}} \mathbb{C}$.

For any $v \in V$, denote by $v^\iota = v \otimes_{(M,\iota)} 1$ the corresponding element of V^ι. Then we have a decomposition

$$v = \sum_{\iota\in\mathrm{Hom}(M,\mathbb{C})} v^\iota \in V \otimes_{\mathbb{Q}} \mathbb{C}.$$

It is exactly the spectral decomposition of the complex vector space $V \otimes_{\mathbb{Q}} \mathbb{C}$ under the action of M.

Chapter Two

Weil Representation and Waldspurger Formula

In this chapter, we will review the theory of Weil representation and its applications to an integral representation of the Rankin–Selberg L-function $L(s, \pi, \chi)$ and to a proof of Waldspurger's central value formula. We will mostly follow Waldspurger's treatment with some modifications including Kudla's construction of incoherent Eisenstein series.

We will start with the classical theory of Weil representation of $O(F) \times SL_2(F)$ on $\mathcal{S}(V)$ for an orthogonal space V over a local field F and its extension to $GO(F) \times GL_2(F)$ on $\mathcal{S}(V \times F^\times)$ by Waldspurger. We then define theta functions, state the Siegel–Weil formula, and define normalized local Shimizu lifting. The main result of this chapter is an integral formula for the L-series $L(s, \pi, \chi)$ using a kernel function $I(s, g, \chi, \Phi)$. This kernel function is a mixed Eisenstein and theta series attached to each $\Phi \in \mathcal{S}(\mathbb{V} \times \mathbb{A}^\times)$ for $\mathbb{V}$, an orthogonal space obtained from a quaternion algebra over $\mathbb{A}$. The Waldspurger formula is a direct consequence of the Siegel–Weil formula.

After the proof of Waldspurger formula, we list some computational results on three types of incoherent Eisenstein series in §2.5. These Eisenstein series will be used in the remaining chapters of this book (for different purposes).

2.1 WEIL REPRESENTATION

Let us start with some basic setup on Weil representation. We follow closely Waldspurger [Wa].

2.1.1 Non-archimedean case

Let k be a non-archimedean local field and (V, q) a quadratic space over k. Let $O = O(V, q)$ denote the orthogonal group of (V, q) and $\widetilde{SL}_2(k)$ the metaplectic double cover of $SL_2(k)$.

Recall that $\widetilde{SL}_2(k)$ is a central extension of $SL_2(k)$ by $\{\pm 1\}$. One can write $\widetilde{SL}_2(k) = SL_2(k) \times \{\pm 1\}$ with group law given by

$$(g_1, \epsilon_1) \cdot (g_2, \epsilon_2) = (g_1 g_2, \epsilon_1 \epsilon_2 \beta(g_1, g_2)).$$

Here $\beta : SL_2(k) \times SL_2(k) \to \{\pm 1\}$ is a cocycle in $H^2(SL_2(k), \{\pm 1\})$.

Denote by $\mathcal{S}(V)$ the space of locally constant and compactly supported complex-valued functions on V. Such functions are also called Schwartz–Bruhat functions on V. The Weil representation is an action r of the group $\widetilde{SL}_2(k) \times O(k)$ on V. For any $\Phi \in \mathcal{S}(V)$, the action is given as follows:

- $r(h)\Phi(x) = \Phi(h^{-1}x)$, $\quad h \in O(k)$;

- $r(m(a))\Phi(x) = \chi_{(V,q)}(a)|a|^{\dim V/2}\Phi(ax)$, $\quad a \in k^\times$;

- $r(n(b))\Phi(x) = \psi(bq(x))\Phi(x)$, $\quad b \in k$;

- $r(w,\epsilon)\Phi = \epsilon^{\dim V}\gamma(V,q)\widehat{\Phi}$, $\quad w = \begin{pmatrix} & 1 \\ -1 & \end{pmatrix}$, $\epsilon \in \{\pm 1\}$.

Several notations need to be explained. The representation depends on the choice of a character $\psi : k \to \mathbb{C}^\times$. We always take ψ to be the standard one described in §1.6. As usual, $m(a)$ and $n(b)$ denote the element $(m(a), 1)$ and $(n(b), 1)$ in $\widetilde{SL}_2(k)$. The character $\chi_{(V,q)} : k^\times \to \mathbb{C}^\times$ takes values in $\{\pm 1, \pm i\}$. The constant $\gamma(V,q)$, an 8-th root of unity, is the Weil index. Finally, $\widehat{\Phi}$ denotes the Fourier transform given by

$$\widehat{\Phi}(x) := \int_V \Phi(y)\psi(\langle x, y \rangle)dy.$$

Here $\langle x, y \rangle := q(x+y) - q(x) - q(y)$ is the corresponding inner product, and the integral uses the self-dual measure.

There are many simplifications if $\dim V$ is even. In that case, r is trivial on the subgroup $\{\pm 1\}$ of $\widetilde{SL}_2(k)$, and thus the Weil representation descends to a representation of $SL_2(k) \times O(k)$ on $\mathcal{S}(V)$. The Weil index $\gamma(V,q)$ is a 4-th root of unity, and the character $\chi_{(V,q)}$ becomes the quadratic character associated to the quadratic space (V,q). Namely,

$$\chi_{(V,q)}(a) = (a, (-1)^{\frac{\dim V}{2}} \det(V,q)), \quad a \in k^\times.$$

Here the right-hand side denotes the Hilbert symbol, and $\det(V,q) \in k^\times/(k^\times)^2$ denotes the image in $k^\times/(k^\times)^2$ of the determinant of the moment matrix $\{(x_i, x_j)\}_{1 \le i,j \le \dim V}$ for any basis $\{x_i\}_{1 \le i \le \dim V}$ of V over k. It is independent of the choice of the basis.

2.1.2 Archimedean case

Now let (V,q) be a quadratic space over $\mathbb{R}$. A Schwartz function Φ on V is an infinitely differentiable complex-valued function on V such that all the partial derivatives of any order are of rapid decay. We explain it as follows. Choose a basis of V, and view V as a Euclidean space of dimension $d = \dim_{\mathbb{R}} V$. Then we have coordinate functions $t_1, \cdots, t_d$. The requirement for Φ is that, for any d-tuples $(e_1, \cdots, e_d)$ and $(e_1', \cdots, e_d')$ of non-negative integers, the partial derivative $\partial\Phi/(\partial^{e_1}t_1 \cdots \partial^{e_d}t_d)$ exists everywhere, and

$$\sup_{(t_1, \cdots, t_d)} \left| t_1^{e_1'} \cdots t_d^{e_d'} \frac{\partial\Phi}{\partial^{e_1}t_1 \cdots \partial^{e_d}t_d} \right| < \infty.$$

The definition does not depend on the choice of the coordinate functions.

Denote by $S(V)$ the space of Schwartz functions on V. The Weil representation r of the group $\widetilde{SL}_2(\mathbb{R}) \times O(\mathbb{R})$ on V is defined by the same formulae as in the non-archimedean case.

It is usually more convenient to consider a smaller space called the Fock model. The theory depends on an orthogonal decomposition $V = V^+ + V^-$ such that the restrictions of q on $V^\pm$ are positive and negative definite respectively, which we fix once for all.

The Fock model $\mathcal{S}(V)$ is the space of functions on V of the form

$$\Phi(x) = P(x)e^{-2\pi(q(x^+)-q(x^-))}, \qquad x = x^+ + x^-,\, x^\pm \in V^\pm.$$

Here P can be any polynomial function on V (via any coordinate functions). It is a linear subspace of $S(V)$, and it is obviously not stable under the action of $SL_2(\mathbb{R})$. However, it has an action of $(\mathcal{G}, K)$ in the sense of Harish-Chandra.

Recall the Lie algebra $\mathcal{G} = sl_2(\mathbb{R}) \times o(V)$ and the maximal compact subgroup $K = \widetilde{SO}_2(\mathbb{R}) \times K^0$. The group $\widetilde{SO}_2(\mathbb{R})$ is the preimage of $SO_2(\mathbb{R})$ in $\widetilde{SL}_2(\mathbb{R})$, and the group $K^0 = O(V^+) \times O(V^-)$ is the stabilizer of the orthogonal decomposition $V = V^+ + V^-$.

By the formulae of the Weil representation above, for any $\Phi \in \mathcal{S}(V)$ and $(g, h) \in \widetilde{SL}_2(\mathbb{R}) \times O(V)$, the action $r(g, h)\Phi$ gives a smooth (rapid-decay) function on V. This function lies in $\mathcal{S}(V)$ for $(g, h) \in K$, but it is not necessarily true for general (g, h). However, it is easy to check that the action of any $(\partial_1, \partial_2) \in sl_2(\mathbb{R}) \times o(V)$ defined by

$$r(\partial_1, \partial_2)\Phi := \frac{d}{dt_1}\Big|_{t_1=0} \frac{d}{dt_2}\Big|_{t_2=0} r(e^{t_1\partial_1}, e^{t_2\partial_2})\Phi$$

still gives an element of $\mathcal{S}(V)$. Therefore, we get a well-defined action of $(\mathcal{G}, K)$ on $\mathcal{S}(V)$.

If (V, q) is (positive or negative) definite, then $O(V)$ is compact and $K^0 = O(V)$, and the Weil representation gives a full action of $O(V)$ on $\mathcal{S}(V)$. Functions in $\mathcal{S}(V)$ are of the form

$$\Phi(x) = P(x)e^{-2\pi|q(x)|}, \qquad x \in V.$$

The distinguished element given by $P(x) = 1$ is just the usual *Gaussian*, and also called *the standard Schwartz function*.

The indefinite case is used at few places in this book. We will omit the case $k \simeq \mathbb{C}$ since it is also only used in Chapter 2. Readers interested in that case may find details in [Wa].

2.1.3 Extension to groups of similitudes

Assume that $\dim V$ is even for simplicity. Following Waldspurger [Wa], we extend this action to an action r of $GL_2(k) \times GO(V)$ in the non-archimedean case and an action of $(\mathcal{G}, K)$ in the real case. In §4.1, we will see a slightly different space of Schwartz functions in the archimedean case.

If k is non-archimedean, let $\mathcal{S}(V \times k^\times)$ be the space of Schwartz–Bruhat functions, i.e., complex-valued locally constant and compactly supported functions on $V \times k^\times$. We also write it as $\overline{\mathcal{S}}(V \times k^\times)$ in the non-archimedean case. The Weil representation is extended by the following formulae:

- $r(h)\Phi(x, u) = \Phi(h^{-1}x, \nu(h)u), \qquad h \in \mathrm{GO}(k);$

- $r(g)\Phi(x, u) = r_u(g)\Phi(x, u), \qquad g \in \mathrm{SL}_2(k);$

- $r(d(a))\Phi(x, u) = \Phi(x, a^{-1}u)|a|^{-\dim V/4}, \qquad a \in k^\times.$

Here $\nu : \mathrm{GO}(k) \to k^\times$ denotes the similitude map. In the right-hand side of the second formula $\Phi(x, u)$ is viewed as a function of x by fixing u, and r_u is the Weil representation on V with new norm uq.

If $k = \mathbb{R}$ is real, let $S(V \times \mathbb{R}^\times)$ be the space of Schwartz functions on $V \times \mathbb{R}^\times$, i.e., infinitely differentiable complex-valued functions $\Phi(x, u)$ which have compact support for u in $\mathbb{R}^\times$ and give Schwartz functions on $x \in V$ for any fixed u. The Weil representation r of $\mathrm{GL}_2(k) \times \mathrm{GO}(V)$ on $S(V \times \mathbb{R}^\times)$ is defined by similar formulae.

The Fock model $\mathcal{S}(V \times \mathbb{R}^\times)$ is the space of finite linear combinations of functions of the form

$$H(u)P(x)e^{-2\pi|u|(q(x^+)-q(x^-))}$$

where $P : V \to \mathbb{C}$ is any polynomial function on V, and H is any compactly supported smooth function on $\mathbb{R}^\times$. Similar to the case of $\mathcal{S}(V)$, the formulae give a smooth function $r(g, h)\Phi$ on $V \times \mathbb{R}^\times$ for any $\Phi \in \mathcal{S}(V \times \mathbb{R}^\times)$ and $(g, h) \in \mathrm{GL}_2(\mathbb{R}) \times \mathrm{GO}(\mathbb{R})$, and induce an action of $(\mathfrak{gl}_2(\mathbb{R}), \mathfrak{go}(\mathbb{R})) \times (O_2(\mathbb{R}), K^0)$ on $\mathcal{S}(V \times \mathbb{R}^\times)$. In this book we always consider the Fock model instead of the whole space of Schwartz functions.

By the action of $\mathrm{GO}(k)$ above, we introduce an action of $\mathrm{GO}(k)$ on $V \times k^\times$ given by

$$h \circ (x, u) := (hx, \nu(h)^{-1}u).$$

This action stabilizes the subset

$$(V \times k^\times)_a := \{(x, u) \in V \times k^\times : uq(x) = a\}.$$

2.1.4　Global case

Now we assume that F is a number field and that (V, q) is a quadratic space over F. Then we can define a Weil representation r on $\mathcal{S}(V_\mathbb{A})$ (which actually depends on ψ) of $\widetilde{\mathrm{SL}}_2(\mathbb{A}) \times O(V_\mathbb{A})$. When $\dim V$ is even, we can define an action r of $\mathrm{GL}_2(\mathbb{A}) \times \mathrm{GO}(V_\mathbb{A})$ on $\mathcal{S}(V_\mathbb{A} \times \mathbb{A}^\times)$ which is the restricted tensor product of $\mathcal{S}(V_v \times F_v)$ with spherical element given by the characteristic function of $V_{O_{F_v}} \times O_{F_v}^\times$ once a global lattice is chosen.

Notice that the representation r depends only on the quadratic space $(V_{\mathbb{A}}, q)$ over $\mathbb{A}$. We may define representations directly for a pair $(\mathbb{V}, q)$ of a free $\mathbb{A}$-module $\mathbb{V}$ with non-degenerate quadratic form $q : \mathbb{V} \to \mathbb{A}$. It still makes sense to define $\mathcal{S}(\mathbb{V} \times \mathbb{A}^\times)$ to be the restricted tensor product of $\mathcal{S}(\mathbb{V}_v \times F_v^\times)$. The Weil representation extends in this case.

If $(\mathbb{V}, q)$ is a base change of an orthogonal space over F, then we call this Weil representation *coherent*; otherwise it is called *incoherent*.

2.1.5　Siegel–Weil formula

Let F be a number field, and (V, q) a quadratic space over F. Then for any $\Phi \in \mathcal{S}(V_{\mathbb{A}})$, we can form a theta series as a function on $\mathrm{SL}_2(F)\backslash\widetilde{\mathrm{SL}}_2(\mathbb{A}) \times O(V)\backslash O(V_{\mathbb{A}})$:

$$\theta(g, h, \Phi) = \sum_{x \in V} r(g, h)\Phi(x), \qquad (g, h) \in \widetilde{\mathrm{SL}}_2(\mathbb{A}) \times O(V_{\mathbb{A}}).$$

Similarly, when V has even dimension we can define theta series for $\Phi \in \mathcal{S}(V_{\mathbb{A}} \times \mathbb{A}^\times)$ as an automorphic form on $\mathrm{GL}_2(F)\backslash\mathrm{GL}_2(\mathbb{A}) \times \mathrm{GO}(V)\backslash\mathrm{GO}(V_{\mathbb{A}})$:

$$\theta(g, h, \Phi) = \sum_{(x,u) \in V \times F^\times} r(g, h)\Phi(x, u).$$

Now we introduce the Siegel Eisenstein series. For $\Phi \in \mathcal{S}(V_{\mathbb{A}})$ and $s \in \mathbb{C}$, we have a section

$$g \longmapsto \delta(g)^s r(g)\Phi(0)$$

in the induced representation $\mathrm{Ind}_{P^1}^{\widetilde{\mathrm{SL}}_2}(\chi_V | \cdot |^{s + \frac{\dim V}{2}})$ defined by

$$\left\{ f : \widetilde{\mathrm{SL}}_2(\mathbb{A}) \to \mathbb{C} \ \middle| \ f\left(\begin{pmatrix} a & b \\ & a^{-1} \end{pmatrix} g\right) = |a|^{s + \frac{\dim V}{2}} \chi_V(a) f(g) \right\}.$$

Here δ is the modulo function explained in the introduction. Thus we can form an Eisenstein series

$$E(s, g, \Phi) = \sum_{\gamma \in P^1(F)\backslash\mathrm{SL}_2(F)} \delta(\gamma g)^s r(\gamma g)\Phi(0).$$

It has a meromorphic continuation to $s \in \mathbb{C}$ and a functional equation with center $s = 1 - \frac{m}{2}$. Here we denote $m = \dim V$.

Denote by r the Witt index of V, i.e., the maximal dimension of F-subspaces of V consisting of elements of norms zero. Then we always have $r \leq \frac{m}{2}$. We call V anisotropic if $r = 0$, i.e., $q(x) = 0$ for $x \in V$ if and only if $x = 0$.

THEOREM 2.1 (Siegel–Weil formula). *Assume that (V, q) is anisotropic or $m - r > 2$. Then*

$$E(0, g, \Phi) = \kappa \int_{\mathrm{SO}(V)\backslash\mathrm{SO}(V_{\mathbb{A}})} \theta(g, h, \Phi) dh.$$

Here the integration uses the Haar measure of total volume one, and

$$\kappa = \begin{cases} 2 & \text{if } m = 1, 2; \\ 1 & \text{if } m > 2. \end{cases}$$

The theorem holds for more general reductive pairs. For the current situation, it was first treated by Siegel [Si] and Weil [We1, We2] for the case $m > 4$, and completed by Rallis [Ra] and Kudla–Rallis [KR1, KR2] (for more general symplectic groups). For a more detailed history, we refer to Kudla [Ku4]. Usually the integration is stated over $O(V)\backslash O(V_{\mathbb{A}})$, but it is easy to see that the integration over $SO(V)\backslash SO(V_{\mathbb{A}})$ gives the same result.

The theorem implicitly states that the Eisenstein series $E(s, g, \Phi)$ is analytic at $s = 0$, and the integration on the right-hand side converges absolutely. Notice that both sides of the equality in the theorem define elements in the space

$$\text{Hom}_{SO(V_{\mathbb{A}}) \times \widetilde{SL}_2(\mathbb{A})}(\mathcal{S}(V(\mathbb{A})), C^{\infty}(SL_2(F)\backslash \widetilde{SL}_2(\mathbb{A}))).$$

One can show that this space is one-dimensional. Thus the most important part of the theorem is the constant κ.

Fix an element $u \in F^{\times}$. We are interested in the Siegel–Weil formula in the following cases:

(1) $(V, q_V) = (E, uq)$ with E a nontrivial quadratic field extension of F and $q = N_{E/F}$;

(2) $(V, q_V) = (B_0, uq)$ where B_0 is the subspace of trace-free elements of anonsplit quaternion algebra B over F and q is induced from the reduced norm on B;

(3) $(V, q_V) = (B, uq)$ with B anonsplit quaternion algebra over F and q the reduced norm.

The Siegel–Weil formula is always valid in (1), and it is valid in (2) and (3) if B is not the matrix algebra. The Tamagawa number of $SO(V)$ in the above cases are respectively $2L(1, \eta), 2, 2$. Then it is easy to convert the integration in terms of the integration using the Tamagawa measure.

2.1.6 Local Siegel–Weil formula

Consider the Fourier expansion of both sides of the Siegel–Weil formula of an anisotropic orthogonal space (V, q) over F. Recall that the Siegel–Weil formula asserts that

$$E(0, g, \Phi) = \frac{\kappa}{\text{vol}(SO(F)\backslash SO(\mathbb{A}))} \int_{SO(F)\backslash SO(\mathbb{A})} \theta(g, h, \Phi) dh.$$

Now we write both sides in terms of Fourier series.

The Eisenstein series has a Fourier expansion

$$E(s, g, \Phi) = \delta(g)^s r(g)\Phi(0) + W_0(s, g, \Phi) + \sum_{a \in F^\times} W_a(s, g, \Phi).$$

Here the a-th Whittaker function for any $a \in F$ is defined by

$$W_a(s, g, \Phi) = \int_{\mathbb{A}} \delta(wn(b)g)^s \, r(wn(b)g)\Phi(0) \, \psi(-ab)db.$$

One can define the local Whittaker integrals

$$W_{a,v}(s, g, \Phi_v) = \int_{F_v} \delta_v(wn(b)g)^s \, r(wn(b)g)\Phi_v(0) \, \psi(-ab)db.$$

It has a meromorphic continuation to all $s \in \mathbb{C}$.

Denote by SO_x the stabilizer of any $x \in V$ in SO as an algebraic group. The right-hand side of the Siegel–Weil formula is equal to

$$\int_{\mathrm{SO}(F)\backslash\mathrm{SO}(\mathbb{A})} \sum_{x \in V} r(g, h)\Phi(x)dh$$

$$= \int_{\mathrm{SO}(F)\backslash\mathrm{SO}(\mathbb{A})} \sum_{x \in \mathrm{SO}(F)\backslash V} \sum_{h' \in \mathrm{SO}_x(F)\backslash\mathrm{SO}(F)} r(g, h)\Phi(h'^{-1}x)dh$$

$$= \sum_{x \in \mathrm{SO}(F)\backslash V} \int_{\mathrm{SO}_x(F)\backslash\mathrm{SO}(\mathbb{A})} r(g, h)\Phi(x)dh$$

$$= \sum_{x \in \mathrm{SO}(F)\backslash V} \mathrm{vol}(\mathrm{SO}_x(F)\backslash\mathrm{SO}_x(\mathbb{A})) \int_{\mathrm{SO}_x(\mathbb{A})\backslash\mathrm{SO}(\mathbb{A})} r(g, h)\Phi(x)dh.$$

For any $a \in F^\times$, the Siegel–Weil formula yields an identity of the a-th Fourier coefficients of the two sides as follows:

$$W_a(0, g, \Phi) = \kappa \frac{\mathrm{vol}(\mathrm{SO}_{x_a}(F)\backslash\mathrm{SO}_{x_a}(\mathbb{A}))}{\mathrm{vol}(\mathrm{SO}(F)\backslash\mathrm{SO}(\mathbb{A}))} \int_{\mathrm{SO}_{x_a}(\mathbb{A})\backslash\mathrm{SO}(\mathbb{A})} r(g, h)\Phi(x_a)dh.$$

Here $x_a \in V$ is any fixed element of norm a. If such an x_a does not exist, the left-hand side is considered to be zero.

Note that both integrals above are products of local integrals. It follows that the global identity induces the local identity

$$W_{a,v}(0, g, \Phi_v) = c_v \int_{\mathrm{SO}_{x_a}(F_v)\backslash\mathrm{SO}(F_v)} r(g, h)\Phi_v(x_a)dh$$

at every place v of F. Here $c_v \neq 0$ is a constant independent of g, Φ_v. Actually Weil [We2] proved the Siegel–Weil formula by first showing such a local version. The delicate part is to determine the constant c_v (after normalizing the Haar measure on the right-hand side). Note that even if the global quadratic space is anisotropic, the local quadratic space can be isotropic.

We state the precise result for any quadratic space (V, q) over any local field k. For any $a \in k$ and $\Phi \in \mathcal{S}(V)$, denote

$$W_a(s, g, \Phi) = \int_k \delta(wn(b)g)^s \, r(wn(b)g)\Phi(0) \, \psi(-ab)db.$$

Denote by $V(a)$ the set of elements of V with norm a. If it is non-empty, then any $x_a \in V(a)$ gives a bijection $V(a) \cong \mathrm{SO}_{x_a}(k)\backslash\mathrm{SO}(k)$. Under this identity, $\mathrm{SO}(k)$-invariant measures of $V(a)$ correspond to Haar measures of $\mathrm{SO}_{x_a}(k)\backslash\mathrm{SO}(k)$. They are unique up to scalar multiples.

THEOREM 2.2 ([We2], local Siegel–Weil). *Let (V, q) be a quadratic space over a local field k. Then the following are true:*

(1) *There is a unique $\mathrm{SO}(k)$-invariant measure $d_a x$ of $V(a)$ for every $a \in k^\times$ such that*

$$\Psi(a) := \int_{V(a)} \Phi(x)d_a x$$

gives a continuous function for $a \in k^\times$, and that

$$\int_k \Psi(a)da = \int_V \Phi(x)dx.$$

Here da and dx are respectively the self-dual measures on k and V with respect to ψ.

(2) *With the above measure,*

$$W_a(0, g, \Phi) = \gamma(V, q) \int_{V(a)} r(g)\Phi(x)d_a x, \quad \forall a \in k^\times, \Phi \in \mathcal{S}(V).$$

The right-hand side is considered to be zero if $V(a)$ is empty.

The measure $d_a x$ is easy to determine for small groups in practice. The case $a = 0$ may be obtained by carefully taking the limit $a \to 0$. The result is similar but more complicated. We omit it here.

Remark. For each $a \in k^\times$ representable by V, both sides of the equality in the second part of the theorem define nonzero elements in the space

$$\mathrm{Hom}_{\mathrm{SO}(V)\times\widetilde{\mathrm{SL}}_2(F)}(\mathcal{S}(V), C^\infty(N(F)\backslash\widetilde{\mathrm{SL}}_2(F), \psi_a)),$$

where $C^\infty(N(F)\backslash\widetilde{\mathrm{SL}}_2(F), \psi_a)$ is the space of smooth functions on $\widetilde{\mathrm{SL}}_2(F)$ with character $\psi_a(x) := \psi(ax)$ under left action of $N \simeq k$. It can be shown that this space is one-dimensional. Thus the significance is still the ratio of these two elements. See [Ra, Proposition 4.2].

2.2   SHIMIZU LIFTING

Let F be a local or global field and B a quaternion algebra over F. Write $V = (B, q)$ as an orthogonal space with quadratic form q defined by the reduced norm on B. Let $B^\times \times B^\times$ act on V by

$$x \mapsto h_1 x h_2^{-1}, \qquad x \in V, \quad h_i \in B^\times.$$

Then we have an exact sequence:

$$1 \longrightarrow F^\times \longrightarrow (B^\times \times B^\times) \rtimes \{1, \iota\} \longrightarrow \mathrm{GO}(V) \longrightarrow 1.$$

Here $\iota : x \mapsto \bar{x}$ is the main involution on $V = B$, and on $B^\times \times B^\times$ by $(h_1, h_2) \mapsto (\bar{h}_2^{-1}, \bar{h}_1^{-1})$, and $F^\times$ is embedded into the group in the middle by $x \mapsto (x, x) \rtimes 1$.

The theta lifting of any representation σ of $\mathrm{GL}_2(F)$ to $\mathrm{GO}(V)$ is induced by the representation $\tilde{\pi} \otimes \pi$ on $B^\times \times B^\times$, where π is the Jacquet–Langlands correspondence of σ. Recall that $\pi \neq 0$ only if $B = M_2(F)$ or σ is discrete. In that case,

$$\dim_{\mathbb{C}} \mathrm{Hom}_{\mathrm{GL}_2(F) \times B^\times \times B^\times}(\sigma \otimes \mathcal{S}(V \times F^\times), \pi \otimes \tilde{\pi}) = 1.$$

In the global case, there is a canonical element in this one-dimensional space given by the theta lifting. We will introduce a normalized form in the local case compatible with the global case.

2.2.1   Global Shimizu lifting

Let F be a number field with $\mathbb{A}$ the adele ring, σ be a cuspidal automorphic representation of $\mathrm{GL}_2(\mathbb{A})$, and B be a quaternion algebra over F. Denote $V = (B, q)$ as above.

For any $\Phi \in \mathcal{S}(V(\mathbb{A}) \times \mathbb{A}^\times)$, we have the theta function

$$\theta(g, h, \Phi) = \sum_{u \in F^\times} \sum_{x \in V} r(g, h)\Phi(x, u), \quad g \in \mathrm{GL}_2(\mathbb{A}), \ h \in B_{\mathbb{A}}^\times \times B_{\mathbb{A}}^\times.$$

For any $\varphi \in \sigma$, define the normalized global Shimizu lifting

$$\theta(\Phi \otimes \varphi)(h) := \frac{\zeta_F(2)}{2L(1, \pi, \mathrm{ad})} \int_{\mathrm{GL}_2(F) \backslash \mathrm{GL}_2(\mathbb{A})} \varphi(g)\theta(g, h, \Phi)dg, \quad h \in B_{\mathbb{A}}^\times \times B_{\mathbb{A}}^\times.$$

$$(2.2.1)$$

It defines an automorphic form $\theta(\Phi \otimes \varphi) \in \pi \otimes \tilde{\pi}$. We will see why we use the normalizing factor.

If ω_σ is the central character of σ, then the central character of π and $\tilde{\pi}$ are respectively ω_σ and ω_σ^{-1}. Let

$$\mathcal{F} : \pi \otimes \tilde{\pi} \to \mathbb{C}$$

be the canonical map defined by the Petersson bilinear pairing

$$(f_1, f_2)_{\text{pet}} = \int_{B^\times \backslash B_\mathbb{A}^\times / Z(\mathbb{A})} f_1(g) f_2(g) dg, \quad f_1 \in \pi, \ f_2 \in \widetilde{\pi}.$$

The integration uses the Tamagawa measure, which gives $\text{vol}(B^\times \backslash B_\mathbb{A}^\times / Z(\mathbb{A})) = 2$.

In the following, let $W_{-1} : \sigma \to W(\sigma, \overline{\psi})$ be the canonical map from σ to its Whittaker model with respect to the additive character $\overline{\psi}$. It is given by the Fourier coefficient. Fix a decomposition $W_{-1} = \otimes_v W_{-1,v}$.

PROPOSITION 2.3 (Waldspurger). *For any $\varphi \in \sigma$ and $\Phi = \otimes_v \Phi_v \in \mathcal{S}(V(\mathbb{A}) \times \mathbb{A}^\times)$, one has*

$$\mathcal{F}\theta(\Phi \otimes \varphi) = \prod_v \frac{\zeta_v(2)}{L(1, \pi_v, \text{ad})} \int_{N(F_v) \backslash \mathrm{GL}_2(F_v)} W_{\varphi, -1, v}(g) r(g) \Phi_v(1, 1) dg.$$

Here the local factor is 1 for almost all places v.

PROOF. Here we only repeat Waldspurger's proof in the case that B is non-split so that the Siegel–Weil formula is applicable. For the split case we refer to the original paper [Wa], which takes a different method without the Siegel–Weil formula.

The pairing between π and $\widetilde{\pi}$ is given by integration on the diagonal of $(B^\times \backslash B_\mathbb{A}^\times / Z(\mathbb{A}))^2$. It follows that

$$\mathcal{F}\theta(\Phi \otimes \varphi) = \frac{\zeta_F(2)}{2L(1, \pi, \text{ad})} \int_{B^\times \backslash B_\mathbb{A}^\times / Z(\mathbb{A})} \int_{\mathrm{GL}_2(F) \backslash \mathrm{GL}_2(\mathbb{A})} \varphi(g) \theta(g, (h, h), \Phi) dg dh.$$

Let B_0 be the subspace of trace-free elements of B. Then $B = F \oplus B_0$ gives an orthogonal decomposition. The diagonal can be identified with $\text{SO}' = \text{SO}(B_0)$. Thus we have globally

$$\mathcal{F}\theta(\Phi \otimes \varphi) = \int_{\text{SO}'(F) \backslash \text{SO}'(\mathbb{A})} dh \int_{\mathrm{GL}_2(F) \backslash \mathrm{GL}_2(\mathbb{A})} \theta(g, h, \Phi) \varphi(g) dg.$$

The Siegel–Weil formula for the orthogonal space B_0 gives

$$\int_{\text{SO}'(F) \backslash \text{SO}'(\mathbb{A})} \theta(g, h, \Phi) dh = 2J(0, g, \Phi). \tag{2.2.2}$$

Here J is a mixed theta-Eisenstein series

$$J(s, g, \Phi) = \sum_{\gamma \in P(F) \backslash \mathrm{GL}_2(F)} \delta(\gamma g)^s \sum_{(x, u) \in F \times F^\times} r(\gamma g) \Phi(x, u).$$

To check the truth of (2.2.2), we can assume that $g \in \mathrm{SL}_2(\mathbb{A})$ because both sides depend only on $r(g)\Phi$. By linearity it suffices to consider the case $\Phi = \Phi_F \otimes \Phi_0$ for $\Phi_F \in \mathcal{S}(\mathbb{A} \times \mathbb{A}^\times)$ and $\Phi_0 \in \mathcal{S}(B_0(\mathbb{A}) \times \mathbb{A}^\times)$ in the sense that

$$\Phi(x, u) = \Phi_F(z, u) \Phi_0(x_0, u), \quad x = z \oplus x_0.$$

Then for any $h \in \mathrm{SO}'(\mathbb{A})$ and $g \in \mathrm{SL}_2(\mathbb{A}) \subset \widetilde{\mathrm{SL}}_2(\mathbb{A})$,

$$\theta(g, h, \Phi) = \sum_{u \in F^\times} \theta(g, u, \Phi_F)\theta(g, h, u, \Phi_0),$$

$$J(s, g, \Phi) = \sum_{u \in F^\times} \theta(g, u, \Phi_F)E(s, g, u, \Phi_0).$$

Here

$$\theta(g, u, \Phi_F) = \sum_{z \in F} r(g)\Phi_F(z, u),$$

$$\theta(g, h, u, \Phi_0) = \sum_{x_0 \in B_0} r(g, h)\Phi_0(x_0, u),$$

$$E(s, g, u, \Phi_0) = \sum_{\gamma \in P^1(F) \backslash \mathrm{SL}_2(F)} \delta(\gamma g)^s r(\gamma g)\Phi_0(0, u).$$

The (2.2.2) is reduced to the Siegel-Weil formula

$$\int_{\mathrm{SO}'(F) \backslash \mathrm{SO}'(\mathbb{A})} \theta(g, h, u, \Phi_0)dh = 2E(0, g, u, \Phi_0).$$

Go back to $\mathcal{F}\theta(\Phi \otimes \varphi)$. We have

$$\mathcal{F}\theta(\Phi \otimes \varphi) = \frac{\zeta_F(2)}{L(1, \pi, \mathrm{ad})} \int_{\mathrm{GL}_2(F) \backslash \mathrm{GL}_2(\mathbb{A})} \varphi(g)J(0, g, \Phi)dg.$$

Denote

$$A(s) = \int_{\mathrm{GL}_2(F) \backslash \mathrm{GL}_2(\mathbb{A})} \varphi(g)J(s, g, \Phi)dg.$$

By definition,

$$A(s) = \int_{\mathrm{GL}_2(F) \backslash \mathrm{GL}_2(\mathbb{A})} \varphi(g) \sum_{\gamma \in P(F) \backslash \mathrm{GL}_2(F)} \sum_{(z,u) \in F \times F^\times} \delta(\gamma g)^s r(\gamma g)\Phi(z, u)dg$$

$$= \int_{P(F) \backslash \mathrm{GL}_2(\mathbb{A})} \delta(g)^s \varphi(g) \sum_{(z,u) \in F \times F^\times} r(g)\Phi(z, u)dg$$

$$= A_1(s) + A_2(s).$$

Here

$$A_1(s) = \int_{P(F) \backslash \mathrm{GL}_2(\mathbb{A})} \delta(g)^s \varphi(g) \sum_{(z,u) \in F^\times \times F^\times} r(g)\Phi(z, u)dg,$$

$$A_2(s) = \int_{P(F) \backslash \mathrm{GL}_2(\mathbb{A})} \delta(g)^s \varphi(g) \sum_{u \in F^\times} r(g)\Phi(0, u)dg.$$

It is easy to see that $A_2(s) = 0$. In fact, since $\delta(g)^s r(g) \Phi(0, u)$ is invariant under the left action of $N(\mathbb{A})$, the vanishing of $A_2(s)$ is a consequence of the cuspidality property

$$\int_{N(F) \backslash N(\mathbb{A})} \varphi(ng) dn = 0.$$

The summation in $A_1(s)$ is a single orbit over the diagonal group of $\mathrm{GL}_2(F)$. Thus

$$A(s) = \int_{N(F) \backslash \mathrm{GL}_2(\mathbb{A})} \delta(g)^s \varphi(g) r(g) \Phi(1, 1) dg$$

$$= \int_{N(\mathbb{A}) \backslash \mathrm{GL}_2(\mathbb{A})} \delta(g)^s W_{\varphi, -1}(g) r(g) \Phi(1, 1) dg.$$

It is easy to check that for almost all v,

$$\int_{N(F_v) \backslash \mathrm{GL}_2(F_v)} \delta(g)^s W_{\varphi, -1, v}(g) r(g) \Phi_v(1, 1) dg = \frac{L(s + 1, \pi_v, \mathrm{ad})}{\zeta_v(2s + 2)}.$$

Thus the result follows. $\qquad\qquad\qquad\qquad\qquad\qquad\qquad\qquad\qquad\qquad\qquad$ $\square$

2.2.2 Normalization of the local Shimizu lifting

Now we go back to local situation. Let F be a local field and σ be an irreducible infinite representation of $\mathrm{GL}_2(F)$. We realize σ as a subspace $\mathcal{W}(\sigma, \overline{\psi})$ of smooth functions on $\mathrm{GL}_2(F)$ with character $\overline{\psi}$ under the left translation of $N(F)$.

Let B be any quaternion algebra over F which gives a quadratic space. Now we normalize the local theta lifting

$$\theta : \mathcal{W}(\sigma, \overline{\psi}) \otimes \mathcal{S}(B \otimes F^\times) \longrightarrow \pi \otimes \widetilde{\pi}$$

by

$$\mathcal{F}\theta(W \otimes \Phi) = \frac{\zeta_F(2)}{L(1, \pi, \mathrm{ad})} \int_{N(F) \backslash \mathrm{GL}_2(F)} W(g) \ r(g) \Phi(1, 1) dg. \qquad (2.2.3)$$

Here $\mathcal{F} : \pi \otimes \widetilde{\pi} \to \mathbb{C}$ is the canonical contraction.

In the global case $V = B$ over a number field, we will have a compatibility of the local and global Shimizu liftings. Note that globally $W_{-1} : \sigma \to \mathcal{W}(\sigma, \overline{\psi})$ is uniquely determined as taking the first Fourier coefficients and $\mathcal{F} : \pi \otimes \widetilde{\pi} \to \mathbb{C}$ is defined in terms of the Petersson pairing. Fix compatible decompositions $\sigma = \otimes_v \sigma_v$ and $W_{-1} = \otimes_v W_{-1, v}$. Fix a decomposition $\mathcal{F} = \otimes_v \mathcal{F}_v$ compatible with $\pi = \otimes_v \pi_v$ and $\widetilde{\pi} = \otimes_v \widetilde{\pi}_v$. Normalize the global θ by (2.2.1), and normalize the local θ_v by (2.2.3). Then we get the following compatibility.

COROLLARY 2.4. *We have a decomposition* $\theta = \otimes \theta_v$ *in*

$$\mathrm{Hom}_{\mathrm{GL}_2(\mathbb{A}) \times B_\mathbb{A}^\times \times B_\mathbb{A}^\times}(\sigma \otimes \mathcal{S}(V(\mathbb{A}) \times \mathbb{A}^\times), \pi \otimes \widetilde{\pi}).$$

2.3   INTEGRAL REPRESENTATIONS OF THE *L*-FUNCTION

In the following we want to describe an integral representation of the L-function $L(s, \pi, \chi)$. Let F be a number field with ring of adeles $\mathbb{A}$. Let $\mathbb{B}$ be a quaternion algebra with ramification set Σ. Fix an embedding $E_\mathbb{A} \hookrightarrow \mathbb{B}$. We have an orthogonal decomposition

$$\mathbb{B} = E_\mathbb{A} + E_\mathbb{A}\mathbf{j}, \qquad \mathbf{j}^2 \in \mathbb{A}^\times.$$

Write $\mathbb{V}$ for the orthogonal space $\mathbb{B}$ with reduced norm q, and $\mathbb{V}_1 = E_\mathbb{A}$ and $\mathbb{V}_2 = E_\mathbb{A}\mathbf{j}$ as subspaces of $\mathbb{V}$. Then $\mathbb{V}_1$ is coherent and it is the adelization of the F-space $V_1 := (E, q)$, and $\mathbb{V}_2$ is coherent if and only if Σ is even.

For $\Phi \in \mathcal{S}(\mathbb{V} \times \mathbb{A}^\times)$, we can form a mixed Eisenstein–theta series

$$I(s, g, \Phi) = \sum_{\gamma \in P(F)\backslash \mathrm{GL}_2(F)} \delta(\gamma g)^s \sum_{(x_1, u) \in V_1 \times F^\times} r(\gamma g)\Phi(x_1, u).$$

Define its χ-component:

$$I(s, g, \chi, \Phi) = \int_{T(F)\backslash T(\mathbb{A})} \chi(t)I(s, g, r(t, 1)\Phi)dt.$$

For any $\varphi \in \sigma$, we introduce the Petersson pairing

$$P(s, \chi, \Phi, \varphi) = \int_{Z(\mathbb{A})\mathrm{GL}_2(F)\backslash \mathrm{GL}_2(\mathbb{A})} \varphi(g)I(s, g, \chi, \Phi)dg. \qquad (2.3.1)$$

Here the integration uses the Tamagawa measure.

PROPOSITION 2.5 (Waldspurger). *If $\Phi = \otimes\Phi_v$ and $\varphi = \otimes\varphi_v$ are decomposable, then*

$$P(s, \chi, \Phi, \varphi) = \prod_v P_v(s, \chi_v, \Phi_v, \varphi_v)$$

where

$$\begin{aligned} &P_v(s, \chi_v, \Phi_v, \varphi_v) \\ &= \int_{Z(F_v)\backslash T(F_v)} \chi(t)dt \int_{N(F_v)\backslash \mathrm{GL}_2(F_v)} \delta(g)^s W_{-1,v}(g)r(g)\Phi_v(t^{-1}, q(t))dg. \end{aligned}$$

Here W_{-1} denotes the Whittaker function of φ with respect to $\bar{\psi}$.

PROOF. Bring the definition formula of $I(s, g, \chi, \Phi)$ to obtain an expression for $P(s, \chi, \Phi, \varphi)$:

$$\int_{Z(\mathbb{A})P(F)\backslash \mathrm{GL}_2(\mathbb{A})} \varphi(g)\delta(g)^s \int_{T(F)\backslash T(\mathbb{A})} \chi(t) \sum_{(x, u) \in V_1 \times F^\times} r(g, (t, 1))\Phi(x, u)dtdg.$$

We decompose the first integral as a double integral

$$\int_{Z(\mathbb{A})P(F)\backslash\mathrm{GL}_2(\mathbb{A})} dg = \int_{Z(\mathbb{A})N(\mathbb{A})P(F)\backslash\mathrm{GL}_2(\mathbb{A})} \int_{N(F)\backslash N(\mathbb{A})} dndg,$$

and perform the integral on $N(F)\backslash N(\mathbb{A})$ to obtain

$$\int_{Z(\mathbb{A})N(\mathbb{A})P(F)\backslash\mathrm{GL}_2(\mathbb{A})} \delta(g)^s dg \int_{T(F)\backslash T(\mathbb{A})} \chi(t)$$
$$\sum_{(x,u)\in V_1\times F^\times} W_{-q(x)u}(g)r(g,(t,1))\Phi(x,u)dt.$$

Here, as φ is cuspidal, the term $x = 0$ has no contribution to the integral. In this way, we may change variable $(x,u) \mapsto (x, q(x^{-1})u)$ to obtain the following expression of the sum:

$$\sum_{(x,u)\in E^\times\times F^\times} W_{-u}(g)r(g,(t,1))\Phi(x,q(x^{-1})u)$$
$$= \sum_{(x,u)\in E^\times\times F^\times} W_{-u}(g)r(g,(tx,1))\Phi(1,u).$$

The sum over $x \in E^\times$ collapses with quotient $T(F) = E^\times$. Thus the integral becomes

$$\int_{Z(\mathbb{A})N(\mathbb{A})P(F)\backslash\mathrm{GL}_2(\mathbb{A})} \delta(g)^s dg \int_{T(\mathbb{A})} \chi(t) \sum_{u\in F^\times} W_{-u}(g)r(g,(t,1))\Phi(1,u)dt.$$

The expression does not change if we make the substitution

$$(g, au) \mapsto (gd(a)^{-1}, u).$$

Thus we have

$$\int_{Z(\mathbb{A})N(\mathbb{A})P(F)\backslash\mathrm{GL}_2(\mathbb{A})} \delta(g)^s dg \int_{T(\mathbb{A})} \chi(t)$$
$$\sum_{u\in F^\times} W_{-u}(d(u^{-1})g)r(d(u^{-1})g,(t,1))\Phi(1,u)dt.$$

The sum over $u \in F^\times$ collapses with quotient $P(F)$, so we obtain the following expression:

$$P(s,\chi,\Phi,\varphi) = \int_{Z(\mathbb{A})N(\mathbb{A})\backslash\mathrm{GL}_2(\mathbb{A})} \delta(g)^s dg \int_{T(\mathbb{A})} \chi(t)W_{-1}(g)r(g)\Phi(t^{-1},q(t))dt.$$

We may decompose the inside integral as

$$\int_{Z(\mathbb{A})\backslash T(\mathbb{A})} \int_{Z(\mathbb{A})}$$

and move the integral $\int_{Z(\mathbb{A})\backslash T(\mathbb{A})}$ to the outside. Then we use the fact that $\omega_\sigma \cdot \chi|_{\mathbb{A}^\times} = 1$ to obtain

$$P(s, \chi, \Phi, \varphi) = \int_{Z(\mathbb{A})\backslash T(\mathbb{A})} \chi(t) dt \int_{N(\mathbb{A})\backslash \mathrm{GL}_2(\mathbb{A})} \delta(g)^s W_{-1}(g) r(g) \Phi(t^{-1}, q(t)) dg.$$

$\square$

When everything is unramified, Waldspurger has computed these integrals and gotten

$$P_v(s, \chi_v, \Phi_v, \varphi_v) = \frac{L((s+1)/2, \pi_v, \chi_v)}{L(s+1, \eta_v)}.$$

Thus we may define a normalized integral P_v° by

$$P_v(s, \chi_v, \Phi_v, \varphi_v) = \frac{L((s+1)/2, \pi_v, \chi_v)}{L(s+1, \eta_v)} P_v^\circ(s, \chi_v, \Phi_v, \varphi_v).$$

This normalized local integral P_v° will be regular at $s = 0$ and equal to

$$\frac{L(1, \eta_v) L(1, \pi_v, \mathrm{ad})}{\zeta_v(2) L(1/2, \pi_v, \chi_v)} \int_{Z(F_v)\backslash T(F_v)} \chi_v(t) \mathcal{F}(\pi(t)\theta(\Phi_v \otimes \varphi_v)) \, dt.$$

We may write this as $\alpha_v(\theta(\Phi_v \otimes \varphi_v))$ with

$$\alpha_v \in \mathrm{Hom}(\pi_v \otimes \widetilde{\pi}_v, \mathbb{C})$$

given by the integration of matrix coefficients:

$$\alpha_v(f_1 \otimes f_2) = \frac{L(1, \eta_v) L(1, \pi_v, \mathrm{ad})}{\zeta_v(2) L(1/2, \pi_v, \chi_v)} \int_{Z(F_v)\backslash T(F_v)} \chi_v(t)(\pi(t)f_1, f_2) dt.$$

We define the global element $\alpha := \otimes_v \alpha_v$ in $\mathrm{Hom}(\pi \otimes \widetilde{\pi}, \mathbb{C})$.

We now take value or derivative at $s = 0$ to obtain

PROPOSITION 2.6.

$$P(0, \chi, \Phi, \varphi) = \frac{L(1/2, \pi, \chi)}{L(1, \eta)} \prod_v \alpha_v(\theta(\Phi_v \otimes \varphi_v)).$$

If Σ is odd, then $L(1/2, \pi, \chi) = 0$, and

$$P'(0, \chi, \Phi, \varphi) = \frac{L'(1/2, \pi, \chi)}{2L(1, \eta)} \prod_v \alpha_v(\theta(\Phi_v \otimes \varphi_v)).$$

Remark. Let $\mathcal{A}_\Sigma(\mathrm{GL}_2, \chi)$ denote the direct sum of cuspidal automorphic representations σ on $\mathrm{GL}_2(\mathbb{A})$ such that $\Sigma(\sigma, \chi) = \Sigma$. If Σ is even, let $\mathcal{I}(g, \chi, \Phi)$ be the projection of $I(0, g, \chi, \Phi)$ on $\mathcal{A}_\Sigma(\mathrm{GL}_2, \chi^{-1})$. If Σ is odd, let $\mathcal{I}'(g, \chi, \Phi)$

denote the projection of $I'(0, g, \chi, \Phi)$ on $\mathcal{A}_\Sigma(\mathrm{GL}_2, \chi^{-1})$. Then we have shown that $\mathcal{I}(g, \chi, \Phi)$ and $\mathcal{I}'(g, \chi, \Phi)$ represent the functionals

$$\varphi \mapsto \frac{L(1/2, \pi, \chi)}{L(1, \eta)} \alpha(\theta(\Phi \otimes \varphi)) \qquad \text{if } \Sigma \text{ is even,}$$

$$\varphi \mapsto \frac{L'(1/2, \pi, \chi)}{2L(1, \eta)} \alpha(\theta(\Phi \otimes \varphi)) \qquad \text{if } \Sigma \text{ is odd,}$$

on σ respectively.

2.4   PROOF OF WALDSPURGER FORMULA

Assume now Σ is even. We are going to sketch Waldspurger's proof of his central value formula in Theorem 1.4. Now the space $\mathbb{V} = V(\mathbb{A})$ is coming from a global $V = B$ over F.

Recall that the Shimizu lifting $\theta(\Phi \otimes \varphi) \in \pi \otimes \widetilde{\pi}$, for any $\Phi \in \mathcal{S}(\mathbb{V} \times \mathbb{A}^\times)$ and $\varphi \in \sigma$, is defined as

$$\theta(\Phi \otimes \varphi)(h) = \frac{\zeta_F(2)}{2L(1, \pi, \mathrm{ad})} \int_{\mathrm{GL}_2(F)\backslash \mathrm{GL}_2(\mathbb{A})} \varphi(g)\theta(g, h, \Phi)dg, \quad h \in B_\mathbb{A}^\times \times B_\mathbb{A}^\times.$$

We are going to compute

$$4L(1, \eta)^2 \cdot (P_{\pi,\chi} \otimes P_{\widetilde{\pi},\chi^{-1}})(\theta(\Phi \otimes \varphi))$$

$$= \int_{(T(F)Z(\mathbb{A})\backslash T(\mathbb{A}))^2} \theta(\Phi \otimes \varphi)(t_1, t_2)\chi(t_1)\chi^{-1}(t_2)dt_1 dt_2.$$

Here the factor $4L(1, \eta)^2$ comes from the Tamagawa measure

$$\mathrm{vol}(T(F)Z(\mathbb{A})\backslash T(\mathbb{A})) = 2L(1, \eta).$$

By definition, the right-hand side equals

$$\frac{\zeta_F(2)}{2L(1, \pi, \mathrm{ad})} \int_{Z(\mathbb{A})\mathrm{GL}_2(F)\backslash \mathrm{GL}_2(\mathbb{A})} \varphi(g)\theta(g, \Phi, \chi)dg$$

where

$$\theta(g, \chi, \Phi) = \int_{Z^\Delta(\mathbb{A})T(F)^2\backslash T(\mathbb{A})^2} \theta(g, (t_1, t_2), \Phi)\chi(t_1 t_2^{-1})dt_1 dt_2.$$

Here Z^Δ is the image of the diagonal embedding $F^\times \hookrightarrow (B^\times)^2$. We change the variables by $t_1 = tt_2$ to get a double integral

$$\theta(g, \chi, \Phi) = \int_{T(F)\backslash T(\mathbb{A})} \chi(t)dt \int_{T(F)Z(\mathbb{A})\backslash T(\mathbb{A})} \theta(g, (tt_2, t_2), \Phi)dt_2.$$

Notice that the diagonal embedding of $Z\backslash T$ in $B^\times \times B^\times$ can be realized as $\mathrm{SO}(Ej, q)$ in the decomposition $B = E + Ej$. See §1.6 for more details. Apply the Siegel–Weil formula in Theorem 2.1 to the space (Ej, uq) to obtain

$$\theta(g, \chi, \Phi) = L(1, \eta)I(0, g, \chi, \Phi).$$

Here we recall that

$$I(s, g, \chi, \Phi) = \int_{T(F)\backslash T(\mathbb{A})} \chi(t)I(s, g, r(t, 1)\Phi)dt$$

with

$$I(s, g, \Phi) = \sum_{\gamma \in P(F)\backslash \mathrm{GL}_2(F)} \delta(\gamma g)^s \sum_{(x,u) \in E \times F^\times} r(\gamma g)\Phi(x, u).$$

Combining with Proposition 2.6, we have

$$\int_{(T(F)Z(\mathbb{A})\backslash T(\mathbb{A}))^2} \theta(\Phi \otimes \varphi)(t_1, t_2)\chi(t_1)\chi^{-1}(t_2)dt_1 dt_2$$

$$= \frac{\zeta_F(2)L(1/2, \pi, \chi)}{2L(1, \pi, \mathrm{ad})}\alpha(\theta(\Phi \otimes \varphi)).$$

Thus we have obtained

$$4L(1, \eta)^2 \cdot (P_{\pi, \chi} \otimes P_{\widetilde{\pi}, \chi^{-1}})(\theta(\Phi \otimes \varphi)) = \frac{\zeta_F(2)L(1/2, \pi, \chi)}{2L(1, \pi, \mathrm{ad})}\alpha(\theta(\Phi \otimes \varphi)).$$

In terms of functionals on $\pi \otimes \widetilde{\pi}$, the above identity can be written as

$$4L(1, \eta)^2 \cdot P_{\pi, \chi} \otimes P_{\widetilde{\pi}, \chi^{-1}} = \frac{\zeta_F(2)L(1/2, \pi, \chi)}{2L(1, \pi, \mathrm{ad})} \cdot \alpha.$$

It proves Theorem 1.4.

2.5 INCOHERENT EISENSTEIN SERIES

Let $(\mathbb{V}, q)$ be an orthogonal space over $\mathbb{A}$, and let $\Phi \in \mathcal{S}(\mathbb{V})$ be a Schwartz function. Recall the associated Siegel–Eisenstein series

$$E(s, g, \Phi) = \sum_{\gamma \in P^1(F)\backslash \mathrm{SL}_2(F)} \delta(\gamma g)^s r(\gamma g)\Phi(0), \quad g \in \widetilde{\mathrm{SL}}_2(\mathbb{A}).$$

It has a meromorphic continuation to $s \in \mathbb{C}$ and a functional equation with center $s = 1 - \frac{\dim V}{2}$.

By the standard theory, we have

$$E(s, g, \Phi) = \delta(g)^s r(g)\Phi(0) + W_0(s, g, \Phi) + \sum_{a \in F^\times} W_a(s, g, \Phi).$$

Here the a-th Whittaker function for any $a \in F$ is defined by

$$W_a(s, g, \Phi) = \int_{\mathbb{A}} \delta(wn(b)g)^s \, r(wn(b)g)\Phi(0) \, \psi(-ab)db.$$

One can define the local Whittaker integrals

$$W_{a,v}(s, g, \Phi_v) = \int_{F_v} \delta_v(wn(b)g)^s \, r(wn(b)g)\Phi_v(0) \, \psi(-ab)db.$$

It has a meromorphic continuation to all $s \in \mathbb{C}$.

If $(\mathbb{V}, q)$ is coherent, then the Siegel–Weil formula (in most convergent cases) gives an expression of $E(0, g, \Phi)$ in terms of the integral of the corresponding theta series.

If $(\mathbb{V}, q)$ is incoherent, then there is no theta series available. However, we can still compute (the Fourier coefficients of) $E(0, g, \Phi)$ in terms of the local Siegel–Weil formula. In fact, Theorem 2.2 asserts a way to write the local Whittaker functions $W_{a,v}(0, g, \Phi_v)$ in terms of averages of the Schwartz functions. Parallel to Theorem 2.1, we are particularly interested in certain subspaces of quaternion algebras.

Let $\mathbb{B}$ be an incoherent quaternion algebra over a totally real number field F. Assume that $\mathbb{B}$ is totally definite at infinity. As usual let q be the reduced norm on $\mathbb{B}$. Fix a number $u \in F^\times$. We have the following three cases:

(1) $(\mathbb{V}, q_{\mathbb{V}}) = (E_{\mathbb{A}}j, uq)$ where $E_{\mathbb{A}}j$ is the orthogonal complement of $E_{\mathbb{A}}$ in $\mathbb{B}$ for some quadratic field extension E of F with a fixed embedding $E_{\mathbb{A}} \hookrightarrow \mathbb{B}$ of $\mathbb{A}$-algebras;

(2) $(\mathbb{V}, q_{\mathbb{V}}) = (\mathbb{B}_0, uq)$ where $\mathbb{B}_0$ is the subspace of trace-free elements of $\mathbb{B}$;

(3) $(\mathbb{V}, q_{\mathbb{V}}) = (\mathbb{B}, uq)$.

We will see that the Eisenstein series $E(s, g, \Phi)$ are always holomorphic at $s = 0$ in these cases. If we take Φ_∞ to be the standard Gaussian, the related Eisenstein series in these cases have weights 1, 3/2 and 2 respectively. So we refer these cases as weight 1, weight 3/2 and weight 2, though our results are true for general Φ_∞ and the weights are different from these numbers in general.

The global Weil index $\gamma(\mathbb{V}, q_{\mathbb{V}}) = -1$ in all three cases. In fact, we first observe that it does not depend on u since it lies in $F^\times$. Then we see that $\gamma(\mathbb{B}, uq) = \varepsilon(\mathbb{B}) = -1$ by definition. Here $\varepsilon(\mathbb{B})$ is the Hasse invariant. In cases (1) and (2), the orthogonal complement $(\mathbb{V}^\perp, q_{\mathbb{V}})$ of $(\mathbb{V}, q_{\mathbb{V}})$ in $(\mathbb{B}, uq)$ is coherent, and thus $\gamma(\mathbb{V}^\perp, q_{\mathbb{V}}) = 1$. It follows that

$$\gamma(\mathbb{V}, q_{\mathbb{V}}) = \gamma(\mathbb{B}, uq)/\gamma(\mathbb{V}^\perp, q_{\mathbb{V}}) = -1.$$

2.5.1   Eisenstein series of weight one

Consider the first case $(\mathbb{V}, q_{\mathbb{V}}) = (E_{\mathbb{A}}\mathfrak{j}, uq)$. The quadratic space is isomorphic to $(E_{\mathbb{A}}, uq(\mathfrak{j})q)$, but we will still write everything based on the space $\mathbb{V}_2 = E_{\mathbb{A}}\mathfrak{j}$.

The center of symmetry of the Eisenstein series $E(s, g, \Phi)$ is exactly the point $s = 0$. We consider the (local) Siegel–Weil formula. It is the most important case in this book. We will see that $E(0, g, \Phi) = 0$ by some essentially local reason. Then it will be more interesting for us to take the derivative $E'(0, g, \Phi)$. See §6.1 for the formula of the derivative, and the context after that for the relation of the derivative with height pairings of CM points.

Recall that

$$W_{a,v}(s, g, \Phi_v) = \int_{F_v} \delta_v(wn(b)g)^s \; r(wn(b)g)\Phi_v(0) \; \psi(-ab)db, \quad a \in F_v.$$

For the case $a = 0$, we use the normalization

$$W_{0,v}^{\circ}(s, g, \Phi_v) \;\; = \;\; \frac{L(s+1, \eta_v)}{L(s, \eta_v)} W_{0,v}(s, g, \Phi_v).$$

Notice that the normalizing factor $\dfrac{L(s+1, \eta_v)}{L(s, \eta_v)}$ has a zero at $s = 0$ when E_v is split, and is equal to π^{-1} at $s = 0$ when v is archimedean. Now we list the precise local Siegel–Weil formula for these local Whittaker functions. We use the convention that $|D_v| = |d_v| = 1$ if v is archimedean.

PROPOSITION 2.7.   *(1) In the sense of analytic continuation for $s \in \mathbb{C}$,*

$$W_{0,v}^{\circ}(0, g, \Phi_v) = |D_v|^{\frac{1}{2}}|d_v|^{\frac{1}{2}} \; \gamma(\mathbb{V}_{2,v}, uq) \; r(g)\Phi_v(0).$$

Therefore,
$$W_0(0, g, \Phi) = -r(g)\Phi(0).$$

Furthermore, for almost all places v,

$$W_{0,v}^{\circ}(s, g, \Phi_v) = \delta_v(g)^{-s} r(g)\Phi_v(0).$$

(2) Assume $a \in F_v^{\times}$.

(a) If a is not represented by $(\mathbb{V}_{2,v}, uq)$, then $W_{a,v}(0, g, \Phi_v) = 0$.

(b) Assume that there exists $x_a \in \mathbb{V}_{2,v}$ satisfying $uq(x_a) = a$. Then

$$W_{a,v}(0, g, \Phi_v) = \frac{\gamma(\mathbb{V}_{2,v}, uq)}{L(1, \eta_v)} \int_{E_v^1} r(g, h)\Phi_v(x_a)dh.$$

Here the integration uses the Haar measure on E_v^1 normalized in §1.6.2.

PROOF. We immediately know that all the results at $s = 0$ are true up to constant multiples independent of g and Φ_v by Theorem 2.2. To determine the constant, one can trace back the measures in Theorem 2.2. Alternatively, we only need to compute the case $g = 1$ for some well-chosen function Φ_v. See [KRY1] for an example.

As for $W_{0,v}^\circ(s, g, \Phi_v)$, it is the image of $\delta(g)^s r_2(g)\Phi_v(0)$ under the normalized intertwining operator for $g \in \mathrm{SL}_2(F_v)$. Hence we know the equality for almost all places.   □

PROPOSITION 2.8. *For any* $\Phi \in \mathcal{S}(\mathbb{V}_2)$, *one has* $E(0, g, \Phi) = 0$.

PROOF. The incoherence theory of Kudla–Rallis [KR3] shows $E(0, g, \Phi) = 0$ in a more general incoherent case. We can also check it directly by our expression of Whittaker functions above.

It is immediate that the constant term $E(0, g, \Phi) = 0$. Now we consider the Whittaker function

$$W_a(0, g, \Phi) = \prod_v W_{a,v}(0, g, \Phi_v)$$

for $a \in F^\times$. We know that $W_{a,v}(0, g, \Phi_v) \neq 0$ only if au^{-1} is represented by $(\mathbb{V}_{2,v}, q)$. We claim that au^{-1} cannot be represented by $(\mathbb{V}_{2,v}, q)$ for all places v.

Denote by B the quaternion algebra over F generated by E and j with relations

$$j^2 = -au^{-1}, \quad jt = \bar{t}j, \ \forall \, t \in E.$$

If au^{-1} is represented by some element x_v of $(\mathbb{V}_{2,v}, q)$ for all places v, then the map $j \mapsto x_v$ gives an isomorphism $B_{\mathbb{A}} \cong \mathbb{B}$. It contradicts the incoherence assumption on $\mathbb{B}$.   □

2.5.2   Eisenstein series of weight two

Now we consider the third case where the Eisenstein series $E(s, g, \Phi)$ is defined based on the quadratic space $(\mathbb{V}, q_{\mathbb{V}}) = (\mathbb{B}, uq)$. For the intertwining part, we still need a normalization

$$W_{0,v}^\circ(s, g, \Phi_v) \quad = \quad \frac{1}{\zeta_v(s+1)} W_{0,v}(s, g, \Phi_v).$$

The following result uses the Haar measures on F_v and $\mathbb{B}_v^1$ normalized in §1.6.1 and §1.6.2.

PROPOSITION 2.9. *The following are true at* $s = 0$:

(1)   (a) If $\mathbb{B}_v$ *is non-split, then* $W_{0,v}^\circ(0, g, \Phi_v) = 0$ *identically.*

(b) If $\mathbb{B}_v$ is split, then under the identification $\mathbb{B}_v = M_2(F_v)$,

$$W_{0,v}^\circ(0, g, \Phi_v)$$

$$= |ud_v^{-1}|_v \int_{\mathrm{SL}_2(O_{F_v})} \int_{F_v} \int_{F_v} r(g)\Phi_v\left(h\begin{pmatrix} y & x \\ 0 & 0 \end{pmatrix}\right) dx\,dy\,dh$$

$$= |ud_v^{-1}|_v \int_{\mathrm{SL}_2(O_{F_v})} \int_{F_v} \int_{F_v} r(g)\Phi_v\left(\begin{pmatrix} 0 & x \\ 0 & y \end{pmatrix} h\right) dx\,dy\,dh.$$

(2) *Assume $a \in F_v^\times$.*

    (a) If a is not represented by $(\mathbb{B}_v, uq)$, then $W_{a,v}(0, g, \Phi_v) = 0$.

    (b) Assume that there exists $x_a \in \mathbb{B}_v$ satisfying $uq(x_a) = a$. Then

$$W_{a,v}(0, g, \Phi_v) = \gamma(\mathbb{B}_v, uq)\, |a|_v \int_{\mathbb{B}_v^1} r(g)\Phi_v(hx_a)dh.$$

    (c) If furthermore $\mathbb{B}_2 = M_2(F_v)$, then (b) has the simplification

$$W_{a,v}(0, g, \Phi_v)$$

$$= |ud_v^{-1}|_v\, \zeta_v(1) \int_{\mathrm{SL}_2(O_{F_v})} \int_{F_v} \int_{F_v}$$

$$r(g)\Phi_v\left(h\begin{pmatrix} y & x \\ & au^{-1}y^{-1} \end{pmatrix}\right) dx\,dy\,dh$$

$$= |ud_v^{-1}|_v\, \zeta_v(1) \int_{\mathrm{SL}_2(O_{F_v})} \int_{F_v} \int_{F_v}$$

$$r(g)\Phi_v\left(\begin{pmatrix} au^{-1}y^{-1} & x \\ & y \end{pmatrix} h\right) dx\,dy\,dh.$$

Here both the measures dx and dy are the Haar measure on F_v.

(3) *The Eisenstein series $E(s, g, \Phi)$ is holomorphic at $s = 0$ with critical value*

$$E(0, g, \Phi) = r(g)\Phi(0) + W_0(0, g, \Phi) - \sum_{a \in F^\times} \int_{\mathbb{B}^1} r(g)\Phi(hx_a)dh.$$

Here $x_a \in \mathbb{B}$ is any element with $uq(x_a) = a$, and the integration is considered to be zero if such x_a does not exist. Furthermore, $W_0(0, g, \Phi) \neq 0$ only if $F = \mathbb{Q}$ and $\Sigma = \{\infty\}$. In that case,

$$W_0(0, g, \Phi) = W'_{0,\infty}(0, g, \Phi_\infty) \prod_{v \neq \infty} W_{0,v}^\circ(0, g, \Phi_v).$$

PROOF. The results of (1) follow from (2) by

$$W_{0,v}(0, g, \Phi_v) = \lim_{a \to 0} W_{a,v}(0, g, \Phi_v).$$

In particular, it is immediate that $W_{0,v}(0, g, \Phi_v) = 0$ if $\mathbb{B}_v$ is non-split.

The proof of (2) is similar to Proposition 2.7. See [KY] for some explicit results which would give (2)(a) and (2)(b). Here we deduce (2)(c) from (2)(b).

If $\mathbb{B}_v = M_2(F_v)$, then $\gamma(\mathbb{B}_v, uq) = 1$. We have

$$W_{a,v}(0, g, \Phi_v) = |a|_v \int_{\mathbb{B}_v^1} r(g)\Phi_v \left(h \begin{pmatrix} 1 & \\ & au^{-1} \end{pmatrix} \right) dh.$$

The Iwasawa decomposition gives the general Haar measure identity

$$\int_{SL_2(F_v)} \Psi(h) dh = |d_v|_v^{-1} \int_{SL_2(O_{F_v})} \int_{F_v^\times} \int_{F_v} \Psi \left(h \begin{pmatrix} 1 & x \\ & 1 \end{pmatrix} \begin{pmatrix} y & \\ & y^{-1} \end{pmatrix} \right) dx d^\times y dh.$$

It yields that

$$W_{a,v}(0, g, \Phi_v)$$
$$= |ad_v^{-1}|_v \int_{SL_2(O_{F_v})} \int_{F_v^\times} \int_{F_v} r(g)\Phi_v \left(h \begin{pmatrix} 1 & x \\ & 1 \end{pmatrix} \begin{pmatrix} y & \\ & y^{-1} \end{pmatrix} \begin{pmatrix} 1 & \\ & au^{-1} \end{pmatrix} \right) dx d^\times y dh$$
$$= |ad_v^{-1}|_v \int_{SL_2(O_{F_v})} \int_{F_v^\times} \int_{F_v} r(g)\Phi_v \left(h \begin{pmatrix} y & au^{-1}y^{-1}x \\ & au^{-1}y^{-1} \end{pmatrix} \right) dx d^\times y dh.$$

By a change of variable $x \to a^{-1}uyx$, we have

$$W_{a,v}(0, g, \Phi_v) = |ud_v^{-1}|_v \zeta_v(1) \int_{SL_2(O_{F_v})} \int_{F_v} \int_{F_v} r(g)\Phi_v \left(h \begin{pmatrix} y & x \\ & au^{-1}y^{-1} \end{pmatrix} \right) dx dy dh.$$

It proves the first identity of (2)(c). The second identity is proved in a similar way by

$$W_{a,v}(0, g, \Phi_v) = |a|_v \int_{\mathbb{B}_v^1} r(g)\Phi_v \left(\begin{pmatrix} au^{-1} & \\ & 1 \end{pmatrix} h \right) dh$$

and the Haar measure identity

$$\int_{SL_2(F_v)} \Psi(h) dh = |d_v|_v^{-1} \int_{SL_2(O_{F_v})} \int_{F_v^\times} \int_{F_v} \Psi \left(\begin{pmatrix} y^{-1} & \\ & y \end{pmatrix} \begin{pmatrix} 1 & x \\ & 1 \end{pmatrix} h \right) dx d^\times y dh.$$

As for (3), we only need to verify the related results on $W_0(s, g, \Phi)$. By definition,

$$W_0(s, g, \Phi) = \zeta_F(s+1) \prod_v W_{0,v}^\circ(s, g, \Phi_v).$$

The zeta functions contribute a simple pole at $s = 0$, and the product of $W_{0,v}^\circ$ contribute a zero at $s = 0$ of multiplicity at least $\#\Sigma$ by (1). It follows that $W_0(s, g, \Phi)$ is always holomorphic at $s = 0$, and vanishes at $s = 0$ if $\#\Sigma > 1$.

If $\#\Sigma = 1$, then $F = \mathbb{Q}$ and $\Sigma = \{\infty\}$. One has

$$W_0(0, g, \Phi) = W_{0,\infty}^{\circ\,\prime}(0, g, \Phi_\infty) \prod_{v \neq \infty} W_{0,v}^\circ(0, g, \Phi_v).$$

It is easy to see $W_{0,\infty}^{\circ\,\prime}(0, g, \Phi_\infty) = W_{0,\infty}'(0, g, \Phi_\infty)$ by $\zeta_\infty(1) = 1$.   □

Remark. The intertwining part $W_0(0, g, \Phi)$ (if nonzero) is the only nonholomorphic part in $E(0, g, \Phi)$.

2.5.3 Eisenstein series of weight 3/2

Now we consider the second case $(\mathbb{V}, q_\mathbb{V}) = (\mathbb{B}_0, uq)$. Let $\Phi \in \mathcal{S}(\mathbb{B}_0)$ and thus we have the Eisenstein series $E(s, g, \Phi)$ and the Whittaker function $W_a(s, g, \Phi)$ and its local component.

To have a better description of the non-holomorphic part of $E(0, g, \Phi)$, we introduce a notation. Let v be a non-archimedean place, and define a linear map

$$\ell_0 : \mathcal{S}(M_2(F_v)_0) \longrightarrow \mathcal{S}(F_v)$$

by sending any $\Phi \in \mathcal{S}(M_2(F_v)_0)$ to

$$(\ell_0 \Phi)(z) = |u|_v^{\frac{1}{2}} \int_{\mathrm{GL}_2(O_{F_v})} \int_{F_v} \Phi_v \left(h^{-1} \begin{pmatrix} z & x \\ & -z \end{pmatrix} h \right) dx dh, \quad z \in F_v.$$

Endow F_v with the quadratic norm $uq^-(z) = -uz^2$ to get a quadratic space (F_v, uq^-). It is standard that the map ℓ_0 is equivariant under the action of $\widetilde{\mathrm{SL}}_2(F_v)$ via the Weil representation. For a proof we refer to Proposition 4.13. By taking the product, we obtain an $\widetilde{\mathrm{SL}}_2(\mathbb{A}_f)$-equivariant map $\ell_0 : \mathcal{S}(M_2(\mathbb{A}_f)_0) \to \mathcal{S}(\mathbb{A}_f)$.

Go back to the Eisenstein series $E(s, g, \Phi)$. For each place v of F, normalize

$$W_{0,v}^\circ(s, g_v, \Phi_v) = \frac{1}{\zeta_v(2s+1)} W_{0,v}(s, g_v, \Phi_v).$$

For each $a \in F^\times$, normalize

$$W_{a,v}^\circ(s, g_v, \Phi_v) = \frac{1}{L_v(s+1, \eta_{-ua})} W_{a,v}(s, g_v, \Phi_v).$$

Here $\eta_{-ua} : \mathbb{A}^\times \to \{\pm 1\}$ denotes the quadratic field associated to the quadratic extension $F(\sqrt{-ua})$ of F. If $-ua$ is a square in F, take the convention that $F(\sqrt{-ua}) = F \oplus F$ and $\eta_{-ua} = 1$.

Now we normalize some Haar measures. For each element $y \in \mathbb{B}_v$ with $q(y) \in F_v^\times$, denote by $\mathbb{B}_{v,y}$ the centralizer of y in $\mathbb{B}_v$. Then $\mathbb{B}_{v,y} = F_v[y] = F_v + F_v y$ is a (possibly split) quadratic extension of F_v. Endow $\mathbb{B}_{v,y}^\times$ and $\mathbb{B}_v^\times$ with the Haar measures normalized in §1.6.2. Endow $\mathbb{B}_{v,y}^\times \backslash \mathbb{B}_v^\times$ with the quotient Haar measure. It is easy to see that $\mathrm{vol}(\mathbb{B}_{v,y}^\times \backslash \mathbb{B}_v^\times) = 2\pi^2$ if v is archimedean.

PROPOSITION 2.10. *Write* $\gamma_v = \gamma(\mathbb{B}_{0,v}, uq)$. *The following are true at* $s = 0$:

(1) (a) *If* $\mathbb{B}_v$ *is non-split, then* $W_{0,v}^\circ(0, g, \Phi_v) = 0$ *identically.*

 (b) *If* $\mathbb{B}_v$ *is split, then under the identification* $\mathbb{B}_v = M_2(F_v)$,

$$W_{0,v}^\circ(0, g, \Phi_v) = \gamma_v \, |2d_v^{-3}|_v^{\frac{1}{2}} \, r(g)(\ell_0 \Phi)(0).$$

(2) *Assume* $a \in F_v^\times$.

 (a) *If* a *is not represented by* $(\mathbb{B}_{0,v}, uq)$, *then* $W_{a,v}(0, g, \Phi_v) = 0$.

 (b) *Assume that there exists* $x_a \in \mathbb{B}_{0,v}$ *satisfying* $uq(x_a) = a$. *Then*

$$W_{a,v}^\circ(0, g, \Phi_v) = \gamma_v \, |8a|_v^{\frac{1}{2}} \int_{\mathbb{B}_{v,x_a}^\times \backslash \mathbb{B}_v^\times} r(g, (h, h))\Phi_v(x_a) dh.$$

 (c) *If furthermore in* (b) *one has* $\mathbb{B}_v = M_2(F_v)$ *and* $a = -uz^2$ *for some* $z \in F_v^\times$, *then we have a simplification*

$$W_{a,v}^\circ(0, g, \Phi_v) = \gamma_v \, |2d_v^{-3}|_v^{\frac{1}{2}} \, r(g)(\ell_0 \Phi_v)(z).$$

(3) *The Eisenstein series* $E(s, g, \Phi)$ *is holomorphic at* $s = 0$ *with critical value as follows:*

 (a) *If* $\#\Sigma > 1$, *then*

$$E(0, g, \Phi) = r(g)\Phi(0)$$
$$- \sum_{a \in F^\times} L(1, \eta_{-ua}) \int_{\mathbb{B}_{x_a}^\times \backslash \mathbb{B}^\times} r(g, (h, h))\Phi(x_a) dh.$$

 Here $x_a \in \mathbb{B}$ *is any element with* $uq(x_a) = a$, *and the integration is considered to be zero if such* x_a *does not exist.*

 (b) *If* $F = \mathbb{Q}$ *and* $\Sigma = \{\infty\}$, *identify* $\mathbb{B}_f = M_2(\mathbb{A}_f)$ *and thus* $\mathbb{B}_f^\times = \mathrm{GL}_2(\mathbb{A}_f)$. *Then*

$$E(0, g, \Phi) = E^{\mathrm{hol}}(0, g, \Phi) + E^{\mathrm{nhol}}(0, g, \Phi),$$

where the holomorphic part and the non-holomorphic part are respectively

$$E^{\mathrm{hol}}(0, g, \Phi)$$

$$= r(g)\Phi(0) - \sum_{a \in \mathbb{Q}^\times} L(1, \eta_{-ua}) \int_{\mathbb{B}_{x_a}^\times \backslash \mathbb{B}^\times} r(g, (h, h))\Phi(x_a)dh,$$

$$E^{\mathrm{nhol}}(0, g, \Phi)$$

$$= -\sum_{z \in \mathbb{Q}} \frac{1}{2\sqrt{2}\gamma_\infty} W'_{-uz^2, \infty}(0, g_\infty, \Phi_\infty)r(g_f)(\ell_0 \Phi_f)(z).$$

PROOF. The proof is similar to Proposition 2.7 and Proposition 2.9. It is easy to see that (1) is a consequence of (2)(a) and (2)(c) by the limit

$$W_{0,v}(0, g, \Phi_v) = \lim_{a \to 0} W_{a,v}(0, g, \Phi_v).$$

See [KRY3] for some explicit results on (2), which could be used to check the explicit constants in (2). Here we only sketch how (2)(b) implies (2)(c). By $a = -uz^2$, we can take $x_a = \begin{pmatrix} z & \\ & -z \end{pmatrix}$ whose norm represents a. The centralizer $\mathbb{B}_{v,x_a}$ of x_a in $\mathbb{B}_v^\times$ is exactly $A(F_v)$, the group of diagonal matrices. It follows that

$$W_{a,v}(0, g, \Phi_v) = \gamma_v |8a|_v^{\frac{1}{2}} \int_{A(F_v) \backslash \mathrm{GL}_2(F_v)} r(g, (h, h))\Phi_v(x_a)dh.$$

By the Iwasawa decomposition, one has a Haar measure identity

$$\int_{A(F_v) \backslash \mathrm{GL}_2(F_v)} \Psi(h)dh = |d_v|_v^{-\frac{3}{2}} \int_{\mathrm{GL}_2(O_{F_v})} \int_{F_v} \Psi(n(b)h)dbdh.$$

It follows that

$$W_{a,v}(0, g, \Phi_v)$$

$$= \gamma_v |8ad_v^{-3}|_v^{\frac{1}{2}} \int_{\mathrm{GL}_2(O_{F_v})} \int_{F_v}$$

$$r(g, (h, h))\Phi_v\left(\begin{pmatrix} 1 & -x \\ & 1 \end{pmatrix}\begin{pmatrix} z & \\ & -z \end{pmatrix}\begin{pmatrix} 1 & x \\ & 1 \end{pmatrix}\right) dxdh$$

$$= \gamma_v |8ad_v^{-3}|_v^{\frac{1}{2}} \int_{\mathrm{GL}_2(O_{F_v})} \int_{F_v} r(g, (h, h))\Phi_v\begin{pmatrix} z & 2xz \\ & -z \end{pmatrix} dxdh$$

$$= \gamma_v |2ud_v^{-3}|_v^{\frac{1}{2}} \int_{\mathrm{GL}_2(O_{F_v})} \int_{F_v} r(g, (h, h))\Phi_v\begin{pmatrix} z & x \\ & -z \end{pmatrix} dxdh.$$

Now we consider (3). It is similar to the result for $W_0(s, g, \Phi)$ in Proposition 2.9, but slightly more complicated. We remark that in both summations on

a, the existence of x_a implies that $ua > 0$ at archimedean places, so η_{-ua} is a nontrivial character and $L(1, \eta_{-ua})$ is a well-defined number.

By definition,

$$W_0(s, g, \Phi) = \zeta_F(2s+1) \prod_v W^\circ_{0,v}(s, g, \Phi_v).$$

The zeta functions contribute a simple pole at $s = 0$, and the product of $W^\circ_{0,v}$ contributes a zero at $s = 0$ of multiplicity at least $\#\Sigma$. It follows that $W_0(s, g, \Phi)$ is always holomorphic at $s = 0$, and it vanishes at $s = 0$ if $\#\Sigma > 1$.

Now let $a \in F^\times$ and consider

$$W_a(s, g, \Phi) = L(s+1, \eta_{-ua}) \prod_v W^\circ_{a,v}(s, g_v, \Phi_v).$$

The product of $W^\circ_{a,v}$ is holomorphic at $s = 0$. If $-ua$ is not a square in $F^\times$, then

$$W_a(0, g, \Phi) = L(1, \eta_{-ua}) \prod_v W^\circ_{a,v}(0, g_v, \Phi_v).$$

It can be written as integrals by (2).

Assume in the following that $-ua$ is a square in $F^\times$, and thus $a = -uz^2$ for some $z \in F^\times$. Then

$$W_a(s, g, \Phi) = \zeta_F(s+1) \prod_v W^\circ_{a,v}(s, g_v, \Phi_v).$$

We claim that a is not represented by $(\mathbb{B}_{0,v}, uq)$ at any $v \in \Sigma$. In fact, if $a = uq(x)$ for some $x \in \mathbb{B}_{0,v}$, then $q(z+x) = z^2 + q(x) = 0$, which is impossible for $z + x \in \mathbb{B}_v$. It follows that the product of $W^\circ_{a,v}$ contributes a zero at $s = 0$ of multiplicity at least $\#\Sigma$. Then $W_a(s, g, \Phi)$ is always holomorphic at $s = 0$, and it vanishes at $s = 0$ if $\#\Sigma > 1$.

Now assume that $\#\Sigma = 1$, i.e., $F = \mathbb{Q}$ and $\Sigma = \{\infty\}$. Still assume that $a = -uz^2$ for some $z \in \mathbb{Q}^\times$. Then

$$W_a(0, g, \Phi) = \mathrm{Res}_{s=0}\zeta_\mathbb{Q}(s+1) \cdot W^{\circ \, '}_{a,\infty}(0, g_\infty, \Phi_\infty) \prod_{v \nmid \infty} W^\circ_{a,v}(0, g_v, \Phi_v).$$

By $W^\circ_{a,\infty}(0, g_\infty, \Phi_\infty) = 0$, we have

$$W^{\circ \, '}_{a,\infty}(0, g_\infty, \Phi_\infty) = \zeta_\infty(1)^{-1} \, W'_{a,\infty}(0, g_\infty, \Phi_\infty) = W'_{a,\infty}(0, g_\infty, \Phi_\infty).$$

It follows that

$$
\begin{aligned}
W_a(0, g, \Phi) &= W'_{a,\infty}(0, g_\infty, \Phi_\infty) \prod_{v \nmid \infty} \gamma_v \, |2d_v^{-3}|_v^{\frac{1}{2}} \, r(g)(\ell_0 \Phi_v)(z) \\
&= -\frac{1}{\sqrt{2}\gamma_\infty} W'_{a,\infty}(0, g_\infty, \Phi_\infty) \, r(g)(\ell_0 \Phi_f)(z).
\end{aligned}
$$

The summation over all such a becomes half of the summation over all $z \in \mathbb{Q}^\times$, since each a corresponds to two z. Similarly,

$$
\begin{aligned}
W_0(0, g, \Phi) &= \operatorname{Res}_{s=0} \zeta_\mathbb{Q}(2s+1) \cdot W_{0,\infty}^{\circ\,\prime}(0, g_\infty, \Phi_\infty) \prod_{v \nmid \infty} W_{0,v}^\circ(0, g_v, \Phi_v) \\
&= -\frac{1}{2\sqrt{2}\gamma_\infty} W_{0,\infty}'(0, g_\infty, \Phi_\infty) \, r(g)(\ell_0 \Phi_f)(0).
\end{aligned}
$$

The extra factor 2 in the denominator comes from the residue of $\zeta_\mathbb{Q}(2s+1)$. It proves the result. $\square$

Remark. In case (3)(b), if setting $u = 1$, taking ϕ_∞ to be the standard Gaussian, and properly choosing ϕ_f, the Eisenstein series $E(0, g, \Phi)$ recovers Zagier's Eisenstein series

$$
\mathcal{F}(z) = \sum_{n=0}^{\infty} H(n) q^n + y^{-\frac{1}{2}} \sum_{m \in \mathbb{Z}} \beta(4\pi m^2 y) \, q^{-m^2}, \quad z \in \mathcal{H}.
$$

Here $y = \operatorname{Im}(z)$, $q = e^{2\pi i z}$, and $H(n)$ is Hurwitz class number, and

$$
\beta(x) - \frac{1}{16\pi} \int_1^\infty e^{-xt} t^{-\frac{3}{2}} dt, \quad \operatorname{Re}(x) \geq 0.
$$

We refer to [HZ] for more details. See also Corollary 2.12 for an expression of the archimedean part $W_{a,\infty}'(0, g_\infty, \Phi_\infty)$ in terms of $\beta(x)$.

2.5.4   Local Whittaker functions at archimedean places

Assume that (V, q) is a positive definite quadratic space over $\mathbb{R}$ of dimension $d \geq 1$, and let $\Phi = e^{-2\pi q} \in \mathcal{S}(V)$ be the standard Gaussian. Here we are going to compute the value at $s = 0$ of the (local) Whittaker function

$$
W_a(s, g, \Phi) = \int_\mathbb{R} \delta(wn(b)g)^s \, r(wn(b)g)\Phi(0) \, \psi(-ab) db, \quad g \in \widetilde{\mathrm{SL}}_2(\mathbb{R}).
$$

It suffices to treat the case $g = 1$, and the general case can be derived by Iwasawa decomposition. The following result is in the literature, and is expressed in terms of confluent hypergeometric functions. See [Shi, KRY1, KRY3, KY] for example. In the following, the Weil index $\gamma_d = \gamma(V, q) = e^{\frac{2\pi i d}{8}}$.

PROPOSITION 2.11. *The following are true:*

(1) For any $a \in \mathbb{R}$,

$$
W_a(s, 1, \Phi) = \gamma_d \frac{2\pi^{s+\frac{d}{2}}}{\Gamma(\frac{s}{2})\Gamma(\frac{s+d}{2})} \int_{t>0,\, t>2a} e^{-2\pi(t-a)} t^{\frac{s+d}{2}-1}(t - 2a)^{\frac{s}{2}-1} dt.
$$

(2) For any $a > 0$,

$$W_a(0, 1, \Phi) = \gamma_d \frac{(2\pi)^{\frac{d}{2}}}{\Gamma(\frac{d}{2})} a^{\frac{d}{2}-1} e^{-2\pi a}.$$

(3) For any $a < 0$,

$$W_a(0, 1, \Phi) = 0,$$

$$W'_a(0, 1, \Phi) = \gamma_d \frac{\pi^{\frac{d}{2}}}{\Gamma(\frac{d}{2})} \int_0^\infty e^{-2\pi(t-a)} t^{\frac{d}{2}-1} (t - 2a)^{-1} dt.$$

(4) For $a = 0$, one has

$$W_0(0, 1, \Phi) = \begin{cases} 0 & \text{if } d \neq 2, \\ \gamma_d \pi & \text{if } d = 2. \end{cases}$$

If $d \neq 2$, then

$$W'_0(0, 1, \Phi) = \gamma_d \pi \frac{2^{2-\frac{d}{2}}}{d - 2}.$$

PROOF. We first consider (1). By definition,

$$W_a(s, g, \Phi) = \gamma_d \int_{\mathbb{R}} \delta(wn(b)g)^s \int_V r(g)\Phi(x)\psi(bq(x)) dx \ \psi(-ab) db.$$

Take an isometry of quadratic space $(V, q) \simeq (\mathbb{R}^d, \|\cdot\|^2)$. Here $\|\cdot\|$ denotes the standard quadratic norm on $\mathbb{R}^d$. We get

$$W_a(s, 1, \Phi) = \gamma_d \int_{\mathbb{R}} \delta(wn(b))^s \int_{\mathbb{R}^d} e^{-2\pi\|x\|^2} e^{2\pi i b\|x\|^2} dx \ e^{-2\pi i a b} db.$$

The integral in x is just

$$\int_{\mathbb{R}^d} e^{-2\pi(1-ib)\|x\|^2} dx = \left(\int_{\mathbb{R}} e^{-2\pi(1-ib)x^2} dx \right)^d = (1 - ib)^{-\frac{d}{2}}.$$

Note that $\delta(wn(b)) = (1 + b^2)^{-\frac{1}{2}}$. We have

$$\begin{aligned} W_a(s, 1, \Phi) &= \gamma_d \int_{\mathbb{R}} (1 + b^2)^{-\frac{s}{2}} (1 - ib)^{-\frac{d}{2}} e^{-2\pi i a b} db \\ &= \gamma_d \int_{\mathbb{R}} (1 + ib)^{-\frac{s}{2}} (1 - ib)^{-\frac{s+d}{2}} e^{-2\pi i a b} db. \end{aligned}$$

After a standard computation as in [Shi, KRY1, KRY3, KY], we have

$$W_a(s, 1, \Phi) = \gamma_d \frac{2\pi^{s+\frac{d}{2}}}{\Gamma(\frac{s}{2})\Gamma(\frac{s+d}{2})} \int_{t>0, \ t>2a} e^{-2\pi(t-a)} t^{\frac{s+d}{2}-1} (t - 2a)^{\frac{s}{2}-1} dt.$$

Now we consider (2) and (3). We first assume $a > 0$. Then a change of variable gives

$$W_a(s,1,\Phi) = \gamma_d \frac{2\pi^{s+\frac{d}{2}}}{\Gamma(\frac{s}{2})\Gamma(\frac{s+d}{2})} e^{-2\pi a} \int_0^\infty e^{-2\pi t}(t+2a)^{\frac{s+d}{2}-1} t^{\frac{s}{2}-1} dt.$$

Note that $\Gamma(\frac{s}{2})$ has a pole at $s = 0$ which contributes a zero of the factor before the integral, while the integral is not convergent at $s = 0$ due to the singularity of $t^{\frac{s}{2}-1}$ at $t = 0$. The difference

$$\int_0^\infty e^{-2\pi t}(t+2a)^{\frac{s+d}{2}-1} t^{\frac{s}{2}-1} dt - \int_0^\infty e^{-2\pi t}(2a)^{\frac{s+d}{2}-1} t^{\frac{s}{2}-1} dt$$

$$= \int_0^\infty e^{-2\pi t} \frac{(t+2a)^{\frac{s+d}{2}-1} - (2a)^{\frac{s+d}{2}-1}}{t} t^{\frac{s}{2}} dt$$

is convergent and holomorphic at $s = 0$. It follows that

$$W_a(s,1,\Phi) = \gamma_d \lim_{s \to 0} \frac{2\pi^{s+\frac{d}{2}}}{\Gamma(\frac{s}{2})\Gamma(\frac{s+d}{2})} e^{-2\pi a} \int_0^\infty e^{-2\pi t}(2a)^{\frac{s+d}{2}-1} t^{\frac{s}{2}-1} dt.$$

The integral is a Gamma function, and the result is easily obtained.

Now we assume that $a < 0$. Then

$$W_a(s,1,\Phi) = \gamma_d \frac{\pi^{s+\frac{d}{2}}}{\Gamma(\frac{s}{2}+1)\Gamma(\frac{s+d}{2})} s \int_0^\infty e^{-2\pi(t-a)} t^{\frac{s+d}{2}-1}(t-2a)^{\frac{s}{2}-1} dt.$$

The integral is holomorphic at $s = 0$. It follows that $W_a(s,1,\Phi) = 0$ by the zero coming from the factors before the integral. We further have

$$W_a'(0,1,\Phi) = \gamma_d \frac{\pi^{\frac{d}{2}}}{\Gamma(\frac{d}{2})} \int_0^\infty e^{-2\pi(t-a)} t^{\frac{d}{2}-1}(t-2a)^{-1} dt.$$

The case $a = 0$ has an explicit expression for all $s \in \mathbb{C}$. In fact,

$$W_0(s,1,\Phi) = \gamma_d \frac{2\pi^{s+\frac{d}{2}}}{\Gamma(\frac{s}{2})\Gamma(\frac{s+d}{2})} \int_0^\infty e^{-2\pi t} t^{s+\frac{d}{2}-2} dt$$

$$= \gamma_d \pi 2^{-(s+\frac{d}{2}-2)} \frac{\Gamma(s+\frac{d}{2}-1)}{\Gamma(\frac{s}{2})\Gamma(\frac{s+d}{2})}.$$

It is easy to obtain the result.   □

In the case $d = 3$, to match the notation in [HZ], we recall the function

$$\beta(x) = \frac{1}{16\pi} \int_1^\infty e^{-xt} t^{-\frac{3}{2}} dt, \quad \mathrm{Re}(x) \geq 0.$$

It will give the derivative $W_a'(0,1,\Phi)$ a slightly different expression.

COROLLARY 2.12. *Assume $d = 3$. For $a \leq 0$,*

$$W'_{a,\infty}(0, 1, \Phi) = \gamma_3 8\sqrt{2}\pi^2 e^{-2\pi a}\beta(-4\pi a).$$

PROOF. By Proposition 2.11 (3) for the case $d = 3$, we have

$$
\begin{aligned}
W'_{a,\infty}(0, 1, \Phi) &= \gamma_3 \cdot 2\pi \int_0^\infty e^{-2\pi(t-a)} t^{\frac{1}{2}} (t - 2a)^{-1} dt \\
&= \gamma_3 \cdot 2\pi e^{2\pi a}(-2a)^{\frac{1}{2}} \int_0^\infty e^{4\pi at} t^{\frac{1}{2}} (t + 1)^{-1} dt \\
&= \gamma_3 \cdot 2^{-\frac{1}{2}} \pi e^{-2\pi a} \int_1^\infty e^{4\pi at} t^{-\frac{3}{2}} dt.
\end{aligned}
$$

The result is also true for $a = 0$. Here the last equality uses the identity

$$4|a|^{\frac{1}{2}} \int_0^\infty e^{4\pi at} t^{\frac{1}{2}} (t + 1)^{-1} dt = \int_0^\infty e^{4\pi at} (t + 1)^{-\frac{3}{2}} dt.$$

It is the special case $p = q = \frac{3}{2}$ of the functional equation

$$U(p, q, z) = z^{1-q} U(1 + p - q, 2 - q, z)$$

of the confluent hypergeometric function

$$U(p, q, z) = \frac{1}{\Gamma(p)} \int_0^\infty e^{-zt} t^{p-1} (t + 1)^{q-p-1} dt.$$

See [Le], p. 265.   □

Chapter Three

Mordell–Weil Groups and Generating Series

The major goal of this chapter is to introduce Theorem 3.21, an identity between the analytic kernel and the geometric kernel, and describe how it is equivalent to Theorem 1.2. We first define the generating series, and then use it to define the geometric kernel. The analytic kernel is the same as that in the Waldspurger formula, except that we take derivative here. As a bridge between these two theorems, we also introduce Theorem 3.15, an identity formulated in terms of projectors.

In §3.1, we review some basic notations and results on Shimura curves.

In §3.2, we will review the Eichler–Shimura theory for abelian varieties parametrized by Shimura curves and give more details on the main result in Theorem 1.2.

In §3.3, we state our main result using the projector $T(f_1 \otimes f_2)$ in Theorem 3.15, and prove that it is equivalent to Theorem 1.2.

In §3.4, we will define $Z(g, \Phi)$, a generating series with coefficients in $\mathrm{Pic}(X \times X)_{\mathbb{C}}$. It acts on X as an algebraic correspondence. This series is an extension of Kudla's generating series for Shimura varieties of orthogonal type [Ku1].

In §3.5, we define the geometric kernel $Z(g, \Phi, \chi)$ and its normalization. It is essentially a linear combination of height pairings of CM points on X with the image of these points under the action of $Z(g, \Phi)$.

In §3.6, we recall the analytic kernel function and state a kernel identity, and we show how this identity implies the main theorem in the introduction. It is based on an expression of the projector $T(f_1 \otimes f_2)$ in terms of the generating series (Theorem 3.22), which will be proved in the next chapter.

3.1 BASICS ON SHIMURA CURVES

Let F be a totally real number field, and Σ be a finite set of places of F with an odd cardinality and including all the infinite places. It gives a totally definite incoherent quaternion algebra $\mathbb{B}$ over $\mathbb{A}$ with ramification set Σ.

3.1.1 Shimura curves

For each open compact subgroup U of $\mathbb{B}^{\times}$, we have a (compactified) Shimura curve X_U. It is a projective curve over F such that, for any embedding $\tau : F \hookrightarrow \mathbb{C}$, the complex points of X_U at τ have the uniformization

$$X_{U,\tau}(\mathbb{C}) \simeq B(\tau)^{\times} \backslash \mathcal{H}^{\pm} \times \mathbb{B}_f^{\times}/U \cup \{\text{cusps}\}.$$

Here $B(\tau)$ is the unique quaternion algebra over F with ramification set $\Sigma \setminus \{\tau\}$, $\mathbb{B}_f$ is identified with $B(\tau)_{\mathbb{A}_f}$ as an $\mathbb{A}_f$-algebra, and $B(\tau)^\times$ acts on $\mathcal{H}^\pm$ through an isomorphism $B(\tau)_\tau \simeq M_2(\mathbb{R})$. The set $\{\text{cusps}\}$ is non-empty if and only if $F = \mathbb{Q}$ and $\Sigma = \{\infty\}$. We usually use $[z, \beta]_U$ to denote the image of $(z, \beta) \in \mathcal{H}^\pm \times \mathbb{B}_f^\times$ in $X_{U,\tau}(\mathbb{C})$. It is smooth when U is small enough.

For any two open compact subsets $U_1 \subset U_2$ of $\mathbb{B}_f^\times$, one has a natural surjective morphism

$$\pi_{U_1, U_2} : X_{U_1} \to X_{U_2}.$$

Let X be the projective limit of the system $\{X_U\}_U$. It is a regular scheme over F, locally noetherian but not of finite type. In terms of the notation above, it has a uniformization

$$X_\tau(\mathbb{C}) \simeq B(\tau)^\times \backslash \mathcal{H}^\pm \times \mathbb{B}_f^\times / D \cup \{\text{cusps}\}.$$

Here D denotes the closure of $F^\times$ in $\mathbb{A}_f^\times$. If $F = \mathbb{Q}$, then $D = \mathbb{Q}^\times$. In general, D is much larger than $F^\times$.

For any U, the curve X_U is connected but not geometrically connected in general. For any $\alpha \in \pi_0(X_{U,\overline{F}})$, we usually denote by $X_{U,\alpha}$ the corresponding connected component of $X_{U,\overline{F}}$. It gives a decomposition

$$X_{U,\overline{F}} = \coprod_{\alpha \in F_+^\times \backslash \mathbb{A}_f^\times / q(U)} X_{U,\alpha}.$$

The action of $\mathrm{Gal}(\overline{F}/F)$ on $\pi_0(X_{U,\overline{F}})$ factors through the reciprocity law

$$\mathrm{Gal}(\overline{F}/F) \longrightarrow F_+^\times \backslash \mathbb{A}_f^\times / q(U).$$

It actually makes $\pi_0(X_{U,\overline{F}})$ a principal homogeneous space under $F_+^\times \backslash \mathbb{A}_f^\times / q(U)$.

In terms of the complex uniformization, one has

$$\pi_0(X_{U,\tau}(\mathbb{C})) \simeq B(\tau)_+^\times \backslash \mathbb{B}_f^\times / U = F_+^\times \backslash \mathbb{A}_f^\times / q(U).$$

Here $B(\tau)_+^\times$ denotes elements of $B(\tau)^\times$ with totally positive norms. The decomposition into connected components is given by

$$X_{U,\tau}(\mathbb{C}) \simeq \coprod_{h \in B(\tau)_+^\times \backslash \mathbb{B}_f^\times / U} \Gamma_h \backslash \mathcal{H}^*, \quad \Gamma_h = B(\tau)_+^\times \cap hUh^{-1}.$$

Here $\mathcal{H}^* = \mathcal{H} \cup \mathbb{P}^1(\mathbb{Q})$ if $F = \mathbb{Q}$ and $\Sigma = \{\infty\}$; otherwise, $\mathcal{H}^* = \mathcal{H}$. A point of $X_{U,\tau}(\mathbb{C})$ represented by $[z, \beta]_U$ with $z \in \mathcal{H}$ and $\beta \in \mathbb{B}_f^\times$ lies in $X_{U,q(\beta)}$.

3.1.2 CM points

Let E/F be a totally imaginary quadratic extension. Fix an embedding $E_{\mathbb{A}} \hookrightarrow \mathbb{B}$. Then $E_{\mathbb{A}}^\times$ acts on X by the right multiplication via the embedding $E_{\mathbb{A}}^\times \hookrightarrow \mathbb{B}^\times$.

Let $X^{E^{\times}}$ be the subscheme of X of fixed points of X under $E^{\times}$. Up to the translation by $\mathbb{B}^{\times}$, the subscheme $X^{E^{\times}}$ does not depend on the choice of the embedding $E_{\mathbb{A}} \hookrightarrow \mathbb{B}$. The scheme $X^{E^{\times}}$ is defined over F.

Let H be the normalizer of E in $\mathbb{B}^{\times}$. A simple calculation shows that H is a dihedral group generated by $E_{\mathbb{A}}^{\times}$ and an element $\mathrm{j} \in \mathbb{B}^{\times}$ such that $\mathrm{j}x\mathrm{j}^{-1} = \bar{x}$ for all $x \in E_{\mathbb{A}}^{\times}$. Then H acts on $X^{E^{\times}}$, which makes $X^{E^{\times}}(\overline{F})$ a principal homogeneous space over H/D_E, where D_E is the closure of $E_{\infty}^{\times}E^{\times}$ in $E_{\mathbb{A}}^{\times}$. The theory of complex multiplication asserts that every point in $X^{E^{\times}}(\overline{F})$ is defined over E^{ab} and that Galois action is given by the reciprocity map

$$\mathrm{Gal}(E^{\mathrm{ab}}/F) \simeq H/D_E.$$

Fix a point $P \in X^{E^{\times}}(E^{\mathrm{ab}})$ throughout this book. It induces a point $P_U \in X_U(E^{\mathrm{ab}})$ for every U. We can normalize the complex uniformization

$$X_{U,\tau}(\mathbb{C}) = B(\tau)^{\times} \backslash \mathcal{H}^{\pm} \times \mathbb{B}_f^{\times}/U \cup \{\text{cusps}\}$$

so that the point P_U is exactly represented by the double coset of $[z_0, 1]_U$. Here $z_0 \in \mathcal{H}$ is the unique fixed point of $E^{\times}$ in $\mathcal{H}$ via the action induced by the embedding $E \hookrightarrow B(\tau)$. Similar description can be made on X.

If we take the geometrically connected component of P_U as the neutral component, we obtain an identification $\pi_0(X_{U,\overline{F}}) = F_+^{\times} \backslash \mathbb{A}_f^{\times}/q(U)$. A point represented by $[z, h]_U$ with $z \in \mathcal{H}$ and $h \in \mathbb{B}_f^{\times}$ lies in the geometrically connected component represented by $q(h)$.

3.1.3 Hodge classes

The curve X_U has a *Hodge class* $L_U \in \mathrm{Pic}(X_U)_{\mathbb{Q}}$. It is the line bundle for holomorphic modular forms of weight two, and it is essentially the canonical bundle modified by the ramification points. The system $L = \{L_U\}_U$ is compatible with pull-back maps.

The following definition is close to that in [Zh1]. Let $\omega_{X_U/F}$ be the canonical bundle of X_U. Define

$$L_U = \omega_{X_U/F} + \sum_{x \in X_U(\overline{F})} \left(1 - e_x^{-1}\right) x.$$

Here group operation in $\mathrm{Pic}(X_U)_{\mathbb{Q}}$ is written additively, and for each $x \in X_U(\overline{F})$, the ramification index is described as follows.

In terms of the above complex uniformization, the index e_x is just the ramification index of any preimage of x in the quotient map $\mathcal{H}^* \to \Gamma_h \backslash \mathcal{H}^*$. We also give an algebraic description here. If x is a cusp, then define $e_x = \infty$ so that the multiplicity $1 - e_x^{-1} = 1$. If x is not a cusp, e_x is defined to be the ramification index of any preimage of x in the map $X_{U'} \to X_U$ for any sufficiently small open compact subgroup U' of U. Here U' is said to be sufficiently small if each geometrically connected component of $X_{U'}$ is a free quotient of $\mathcal{H}$ under the complex uniformization (for any embedding $\tau : F \hookrightarrow \mathbb{C}$). The index e_x is independent of the choice of U' and the preimage of x, and the Hodge bundle L_U is defined over F.

For each component $\alpha \in \pi_0(X_{U,\overline{F}})$, denote the restriction $L_{U,\alpha} = L_U|_{X_{U,\alpha}}$ to the connected component $X_{U,\alpha}$ of $X_{U,\overline{F}}$ corresponding to α. It is also viewed as a divisor class on X_U via push-forward under $X_{U,\alpha} \to X_U$. Denote by

$$\xi_{U,\alpha} = \frac{1}{\deg L_{U,\alpha}} L_{U,\alpha}$$ the normalized Hodge class on $X_{U,\alpha}$, and by $\xi_U = \sum_\alpha \xi_{U,\alpha}$ the normalized Hodge class on X_U.

We remark that $\deg L_{U,\alpha}$ is independent of α since all geometrically connected components are Galois conjugate to each other. It follows that $\deg L_{U,\alpha} = \deg L_U / |F_+^\times \backslash \mathbb{A}_f^\times / q(U)|$. The degree of L_U can be further expressed as the volume $\mathrm{vol}(X_U)$, defined as the integral of $\dfrac{dx dy}{2\pi y^2}$ on $X_{U,\tau}(\mathbb{C})$ for any embedding $\tau : F \hookrightarrow \mathbb{C}$.

LEMMA 3.1. *The following are true:*

(1) *One always has* $\deg L_U = \mathrm{vol}(X_U)$. *Hence,* $\mathrm{vol}(X_U)$ *is a rational number independent of the choice of* $\tau : F \hookrightarrow \mathbb{C}$.

(2) *For any inclusion* $U_1 \subset U_2$, *the projection* $\pi_{U_1,U_2} : X_{U_1} \to X_{U_2}$ *has degree*

$$\deg(\pi_{U_1,U_2}) = \mathrm{vol}(X_{U_1})/\mathrm{vol}(X_{U_2}).$$

PROOF. It is easy to obtain (2) by the definition. To prove (1), we first introduce the Petersson metric $\| \cdot \|_{\mathrm{Pet}}$ of $L_{U,\tau}(\mathbb{C})$. On any connected component $\Gamma \backslash \mathcal{H}^*$ of $X_{U,\tau}(\mathbb{C})$, a section of $L_{U,\tau}(\mathbb{C})$ is of the form $f(z)dz$, where $f : \mathcal{H} \to \mathbb{C}$ is a holomorphic modular form of weight two with respect to Γ. The Petersson metric on this connected component is given by

$$\|f(z)dz\|_{\mathrm{Pet}} = 4\pi \cdot \mathrm{Im} z \cdot |f(z)|.$$

The result of the lemma follows from Chern's integration formula

$$\deg L_U = \int_{X_{U,\tau}(\mathbb{C})} c_1(L_{U,\tau}(\mathbb{C}), \| \cdot \|_{\mathrm{Pet}}).$$

Here the Chern form (on each connected component) is exactly

$$c_1(L_{U,\tau}(\mathbb{C}), \| \cdot \|_{\mathrm{Pet}}) = \frac{\partial \overline{\partial}}{\pi i} \log(4\pi y) = \frac{dx \wedge dy}{2\pi y^2}.$$

The Petersson metric has a logarithmic singularity at the cusps, but Chern's integration formula still applies here by a simple local computation. $\quad\square$

3.1.4   Hecke correspondences

Infinite level

The Shimura curve X is endowed with an action T_x of $x \in \mathbb{B}^\times$ given by "*the right multiplication by x_f.*" The action T_x is trivial if and only if $x_f \in D$. Each X_U is just the quotient of X by the action of U under this action. In terms of the system $\{X_U\}_U$, the action gives an isomorphism $\mathrm{T}_x : X_{xUx^{-1}} \to X_U$ for each U.

Recall that the Hecke algebra $\mathcal{H} := C_c^\infty(\mathbb{B}_f^\times)$ consists of smooth and compactly supported functions $\phi : \mathbb{B}_f^\times \to \mathbb{C}$. Its multiplication is given by the convolution

$$(\phi_1 * \phi_2)(x) := \int_{\mathbb{B}_f^\times} \phi_1(x')\phi_2(x'^{-1}x)dx'.$$

It has no multiplicative unit. For any admissible representation (V, ρ) of $\mathbb{B}_f^\times$, there is a standard action of $\mathcal{H}$ on V by

$$\mathcal{H} \longrightarrow \mathrm{End}(V), \quad \rho(\phi) : v \longmapsto \int_{\mathbb{B}_f^\times} \phi(x)\rho(x)v.$$

It actually gives an equivalence between the category of admissible representations of $\mathbb{B}_f^\times$ and the category of admissible representations of $\mathcal{H}$.

For each $\phi \in \mathcal{H}$, define a "correspondence" on X with complex coefficients by

$$\mathrm{T}(\phi) := \int_{\mathbb{B}_f^\times} \phi(x)\mathrm{T}_x.$$

The algebra $\mathcal{H}_U$ has an involution

$$\phi \longmapsto (\phi^{\mathrm{t}} : x \mapsto \phi(x^{-1})),$$

which is compatible with the transpose of the algebraic correspondence $\mathrm{T}(\phi)$.

Finite level

Fix an open compact subgroup U of $\mathbb{B}_f^\times$. Let $Z(x) \subset X \times X$ be the graph of T_x. Let $Z(x)_U$ denote the image of $Z(x)$ in $X_U \times X_U$. Then $Z(x)_U$ can also be defined as the image of the morphism

$$(\pi_{U_x,U}, \pi_{U_x,U} \circ \mathrm{T}_x) : \quad X_{U_x} \longrightarrow X_U \times X_U.$$

Here $U_x = U \cap xUx^{-1}$ is an open and compact subgroup of $\mathbb{B}_f^\times$. It is an algebraic correspondence on X_U, and also viewed as an element of $\mathrm{Pic}(X_U \times X_U)$.

In terms of complex uniformization above, the push-forward gives

$$(Z(x)_U)_* : [z, \beta]_U \longmapsto \sum_{y \in UxU/U} [z, \beta y]_U.$$

The transpose $Z(x)_U^t$ of $Z(x)_U$ is simply equal to $Z(x^{-1})_U$.

We can extend the action to the Hecke algebra

$$\mathcal{H}_U = C_c^\infty(U \backslash \mathbb{B}_f^\times / U) = \{f \in \mathcal{H} : f(UxU) = x, \forall x \in \mathbb{B}_f^\times\}$$

of compactly supported functions $\phi : \mathbb{B}_f^\times \to \mathbb{C}$ that are bi-invariant under the action of U. It is a sub-algebra of $\mathcal{H}$, with multiplicative unit $\mathrm{vol}(U)^{-1}1_U$.

For each $\phi \in \mathcal{H}_U$, the linear combination

$$\mathrm{T}(\phi)_U := \sum_{x \in U \backslash \mathbb{B}_f^\times / U} \phi(x)\, Z(x)_U$$

gives an element of $\mathrm{Pic}(X_U \times X_U)_\mathbb{C}$. It is an algebraic correspondence on X_U with complex coefficients. The algebra $\mathcal{H}_U$ has an involution

$$\phi \longmapsto (\phi^t : x \mapsto \phi(x^{-1})),$$

which is compatible with the transpose of algebraic correspondences.

Now we vary U. It is easy to check that, for any $\phi \in \mathcal{H}$, the system

$$\widetilde{\mathrm{T}}(\phi) = \{\mathrm{vol}(X_U)\, \mathrm{T}(\phi)_U\}_U$$

is compatible with pull-back maps and thus defines an element in

$$\mathrm{Pic}(X \times X)_\mathbb{C} := \varinjlim_U \mathrm{Pic}(X_U \times X_U)_\mathbb{C}.$$

In this book, we take the convention that algebraic correspondences act on cohomology groups via pull-back, and act on divisors via push-forward. There may be a few exceptions, which we will specify. We sometimes abbreviate the push-forward D_* as D for a correspondence D.

3.1.5 Differential forms

Denote by $\mathcal{A}(\mathbb{B}^\times)$ the set of (equivalence classes of) irreducible admissible representations π of $\mathbb{B}^\times$ such that the Jacquet–Langlands correspondence $\mathrm{JL}(\pi)$ of π on $\mathrm{GL}_2(\mathbb{A})$ is a cuspidal automorphic representation of $\mathrm{GL}_2(\mathbb{A})$, discrete of parallel weight two at infinity. Note the condition that $\mathrm{JL}(\pi)$ is discrete of parallel weight two at infinity is equivalent to the condition that π_∞ is the trivial representation of $\mathbb{B}_\infty^\times$.

Fix an embedding $\tau : F \hookrightarrow \mathbb{C}$. Then we have a cohomological group

$$H^1(X_{U,\tau}(\mathbb{C}), \mathbb{Q})$$

with Hodge structure

$$H^1(X_{U,\tau}(\mathbb{C}), \mathbb{Q}) \otimes \mathbb{C} = H^{1,0}(X_{U,\tau}(\mathbb{C})) \oplus H^{0,1}(X_{U,\tau}(\mathbb{C})),$$

where

$$H^{1,0}(X_{U,\tau}) \; = \; \Gamma(X_{U,\tau}(\mathbb{C}), \Omega^1_{X_{U,\tau}(\mathbb{C})/\mathbb{C}}),$$
$$H^{0,1}(X_{U,\tau}) \; = \; \Gamma(X_{U,\tau}(\mathbb{C}), \overline{\Omega}^1_{X_{U,\tau}(\mathbb{C})/\mathbb{C}}).$$

By the complex uniformization, these spaces can be identified with the spaces of cusp forms of $B(\tau)^\times_\mathbb{A}$ with weight $(2, 0, \cdots, 0)$ and $(-2, 0, \cdots, 0)$ invariant under U. By the Jacquet–Langlands correspondence, we exactly have

$$H^{1,0}(X_{U,\tau}) \simeq \bigoplus_{\pi \in \mathcal{A}(\mathbb{B}^\times)} \pi^U, \qquad H^{1,0}(X_{U,\tau}) \simeq \bigoplus_{\pi \in \mathcal{A}(\mathbb{B}^\times)} \bar{\pi}^U.$$

Here π^U is the space of vector of π fixed by U, and it is nonzero only for finitely many π. Each π appears with multiplicity one in the above decomposition.

In particular, the decomposition is compatible with the action of the Hecke algebra $\mathcal{H}_U$. Here $\mathcal{H}_U$ acts on $H^{1,0}(X_{U,\tau})$ via the pull-back maps of algebraic correspondences described above, and acts on π^U by the restriction of the action of $\mathcal{H}$ on π.

Taking the direct limit, we obtain decompositions

$$H^{1,0}(X_\tau) \simeq \bigoplus_{\pi \in \mathcal{A}(\mathbb{B}^\times)} \pi, \qquad H^{0,1}(X_\tau) \simeq \bigoplus_{\pi \in \mathcal{A}(\mathbb{B}^\times)} \bar{\pi}$$

as representations of $\mathbb{B}^\times$. Here

$$H^{1,0}(X_\tau) := \varinjlim_U H^{1,0}(X_{U,\tau}), \qquad H^{0,1}(X_\tau) := \varinjlim_U H^{0,1}(X_{U,\tau}).$$

3.1.6 Jacobian variety and its dual

For each level U, denote by J_U the Jacobian variety of X_U. There is a canonical isomorphism $J_U \to J_U^\vee$, and J_U can be considered either as the Albanese variety or as the Picard variety of X_U of line bundles of degree 0 by canonical duality.

In this book, we take the convention that J_U denotes the Albanese variety and $J_U^\vee$ denotes the Picard variety. It seems to be superfluous for fixed U, but makes an essential difference when varying U.

For any natural map $\pi_{U',U} : X_{U'} \to X_U$ given by an inclusion $U' \subset U$, we have an induced algebraic homomorphism $(\pi_{U',U})_* : J_{U'} \to J_U$ given by push-forward of divisors, and an algebraic homomorphism $\pi^*_{U',U} : J_U^\vee \to J_{U'}^\vee$ given by pull-back of line bundles. Thus we obtain a projective system $J := \{J_U\}_U$ and a direct system $J^\vee := \{J_U^\vee\}_U$.

We may think of J as a projective limit and $J^\vee$ as a direct limit, but they are obviously not algebraic varieties. We will only consider their realizations in terms of algebraic points, cohomologies, and homomorphisms.

Algebraic points

For a ring $R = \mathbb{Z}, \mathbb{Q}, \mathbb{R}, \mathbb{C}$, and any field extension F' of F, define

$$J(F')_R : = \varprojlim_U J_U(F') \otimes_\mathbb{Z} R = \varprojlim_U \text{Cl}^0(X_{U,F'}) \otimes_\mathbb{Z} R,$$

$$J^\vee(F')_R : = \varinjlim_U J_U^\vee(F') \otimes_\mathbb{Z} R = \varinjlim_U \text{Pic}^0(X_{U,F'}) \otimes_\mathbb{Z} R.$$

Similarly, we also define

$$\text{Cl}(X_{F'})_R := \varprojlim_U \text{Cl}(X_{U,F'}) \otimes_\mathbb{Z} R, \qquad \text{Pic}(X_{F'})_R := \varinjlim_U \text{Pic}(X_{U,F'}) \otimes_\mathbb{Z} R.$$

When $R = \mathbb{Z}$, we omit the subscript R.

The following are some examples we have just constructed:

- The Hodge bundle $L = \{L_U\}_U$ is an element of $\text{Pic}(X)_\mathbb{Q}$.

- The CM point P fixed above gives an element $P = \{P_U\}_U \in \text{Cl}(X_{\overline{F}})$.

- The normalized Hodge bundle $\xi_P = \{\xi_{U,P}\}_U$ gives an element of $\text{Cl}(X_{\overline{F}})_\mathbb{Q}$. Here $\xi_{U,P} = \xi_{U,\alpha(P)}$ where $\alpha(P) \in \pi_0(X_{U,\overline{F}})$ denotes the component containing P_U.

- The difference $P^\circ := P - \xi_P$ lies in $J(\overline{F})_\mathbb{Q}$.

There is a natural height pairing

$$\langle \cdot, \cdot \rangle_{\text{NT}} : J(\overline{F})_\mathbb{Q} \times J^\vee(\overline{F})_\mathbb{Q} \longrightarrow \mathbb{R}.$$

We extend this pairing to a bilinear pairing

$$\langle \cdot, \cdot \rangle_{\text{NT}} : J(\overline{F})_\mathbb{C} \times J^\vee(\overline{F})_\mathbb{C} \longrightarrow \mathbb{C}.$$

In fact, the usual theory of Néron–Tate heights gives a pairing

$$\langle \cdot, \cdot \rangle_{\text{NT}} : J_U(\overline{F})_\mathbb{Q} \times J_U^\vee(\overline{F})_\mathbb{Q} \longrightarrow \mathbb{R}$$

over F. See §7.1 for example. Now we vary U. Let $U' \subset U$ be any inclusion, and $\pi_{U',U} : X_{U'} \to X_U$ be the natural map as above. By the projection formula,

$$\langle \pi_{U',U}^* \alpha, \beta' \rangle_{\text{NT}} = \langle \alpha, (\pi_{U',U})_* \beta' \rangle_{\text{NT}}, \quad \alpha \in J_U^\vee(\overline{F})_\mathbb{Q}, \ \beta' \in J_U(\overline{F})_\mathbb{Q}.$$

It follows from Proposition 7.3 or the projection formula for the arithmetic intersection theory. Then we can take limits to define the height pairing between J and $J^\vee$.

Cohomology groups

Fix an embedding $\tau : F \hookrightarrow \mathbb{C}$. Consider the cohomology with Hodge structure

$$H^1(Y_{U,\tau}) := H^1(Y_{U,\tau}(\mathbb{C}), \mathbb{Q}), \qquad H^1(Y_{U,\tau}) \otimes_{\mathbb{Q}} \mathbb{C} = H^{1,0}(Y_{U,\tau}) \oplus H^{0,1}(Y_{U,\tau}).$$

Here Y represents X, J or $J^\vee$. Taking limit, we obtain

$$
\begin{aligned}
H^1(J_\tau) : \quad &= \quad \varinjlim_U H^1(J_{U,\tau}), \\
H^1(J_\tau^\vee) : \quad &= \quad \varprojlim_U H^1(J_{U,\tau}^\vee).
\end{aligned}
$$

They are endowed with rational Hodge structures of weight 1 by the limit construction.

There is a canonical isomorphism $H^1(J_{U,\tau}) = H^1(X_{U,\tau})$ keeping the direct system. Thus $H^1(J_\tau)$ is canonically isomorphic to

$$H^1(X_\tau) = \varinjlim_U H^1(X_{U,\tau}).$$

By definition,
$$H^1(X_\tau) \otimes_{\mathbb{Q}} \mathbb{C} = H^{1,0}(X_\tau) \oplus H^{0,1}(X_\tau).$$

Here $H^{1,0}(X_\tau)$ has been considered in §3.1.5, and $H^{0,1}(X_\tau)$ has a similar decomposition.

Write $H^1(X_{U,\tau})' = H^1(X_{U,\tau})$ for each U, but view $\{H^1(X_{U,\tau})'\}_U$ as a projective system by the transition map $(\pi_{U',U})_*$. Denote

$$H^1(X_\tau)' := \varprojlim_U H^1(X_{U,\tau})'.$$

It is canonically isomorphic to $H^1(J_\tau^\vee)$.

Algebraic homomorphisms

By abstract non-sense, we get a direct system $\{\mathrm{Hom}(J_U, J_U^\vee)\}_U$. Here each term denotes the group of algebraic homomorphisms between the two abelian varieties over F. Hence, it is reasonable to define

$$\mathrm{Hom}(J, J^\vee) := \varinjlim_U \mathrm{Hom}(J_U, J_U^\vee).$$

Similarly, define

$$\mathrm{Hom}^0(J, J^\vee) := \varinjlim_U \mathrm{Hom}(J_U, J_U^\vee) \otimes_{\mathbb{Z}} \mathbb{Q}.$$

It is the group of homomorphisms up to isogeny. Define $\mathrm{Hom}^0(J, J^\vee)_{\mathbb{C}}$ to be the base change of $\mathrm{Hom}^0(J, J^\vee)$.

Consider the cohomology groups above. We have a natural injection

$$\mathrm{Hom}^0(J_U, J_U^\vee)_{\mathbb{C}} \longrightarrow \mathrm{Hom}(H^1(X_{U,\tau})', \; H^1(X_{U,\tau}))$$

induced by the pull-back map of $H^{0,1}$. Here the right-hand side denotes the group of homomorphisms between Hodge structures. The limit gives an injection

$$\mathrm{Hom}^0(J, J^\vee)_{\mathbb{C}} \longrightarrow \mathrm{Hom}(H^1(X_\tau)', \; H^1(X_\tau)).$$

The image is actually contained in the smaller space

$$\mathrm{Hom}_{\mathrm{cont}}(H^1(X_\tau)', \; H^1(X_\tau)) := \varinjlim_U \mathrm{Hom}(H^1(X_{U,\tau})', \; H^1(X_{U,\tau})).$$

Conversely, we can define

$$\mathrm{Hom}^0(J^\vee, J) := \varprojlim_U \mathrm{Hom}(J_U^\vee, J_U) \otimes_{\mathbb{Z}} \mathbb{Q}.$$

It has a canonical element $\mathrm{vol}(X)^{-1} = \{\mathrm{vol}(X_U)^{-1}\}_U$ as the limit of the scalar

$$\mathrm{vol}(X_U)^{-1} \in \mathrm{Hom}(J_U^\vee, J_U) \otimes_{\mathbb{Z}} \mathbb{Q}.$$

This space is less used in this book.

Direct system of algebraic correspondences

Consider the group

$$\mathrm{Pic}(X \times X) := \varinjlim_U \mathrm{Pic}(X_U \times X_U).$$

On each level U, we obtain a map

$$\mathrm{Pic}(X_U \times X_U) \longrightarrow \mathrm{Hom}(J_U, J_U^\vee)$$

given by push-forward maps of correspondences. Both sides are direct systems when varying U. We claim that these two direct systems are compatible. Then we obtain a well-defined map

$$\mathrm{Pic}(X \times X) \longrightarrow \mathrm{Hom}(J, J^\vee).$$

The compatibility follows from the following basic result.

LEMMA 3.2. *Let $\varphi : Y' \to Y$ be a finite morphism of two projective curves over any field, and let $D \in \mathrm{Pic}(Y \times Y)$ be a correspondence on Y. View $D' = (\varphi \times \varphi)^* D \in \mathrm{Pic}(Y' \times Y')$ as a correspondence on Y'. Then*

$$D' = \Gamma_\varphi^{\mathrm{t}} \circ D \circ \Gamma_\varphi.$$

Here Γ_φ denotes the graph of φ in $Y \times Y'$ viewed as a correspondence from Y to Y', and Γ_φ^t denotes the transpose of Γ_φ viewed as a correspondence from Y' to Y. Here $\circ$ denotes the usual composition of correspondences. For a proof of the lemma, we refer to [Li, Proposition 1.13]. Here we explain it in a more direct language.

Let $\alpha' \in \mathrm{Pic}(Y')$ be a divisor on Y'. Then the lemma says

$$D'_* \alpha' = \varphi^*(D_* \varphi_* \alpha').$$

Go back to the map

$$\mathrm{Pic}(X_U \times X_U) \longrightarrow \mathrm{Hom}^0(J_U, J_U^\vee).$$

Let $D = \{D_U\}_U$ be an element of $\mathrm{Pic}(X \times X)$, and $\alpha = \{\alpha_U\}_U$ be an element of $J(\overline{F})$. Namely, $\{D_U\}_U$ is compatible with pull-back and $\{\alpha_U\}_U$ is compatible with push-forward. By Lemma 3.2 above, we see that the new system $D_* \alpha = \{(D_U)_* \alpha_U\}_U$ is compatible with pull-back. It gives the compatibility.

3.2 ABELIAN VARIETIES PARAMETRIZED BY SHIMURA CURVES

In this section, we will explain the background of Theorem 1.2.

3.2.1 Abelian varieties of GL(2)-type

Let K be a field of characteristic zero. We will work on the category AV^0 of *abelian varieties over K up to isogeny*. Namely, the objects of AV^0 are abelian varieties over K, and the morphism group of two objects A and B in AV^0 is defined by

$$\mathrm{Hom}^0(A, B) := \mathrm{Hom}_K(A, B) \otimes_{\mathbb{Z}} \mathbb{Q}.$$

We also denote $\mathrm{End}^0(A) = \mathrm{Hom}^0(A, A)$ conventionally.

An abelian variety A over K is said to be *of GL(2)-type over K*, if there exists a commutative subring M of $\mathrm{End}^0(A)$, such that the induced action of M on $\mathrm{Lie}(A)$ makes $\mathrm{Lie}(A)$ a free module of rank one over $M \otimes_{\mathbb{Q}} K$. In that case, we also say that the pair (A, M) is *of GL(2)-type*. Note that the choice M may not be unique. The notion of abelian varieties of GL(2)-type is the simplest generalization of elliptic curves.

We say that a simple abelian variety A over K *is of strict GL(2)-type over K* if $(A, \mathrm{End}^0(A))$ is an abelian variety of GL(2)-type over K. Equivalently, the algebra $\mathrm{End}^0(A)$ is a field of degree equal to $\dim A$. In that case, the only choice for M is $\mathrm{End}^0(A)$. By the positivity of the Rosati involution, M is either a totally real field or a CM field. See [Mu] for example.

In the next subsection, we will restrict to the case that K is a totally real number field. Then GL(2)-type over K is actually equivalent to strict GL(2)-type. In fact, we have the following result:

LEMMA 3.3. *Let A be a simple abelian variety of dimension g over a subfield K of $\mathbb{R}$. Then the degree $[D : \mathbb{Q}]$ of the endomorphism ring $D = \operatorname{End}^0(A)$ divides g.*

PROOF. Consider the representation of D on $V = H^1(A(\mathbb{C}), \mathbb{Q})$ induced by the endomorphisms. The complex conjugation gives an action $c : A(\mathbb{C}) \to A(\mathbb{C})$, viewed as an automorphism of the real Lie group $A(\mathbb{C})$. Then it induces an action c^* on V, which commutes with the action of D on V. Therefore, the decomposition $V = V^+ \oplus V^-$ of V into the two eigenspaces of c^* gives two subrepresentations of V. We claim that both subspaces have dimension g. Once this is true, then V^+ is a vector space over the division algebra D gives

$$[D : \mathbb{Q}] \mid g.$$

For the claim, consider the Hodge decomposition

$$V \otimes_{\mathbb{Q}} \mathbb{C} = \Gamma(A(\mathbb{C}), \Omega^1) \oplus \Gamma(A(\mathbb{C}), \bar{\Omega}).$$

The action c^* switches the factors on the right-hand side due to the condition that A is defined over a field invariant under c. It follows that $\dim V^+ = \dim V^- = g$. □

In the following, assume that K is a number field, and (A, M) is a simple abelian variety of GL(2)-type over K. We will define an L-series $L(s, A, M)$ taking values in $M \otimes_{\mathbb{Q}} \mathbb{C}$ such that the Hasse–Weil L-function is given by

$$L(s, A) = \mathrm{N}_{M \otimes \mathbb{C}/\mathbb{C}} L(s, A, M).$$

We only need to decompose the local L-function.

Fix a non-archimedean place v of K and a prime number ℓ not divisible by v such that $M_\ell := M \otimes \mathbb{Q}_\ell$ is still a field. Then the local L-series at v is defined by

$$L_v(s, A) = P_v(N_v^{-s})^{-1},$$

where N_v is the cardinality of the residue field $k(v)$ of v, and the characteristic polynomial $P_v \in \mathbb{Z}[T]$ is defined by

$$P_v(T) = \det_{\mathbb{Q}_\ell}(1 - \operatorname{Frob}(v)T|V_\ell(A)^{I_v}).$$

Here $V_\ell(A)$ is the ℓ-adic Tate module over $\mathbb{Q}_\ell$, I_v is the inertia subgroup of the decomposition group in $D_v := \operatorname{Gal}(\overline{K}_v/K_v)$, and $\operatorname{Frob}(v)$ is the geometric Frobenius in $D_v/I_v = \operatorname{Gal}(\bar{k}(v)/k(v))$. The polynomial $P_v(T)$ has coefficients in $\mathbb{Z}$ and does not depend on the choice of ℓ.

Since $V_\ell(A)$ has a module structure over M_ℓ, we have a decomposition

$$P_v(T) = \mathrm{N}_{M_\ell/\mathbb{Q}_\ell} P_v(T, M)$$

where $\mathrm{N}_{M_\ell/\mathbb{Q}_\ell} : M_\ell[T] \to \mathbb{Q}_\ell[T]$ denotes the norm map, and

$$P_v(T, M) := \det_{M_\ell}(1 - \operatorname{Frob}(v)|V_\ell(A)^{I_v}).$$

Here $\det_{M_\ell}$ denotes the determinant of the operator on the M_ℓ-space $V_\ell(A)^{I_v}$. The polynomial $P_v(T, M)$ has coefficients in M and does not depend on the choice of ℓ. Its degree is 2 if A has good reduction at v. Otherwise, its degree is 0 or 1. We define a formal Dirichlet series with coefficients in M as follows:

$$L(s, A, M) := \prod_{v \nmid \infty} L_v(s, A, M), \qquad L_v(s, A, M) := P_v(N_v^{-s}, M)^{-1}.$$

For any embedding $\iota : M \hookrightarrow \mathbb{C}$, we obtain the usual (complex) L-functions $L(s, A, \iota)$ and $L_v(s, A, \iota)$ by base change. By Weil's result on the Riemann hypothesis for abelian varieties in positive characteristics, $L(s, A, \iota)$ is absolutely convergent for $\mathrm{Re}\, s > 3/2$. We have the desired decomposition

$$L(s, A) = \prod_\iota L(s, A, \iota).$$

In the next subsection we will discuss the modularity conjecture of A when $F = K$ is totally real in the sense that there is an absolutely irreducible automorphic representation $\sigma_f = \otimes \sigma_v$ of $\mathrm{GL}_2(\mathbb{A}_f)$ over M such that

$$L_v(s, A, M) = L(s - 1/2, \sigma_v).$$

3.2.2   Rational representations on GL(2)

Let F be a totally real number field with adeles $\mathbb{A}$. Let $\sigma_\infty^{(2)}$ be the *complex* admissible representation of $\mathrm{GL}_2(F_\infty)$, discrete of (parallel) weight 2 with trivial central character.

Let σ_f be an irreducible automorphic representation of $\mathrm{GL}_2(\mathbb{A}_f)$ over a $\mathbb{Q}$-vector space. We say that σ_f is *automorphic (resp. cuspidal) of weight* 2 if $\sigma_\infty^{(2)} \otimes_\mathbb{Q} \sigma_f$ is a direct sum of irreducible automorphic (resp. cuspidal) complex representations σ_i of $\mathrm{GL}_2(\mathbb{A})$. We would like to study how σ_f is decomposed into a product of local representations of $\mathrm{GL}_2(F_v)$ and how $\sigma_f \otimes_\mathbb{Q} \mathbb{C}$ is decomposed into irreducible representations.

To do this, it will be more convenient to use the quaternion algebra $\mathbb{B}_0$ over $\mathbb{A}$ whose ramification set is the set Σ_0 of all archimedean places. In this way, we can view σ_f as a representation of $\mathbb{B}_{0,f}^\times$ and extend it to a representation π on $\mathbb{B}^\times$ over $\mathbb{Q}$ so that $\mathbb{B}_{0,\infty}^\times$ acts trivially.

More generally, we will work on an arbitrary quaternion algebra $\mathbb{B}$ over $\mathbb{A}$ whose ramification set Σ contains all archimedean places. It can be either coherent or incoherent. Recall that $\mathcal{A}(\mathbb{B}^\times)$ denotes the set of isomorphism classes of irreducible automorphic representations π of $\mathbb{B}^\times$ of weight 0, i.e., irreducible admissible complex representations π of $\mathbb{B}^\times$ whose Jacquet–Langlands correspondence σ on $\mathrm{GL}_2(\mathbb{A})$ is automorphic cuspidal and discrete of parallel weight 2. Let $\mathcal{A}(\mathbb{B}^\times, \mathbb{Q})$ be the set of isomorphism classes of irreducible representations π of $\mathbb{B}^\times$ over $\mathbb{Q}$ such that $\pi \otimes_\mathbb{Q} \mathbb{C}$ is a direct sum of representations in $\mathcal{A}(\mathbb{B}^\times)$.

THEOREM 3.4. *The following are true:*

(1) *For every $\pi \in \mathcal{A}(\mathbb{B}^\times)$ and $\alpha \in \mathrm{Aut}(\mathbb{C})$, the α-conjugate representation $\pi^\alpha := \pi \otimes_{(\mathbb{C},\alpha)} \mathbb{C}$ is still in $\mathcal{A}(\mathbb{B}^\times)$. The correspondence $(\alpha, \pi) \mapsto \pi^\alpha$ defines an action of $\mathrm{Aut}(\mathbb{C})$ on $\mathcal{A}(\mathbb{B}^\times)$. The stabilizer of every π is $\mathrm{Gal}(\mathbb{C}/M)$ for some number field M, and the orbit of π is indexed by the set $\mathrm{Hom}(M, \mathbb{C})$ of embeddings. Moreover, M is the number field generated by eigenvalues of spherical Hecke operators of π.*

(2) *Let $\{\pi^\iota : \iota \in \mathrm{Hom}(M, \mathbb{C})\}$ be an orbit of $\mathcal{A}(\mathbb{B}^\times)$ under the action of $\mathrm{Aut}(\mathbb{C})$ as above. Then there is a unique $\pi \in \mathcal{A}(\mathbb{B}^\times, \mathbb{Q})$ with $\mathrm{End}_{\mathbb{B}^\times}(\pi) = M$ such that $\pi^\iota = \pi \otimes_{(M,\iota)} \mathbb{C}$. It follows that*

$$\pi \otimes_{\mathbb{Q}} \mathbb{C} = \bigoplus_{\iota \in \mathrm{Hom}(M,\mathbb{C})} \pi^\iota.$$

Moreover, the correspondence $\{\pi^\iota : \iota\} \longmapsto \pi$ gives a bijection

$$\mathcal{A}(\mathbb{B}^\times)/\mathrm{Aut}(\mathbb{C}) \xrightarrow{\sim} \mathcal{A}(\mathbb{B}^\times, \mathbb{Q}).$$

(3) *For every $\pi \in \mathcal{A}(\mathbb{B}^\times, \mathbb{Q})$ with $\mathrm{End}_{\mathbb{B}^\times}(\pi) = M$, we have a unique decomposition*

$$\pi = \bigotimes_v \pi_v.$$

Here π_v is an irreducible admissible representation of $\mathbb{B}_v^\times$ over M, and the tensor product is a restricted tensor product over M. Moreover, M is the number field generated by the spherical Hecke eigenvalues of π.

(4) *(Jacquet–Langlands correspondence) For any $\pi \in \mathcal{A}(\mathbb{B}^\times, \mathbb{Q})$ there is a unique $\sigma \in \mathcal{A}(\mathbb{B}_0^\times, \mathbb{Q})$ with $\mathrm{End}_{\mathbb{B}_0^\times}(\sigma) = \mathrm{End}_{\mathbb{B}^\times}(\pi)$, such that $\pi_v \simeq \sigma_v$ at every place $v \notin \Sigma(\mathbb{B})$. The correspondence $\pi \mapsto \sigma$ gives a bijection between $\mathcal{A}(\mathbb{B}^\times, \mathbb{Q})$ and the subset of $\mathcal{A}(\mathbb{B}_0^\times, \mathbb{Q})$ of elements σ such that $\sigma_v^\iota = \sigma_v \otimes_\iota \mathbb{C}$ is square-integrable for every finite place $v \in \Sigma(\mathbb{B})$ and every embedding $\iota : \mathrm{End}_{\mathbb{B}_0^\times}(\sigma) \hookrightarrow \mathbb{C}$.*

Let $\sigma = \otimes_v \sigma_v \in \mathcal{A}(\mathbb{B}_0^\times, \mathbb{Q})$ with $M = \mathrm{End}_{\mathbb{B}^\times}(\sigma)$. Define the local L-function

$$L(s, \sigma_v) = P_v(q_v^{-s})^{-1} \in M \otimes_{\mathbb{Q}} \mathbb{C}$$

by the classification of the representation σ_v. Here $P_v(T) \in M(T)$ is a polynomial over M of degree at most two. By the above theorem, $L(s, \sigma_v)$ is the unique function of $s \in \mathbb{C}$ valued in $M \otimes \mathbb{C}$ whose component for $\iota : M \hookrightarrow \mathbb{C}$ is given by

$$L(s, \sigma_v, \iota) = L(s, \sigma_v^\iota).$$

We may also define the global L-function

$$L(s, \sigma) = \prod_v L(s, \sigma_v) \in M \otimes_{\mathbb{Q}} \mathbb{C}$$

by putting the usual Γ-factors at infinite places. Then $L(s, \sigma)$ has a holomorphic continuation to whole complex plane with a functional equation

$$L(s, \sigma) = \epsilon(s, \sigma) \, L(1 - s, \widetilde{\sigma})$$

where $\epsilon(s, \sigma)$ is a function of s valued in $(M \otimes_{\mathbb{Q}} \mathbb{C})^{\times}$.

If $\pi \in \mathcal{A}(\mathbb{B}^{\times}, \mathbb{Q})$ with Jacquet–Langlands correspondence $\sigma \in \mathcal{A}(\mathbb{B}_0^{\times}, \mathbb{Q})$, define

$$L(s, \pi_v) = L(s, \sigma_v), \qquad L(s, \pi) = L(s, \sigma).$$

The strong multiplicity one theorem on $\mathcal{A}(\mathbb{B}_0^{\times})$ induces the strong multiplicity one theorem on $\mathcal{A}(\mathbb{B}^{\times})$. Since a local unramified representation is determined by its local L-series, two representations in $\mathcal{A}(\mathbb{B}^{\times}, \mathbb{Q})$ are isomorphic if and only if they have the same local L-series at all but finitely many places.

Let us return to the proof of Theorem 3.4. Recall that we have considered the complex Hecke algebras

$$\begin{aligned}
\mathcal{H} &= \{\text{locally constant and compactly supported } \phi : \mathbb{B}_f^{\times} \to \mathbb{C}\}, \\
\mathcal{H}_U &= \{\phi \in \mathcal{H} : \phi(UxU) = \phi(x), \ \forall x \in \mathbb{B}_f^{\times}\}.
\end{aligned}$$

Now we introduce the rational Hecke sub-algebras

$$\begin{aligned}
\mathcal{H}_{\mathbb{Q}} &= \{\text{locally constant and compactly supported } \phi : \mathbb{B}_f^{\times} \to \mathbb{Q}\}, \\
\mathcal{H}_{U,\mathbb{Q}} &= \{\phi \in \mathcal{H} : \phi(UxU) = \phi(x), \ \forall x \in \mathbb{B}_f^{\times}\}.
\end{aligned}$$

We further introduce the spherical subalgebras

$$\begin{aligned}
\mathcal{T}_U &= \{\phi \in \mathcal{H}_U : \phi = 1_{U_S} \otimes \phi^S, \ \phi^S : \mathbb{B}_f^{S,\times} \to \mathbb{C}\}, \\
\mathcal{T}_{U,\mathbb{Q}} &= \{\phi \in \mathcal{H}_{U,\mathbb{Q}} : \phi = 1_{U_S} \otimes \phi^S, \ \phi^S : \mathbb{B}_f^{S,\times} \to \mathbb{Q}\}.
\end{aligned}$$

Here S is the set of finite places v of F outside $\Sigma(\mathbb{B})$ such that U_v is not maximal. It is well-known that $\mathcal{T}_U$ and $\mathcal{T}_{U,\mathbb{Q}}$ are commutative.

Case where $\mathbb{B}$ is coherent

Assume first that $\mathbb{B}$ is coherent. Then $\mathbb{B} = B \otimes_F \mathbb{A}$ for a totally definite quaternion algebra B over F. Denote by $C^{\infty}(B^{\times}\backslash\mathbb{B}^{\times}/B_{\infty}^{\times}, \mathbb{Q})$ the space of locally constant functions $f : \mathbb{B}^{\times} \to \mathbb{Q}$ left invariant under $B^{\times}$. Since $B_{\infty}^{\times}$ is connected, it acts trivially on this space. Define $C^{\infty}(B^{\times}\backslash\mathbb{B}^{\times}/B_{\infty}^{\times}, \mathbb{C})$ similarly.

By the spectral decomposition,

$$C^{\infty}(B^{\times}\backslash\mathbb{B}^{\times}/B_{\infty}^{\times}, \mathbb{C}) = \bigoplus_{\pi \in \mathcal{A}(\mathbb{B})} \pi.$$

Since the left-hand side has a $\mathbb{Q}$-structure, $\mathrm{Aut}(\mathbb{C})$ acts on $\mathcal{A}(\mathbb{B})$. We need to filter the above spaces in terms of open compact subgroups U of $\mathbb{B}_f^{\times}$. In fact,

$$C^{\infty}(B^{\times}\backslash\mathbb{B}^{\times}/B_{\infty}^{\times}, \mathbb{Q}) = \varinjlim_U C(B^{\times}\backslash\mathbb{B}_f^{\times}/U, \mathbb{Q}),$$

$$C^\infty(B^\times\backslash\mathbb{B}^\times/B^\times_\infty,\ \mathbb{C}) = \varinjlim_U C(B^\times\backslash\mathbb{B}^\times_f/U,\ \mathbb{C}).$$

Here $C(B^\times\backslash\mathbb{B}^\times_f/U,\ \mathbb{Q})$ (resp. $C(B^\times\backslash\mathbb{B}^\times_f/U,\ \mathbb{C})$) denotes the space of maps from the finite set $B^\times\backslash\mathbb{B}^\times_f/U$ to $\mathbb{Q}$ (resp. $\mathbb{C}$).

Fix such a U. Let $\mathbb{H}_{U,\mathbb{Q}}$ (resp. $\mathbb{T}_{U,\mathbb{Q}}$) be the image of $\mathcal{H}_{U,\mathbb{Q}}$ (resp. $\mathcal{T}_{U,\mathbb{Q}}$) in $\mathrm{End}(C(B^\times\backslash\mathbb{B}^\times_f/U,\mathbb{Q}))$. Let $\mathbb{H}_U$ and $\mathbb{T}_U$ be their base changes to $\mathbb{C}$. We have the following decomposition of $\mathbb{H}_U$-modules

$$C(B^\times\backslash\mathbb{B}^\times_f/U,\mathbb{C}) = \bigoplus_{\pi\in\mathcal{A}(\mathbb{B}^\times)} \pi^U.$$

By the strong multiplicity one theorem, $\{\pi^U \neq 0\}$ is a finite set of distinct finite-dimensional irreducible $\mathcal{H}_U$-modules. By the density theorem of Jacobson and Chevalley for semisimple modules, the map

$$\mathcal{H}_U \longrightarrow \bigoplus_{\pi^U\neq 0} \mathrm{End}_\mathbb{C}(\pi^U)$$

is surjective. See Theorem 11.16 of [La] for example.

Hence, we have obtained

$$\mathbb{T}_U = \bigoplus_{\pi^U\neq 0} \mathbb{C}, \qquad \mathbb{H}_U = \bigoplus_{\pi^U\neq 0} \mathrm{End}(\pi^U).$$

This shows that $\mathbb{H}_{U,\mathbb{Q}}$ is semisimple with center $\mathbb{T}_{U,\mathbb{Q}}$, and that the set of π that appears in the sum is indexed by $(\mathrm{Spec}\,\mathbb{T}_{U,\mathbb{Q}})(\mathbb{C})$ with a compatible action by $\mathrm{Aut}(\mathbb{C})$. Thus the orbits of $\{\pi^U \neq 0\}$ under $\mathrm{Aut}(\mathbb{C})$ are indexed by closed points in $\mathrm{Spec}\,\mathbb{T}_{U,\mathbb{Q}}$.

More precisely, decompose $\mathbb{T}_{U,\mathbb{Q}}$ as a direct sum of fields M in the form

$$\mathbb{T}_{U,\mathbb{Q}} = \bigoplus M.$$

Then each M represents a conjugacy class $\{(\pi^U)^\iota : \iota \in \mathrm{Hom}(M,\mathbb{C})\}$ such that the action of $\mathbb{T}_{U,\mathbb{Q}}$ on $(\pi^U)^\iota$ is given by the composition

$$\mathbb{T}_{U,\mathbb{Q}} \longrightarrow M \overset{\iota}{\longrightarrow} \mathbb{C}.$$

Fix an M and define

$$\pi^U = C(B^\times\backslash\mathbb{B}_f/U,\mathbb{Q}) \otimes_{\mathbb{T}_{U,\mathbb{Q}}} M.$$

Then π^U is a module over $\mathbb{H}_{U,\mathbb{Q}}$ such that

$$\pi^U \otimes_{(M,\iota)} \mathbb{C} = (\pi^U)^\iota, \qquad \pi^U \otimes_\mathbb{Q} \mathbb{C} = \bigoplus(\pi^U)^\iota.$$

As $(\pi^U)^\iota$ are non-isomorphic to each other, π^U is irreducible over $\mathbb{Q}$ and geometrically irreducible over M. Thus $\mathrm{End}_{\mathcal{H}_{U,\mathbb{Q}}}(\pi^U) = M$.

If U' is an open compact subgroup of U, then we have a morphism $\mathbb{T}_{U',\mathbb{Q}} \to \mathbb{T}_{U,\mathbb{Q}}$. The strong multiplicity one theorem shows that this is surjective. If M appears in $\mathrm{Spec}\,\mathbb{T}_{U',\mathbb{Q}}$, we can define the $\mathcal{H}_{U,\mathbb{Q}}$-module $\pi^{U'}$ which includes π^U. The direct limit of these spaces forms a representation π such that

$$\pi \otimes_{(M,\iota)} \mathbb{C} = \pi^\iota, \quad \pi \otimes_\mathbb{Q} \mathbb{C} = \oplus \pi^\iota.$$

This proves part (1). The remaining part of part (2) follows from the following:

LEMMA 3.5. *For two irreducible representations π_1 and π_2 in $\mathcal{A}(\mathbb{B}^\times,\mathbb{Q})$,*

$$\mathrm{Hom}_{\mathbb{B}^\times}(\pi_1,\pi_2) \otimes_\mathbb{Q} \mathbb{C} = \mathrm{Hom}_{\mathbb{B}^\times}(\pi_1 \otimes_\mathbb{Q} \mathbb{C}, \pi_2 \otimes_\mathbb{Q} \mathbb{C}).$$

Thus π_1 and π_2 are isomorphic if and only if $\pi_1 \otimes_\mathbb{Q} \mathbb{C}$ and $\pi_2 \otimes_\mathbb{Q} \mathbb{C}$ have a common irreducible component in $\mathcal{A}(\mathbb{B}^\times)$.

PROOF. It is clear that the left-hand side is included in the right-hand side. We need to prove the other direction. Let ϕ be one element in the right-hand side. Decomposing $\pi_i \otimes \mathbb{C}$ into irreducible representations over $\mathbb{C}$, there is an irreducible complex representation σ and surjective morphisms $\alpha_i : \pi_i \otimes \mathbb{C} \longrightarrow \sigma$ such that $\alpha_1 = \alpha_2 \circ \phi$. Let π_0 be a representation satisfying the first part of Theorem 3.4 (2) for the conjugacy class of σ. Thus we have another projection $\alpha_0 : \pi_0 \otimes \mathbb{C} \longrightarrow \sigma$.

It is clear that the restriction of α_i on each $\pi_i \otimes 1$ is injective with images $\alpha_i(\pi_i \otimes 1)$ generating σ respectively. Let U be an open compact subgroup of $\mathbb{B}^\times$ such that σ^U is one-dimensional. Applying the element $1_U \in \mathbb{H}_U$, we have an

$$\sigma^U = \mathbb{C}\alpha_i(\pi_i^U).$$

This implies that $\pi_i^U \neq 0$ for each i. Let $v_i \in \pi_i^U$ be any nonzero elements. Then there is a $c_i \in \mathbb{C}^\times$ such that $c_i\alpha_i(v_i) = \alpha_0(v_0)$ for $i = 1,2$. We see that $c_i\alpha_i(\pi_i) = \alpha_0(\pi_0)$ since both sides are irreducible over $\mathbb{Q}$ with a common element $\alpha_0(v_0)$. This shows that there is an isomorphism $f : \pi_i \longrightarrow \pi_0$ bringing v_i to v_0. Thus we may assume in the lemma that both π_i are equal to π_0. Let M be the field of spherical eigenvalues of σ. Then the right hand is given by:

$$\mathrm{End}(\pi_0 \otimes_\mathbb{Q} \mathbb{C}) = \mathrm{End}(\oplus_{\iota:M\to\mathbb{C}}\sigma^\iota) = \oplus_{\iota:M\longrightarrow\mathbb{C}}\mathbb{C} = M \otimes \mathbb{C}.$$

Since M acts on π_0, the left-hand side includes $M \otimes \mathbb{C}$. Thus we must have the equality.   □

Part (3) and part (4) of the theorem are clear as π is absolutely irreducible over M. The following is a by-product of the proof.

THEOREM 3.6. *Assume that $\mathbb{B}$ is coherent with rational structure B over F. Then*

$$C^\infty(B^\times \backslash \mathbb{B}^\times / B_\infty^\times, \mathbb{Q}) = \bigoplus_{\pi \in \mathcal{A}(\mathbb{B}^\times,\mathbb{Q})} \pi.$$

Case when $\mathbb{B}$ *is incoherent*

Assume $\mathbb{B}$ is incoherent. Then $\mathbb{B}$ defines a Shimura curve $X = \varprojlim_U X_U$. Fix an open and compact subgroup U of $\mathbb{B}_f^{\times}$. Fix an embedding $\tau : F \to \mathbb{C}$. Then we have an action of $\mathbb{B}^{\times}$ on the Hodge structure $H^1(X_{\tau}, \mathbb{Q})$. Let c be the involution on $H^1(X_{\tau}, \mathbb{Q})$ induced by complex conjugation on $X_{\tau}(\mathbb{C})$.

Let $\mathbb{H}_{U,\mathbb{Q}}$ and $\mathbb{T}_{U,\mathbb{Q}}$ be the images of $\mathcal{H}_{U,\mathbb{Q}}$ and $\mathcal{T}_{U,\mathbb{Q}}$ in $\operatorname{End}(H^1(X_{U,\tau}, \mathbb{Q}))$ respectively. Both of them are finite-dimensional since $H^1(X_{U,\tau}, \mathbb{Q})$ is finite-dimensional. The complex conjugation c commutes with the action of $\mathbb{H}_{U,\mathbb{C}}$.

Consider the $\mathbb{C}$-linear actions of

$$\mathbb{T}_{U,\mathbb{C}} := \mathbb{T}_{U,\mathbb{Q}} \otimes_{\mathbb{Q}} \mathbb{C}, \qquad \mathbb{H}_{U,\mathbb{C}} := \mathbb{H}_{U,\mathbb{Q}} \otimes_{\mathbb{Q}} \mathbb{C}, \qquad c$$

on the complex Hodge structure

$$H^1(X_{U,\tau}, \mathbb{C}) = H^{1,0}(X_{U,\tau}) \oplus H^{0,1}(X_{\tau}).$$

The action c switches the last two factors. We call a rational Hodge structure (V, c) of weight 1 with an involution on the underlying $\mathbb{Q}$ vector space a *rational Hodge structure over* $\mathbb{R}$ if c switches the Hodge factors $V^{1,0}$ and $V^{0,1}$.

By the Jacquet–Langlands correspondence, we have an isomorphism of $\mathbb{H}_{U,\mathbb{C}}$-modules:

$$H^1(X_{U,\tau}, \mathbb{C}) = \bigoplus_{\pi \in \mathcal{A}(\mathbb{B}^{\times})} (\pi^U \oplus \bar{\pi}^U),$$

where $\bar{\pi} := \pi \otimes_{(\mathbb{C},c)} \mathbb{C}$ is the complex conjugation of π. The direct sum $\pi^U \oplus \bar{\pi}^U$ gives a Hodge structure over $\mathbb{R}$. Since the left-hand has a rational structure, this shows that $\mathcal{A}(\mathbb{B}^{\times})$ is closed under $\operatorname{Aut}(\mathbb{C})$. Under this isomorphism, we have

$$\mathbb{T}_{U,\mathbb{C}} = \bigoplus_{\pi^U \neq 0} \mathbb{C}, \qquad \mathbb{H}_{U,\mathbb{C}} = \bigoplus_{\pi \in \mathcal{A}(\mathbb{B}^{\times})} \operatorname{End}(\pi^U).$$

Note that the direct sums have only finitely many nonzero terms.

This shows that $\mathbb{H}_{U,\mathbb{Q}}$ is semisimple with center $\mathbb{T}_{U,\mathbb{Q}}$ and the representations appearing in the sum are indexed by $\mathbb{C}$-points of the scheme $\operatorname{Spec}\mathbb{T}_{U,\mathbb{Q}}$. Then we can decompose $\mathbb{T}_{U,\mathbb{Q}}$ as a direct sum of fields M in the form

$$\mathbb{T}_{U,\mathbb{Q}} = \oplus M.$$

Each M represents a conjugacy class $\{\pi^{\iota} : \iota \in \operatorname{Hom}(M, \mathbb{C})\}$ with each $\pi^{\iota} \in \mathcal{A}(\mathbb{B}^{\times})$ such that the action on $\mathbb{T}_{U,\mathbb{Q}}$ on $(\pi^U)^{\iota}$ is given by composition

$$\mathbb{T}_{U,\mathbb{Q}} \longrightarrow M \overset{\iota}{\longrightarrow} \mathbb{C}.$$

Fix M. Define $V^U := H^1(X_{U,\tau}, \mathbb{Q}) \otimes_{\mathbb{T}_{U,\mathbb{Q}}} M$. Then V^U is a rational Hodge structure over $\mathbb{R}$ with an action by $\mathbb{H}_{U,\mathbb{Q}}$. It further has decompositions:

$$V^U \otimes_{(M,\iota)} \mathbb{C} = (\pi^U)^{\iota} \oplus (\bar{\pi}^U)^{\iota}, \qquad V^U \otimes_{\mathbb{Q}} \mathbb{C} = \bigoplus_{\iota:M \longrightarrow \mathbb{C}} ((\pi^U)^{\iota} \oplus (\bar{\pi}^U)^{\bar{\iota}}).$$

Here $\bar{\iota}$ means the composition of ι and the complex conjugation of $\mathbb{C}$. If U' is an open compact subgroup of U, then $\mathrm{Spec}\mathbb{T}_{U,\mathbb{Q}}$ is a subscheme of $\mathrm{Spec}\mathbb{T}_{U',\mathbb{Q}}$. Thus we can similarly construct the space $V^{U'}$, which includes V^U. The direct limit of V^U forms a rational Hodge structure V over $\mathbb{R}$ with an action by M. The Hodge decompositions are

$$V \otimes_{(M,\iota)} \mathbb{C} = \pi^\iota \oplus \pi^{\bar{\iota}}, \qquad V \otimes_{\mathbb{Q}} \mathbb{C} = \bigoplus_{\iota: M \longrightarrow \mathbb{C}} (\pi^\iota \oplus \pi^{\bar{\iota}}).$$

Choose a U such that $(\pi^U)^\iota$ is one-dimensional for all ι. Then V^U has dimension 2 over M, and dimension $2 \dim M$ over $\mathbb{Q}$. It follows that V^U is simple as an M-Hodge structure with

$$\mathrm{End}_{M-\mathrm{Hodge}/\mathbb{R}}(V^U) = M.$$

Now we define

$$\pi := \mathrm{Hom}_{M-\mathrm{Hodge}/\mathbb{R}}(V^U, V)$$

as a representation of $\mathbb{B}^\times$. By construction, we have the following properties:

(a) π has coefficient M, and $\dim_M \pi^U = 1$;

(b) $\pi \otimes \mathbb{C}$ is included into $\mathrm{Hom}_{\mathrm{Hodge}/\mathbb{R}}(V_{\mathbb{C}}^U, V_{\mathbb{C}}) \simeq V_{\mathbb{C}}^{\dim M}$.

By these two properties, we have isomorphisms of $\mathbb{B}^\times$-modules:

$$\pi \otimes_{\mathbb{Q}} \mathbb{C} \simeq \bigoplus_{\iota: M \hookrightarrow \mathbb{C}} \pi^\iota, \qquad \pi \otimes_{(M,\iota)} \mathbb{C} \simeq \pi^\iota.$$

This proved part (1) and the first half of part (2). The other parts can be proved using the same argument as in the coherent case. The following is a by-product of the proof.

THEOREM 3.7. *Assume that $\mathbb{B}$ is incoherent and defines a Shimura curve X over F. Fix an embedding $\tau : F \hookrightarrow \mathbb{C}$. Then we have the following decomposition of rational Hodge structures over $\mathbb{R}$ with an $\mathbb{B}^\times$-action:*

$$H^1(X_\tau(\mathbb{C}), \mathbb{Q}) = \bigoplus_{\pi \in \mathcal{A}(\mathbb{B}^\times, \mathbb{Q})} \pi \otimes_M V(\pi).$$

Here $M = \mathrm{End}_{\mathbb{B}^\times}(\pi)$ is a number field, and $V(\pi)$ is an irreducible rational Hodge structure over $\mathbb{R}$ defined by

$$V(\pi) := \mathrm{Hom}_{\mathbb{B}^\times}(\pi, H^1(X_\tau(\mathbb{C}), \mathbb{Q})).$$

3.2.3 Abelian varieties parametrized by Shimura curves

Let $\mathbb{B}$ be a totally definite incoherent quaternion algebra and let $X = \lim_U X_U$ be the associated Shimura curve.

Let A be a simple abelian variety over F. We say that A *is parametrized by* X if there is a non-constant morphism $f : X_U \to A$ over F for some compact and open subgroup U of $\mathbb{B}_f^\times$. Using the normalized Hodge class ξ_U as a base point, it is equivalent to the existence of a nonzero homomorphism $J_U \to A$ where $J_U = \mathrm{Cl}^0(X_U)$ is the Jacobian variety.

Recall in Chapter 1 we have introduced the key definition

$$\pi_A = \mathrm{Hom}_\xi^0(X, A) = \varinjlim_U \mathrm{Hom}_{\xi_U}^0(X_U, A).$$

Here $\mathrm{Hom}_{\xi_U}^0(X_U, A)$ denotes the $\mathbb{Q}$-vector space of morphisms in

$$\mathrm{Hom}_F(X_U, A) \otimes_{\mathbb{Z}} \mathbb{Q}$$

which maps the Hodge class ξ_U of X_U to zero in A. More precisely,

$$\mathrm{Hom}_{\xi_U}^0(X_U, A)$$

consists of elements $a \otimes f$ for any rational number $a \in \mathbb{Q}$ and any morphism $f : X_U \to A$ such that ξ_U is mapped to zero in the composition

$$\mathrm{Pic}(X_U)_\mathbb{Q} \xrightarrow{f_*} \mathrm{CH}_0(A)_\mathbb{Q} \xrightarrow{\alpha} A(F)_\mathbb{Q}.$$

Here $\mathrm{CH}_0(A)$ is the Chow group of cycles of dimension zero on A, and $\alpha : \mathrm{CH}_0(A) \to A(F)$ is the canonical map sending a formal summation of points (up to linear equivalence) to the same summation of points with respect to the group law on A.

The collection $\{\mathrm{Hom}_{\xi_U}^0(X_U, A)\}_U$ forms a direct system naturally, and thus the direct limit makes sense. Since any morphism $X_U \to A$ factors through the Jacobian J_U of X_U, we have

$$\mathrm{Hom}_{\xi_U}^0(X_U, A) = \mathrm{Hom}^0(J_U, A),$$

and

$$\pi_A = \mathrm{Hom}^0(J, A) = \varinjlim_U \mathrm{Hom}^0(J_U, A).$$

It is easy to see that representation π_A is admissible.

THEOREM 3.8. *Let A be a simple abelian variety over F parametrized by X with $M = \mathrm{End}^0(A)$. Then A is of strict GL(2)-type, $\pi_A \in \mathcal{A}(\mathbb{B}^\times, \mathbb{Q})$, $\mathrm{End}(\pi_A) = M$ and*

$$L_v(s, A, M) = L(s, \pi_{A,v})$$

for each finite place v of F.

Moreover, the map $A \mapsto \pi_A$ defines a bijection

$$\{abelian\ varieties\ parametrized\ by\ X\ up\ to\ isogeny\} \xrightarrow{\sim} \mathcal{A}(\mathbb{B}^\times, \mathbb{Q}).$$

PROOF. Fix an embedding $\tau : F \hookrightarrow \mathbb{R}$. Theorem 3.7 gives a decomposition of the Hodge structure on $H^1(X_\tau(\mathbb{C}), \mathbb{Q})$ over $\mathbb{R}$ into a direct sum of $\pi \otimes_M V(\pi)$ over $\pi \in \mathcal{A}(\mathbb{B}^\times)$, where

$$M = \text{End}(\pi), \qquad V(\pi) = \text{Hom}_{\mathbb{B}^\times}(\pi, H^1(X_\tau(\mathbb{C}), \mathbb{Q})).$$

By Riemann's theorem, $V(\pi)$ defines a simple abelian variety $A_{\pi,\tau}$ over $\mathbb{R}$ such that $H^1(A_{\pi,\tau}(\mathbb{C}), \mathbb{Q}) = V(\pi)$ as a rational Hodge structure over $\mathbb{R}$. Moreover, the embedding

$$\pi \longrightarrow \text{Hom}_{\text{Hodge}/\mathbb{R}}(V(\pi), H^1(X_\tau(\mathbb{C}), \mathbb{Q}))$$

defines an embedding

$$\pi \longrightarrow \text{Hom}_{\mathbb{R}}^0(J_\tau, A_{\pi,\tau}).$$

LEMMA 3.9. *The following are true:*

(1) Every $A_{\pi,\tau}$ has a unique model A_π over F such that the image of π in $\text{Hom}^0(J_\tau, A_{\pi,\tau})$ is in $\text{Hom}^0(J, A_\pi)$;

(2) Each A_π has the endomorphism ring equal to $\text{End}(\pi)$ and the local L-series of A_π is given by the local L-series of π.

The lemma shows that each J_U has a decomposition in the category of abelian varieties up to isogeny by

$$J_U \simeq \bigoplus_{\pi \in \mathcal{A}(\mathbb{B}^\times, \mathbb{Q})} \pi^U \otimes_{\text{End}(\pi)} A_\pi.$$

Here A_π is simple and its local L-series is equal to those of π. This implies in particular that A_π is not isogenous to each other. Thus we have

$$\text{Hom}^0(J_U, A_\pi) \simeq \pi^U.$$

Consequently, every abelian varieties parametrized by X must be isogenous to one of A_π and then $\pi_A \simeq \pi$. The theorem follows.

It remains to prove the lemma. We need only check that in the proof of Theorem 3.7, the action of $\mathcal{H}_U$ on $H^1(J, \mathbb{Q})$ can be realized as algebraic correspondences on J. In fact, the characteristic function of the double coset UxU is given by the Hecke correspondence $Z(x)_U$ on the Jacobian J_U of X_U via push-forward. It gives an action of $\mathcal{H}_{U,\mathbb{Q}}$ on J_U (in the category of abelian varieties up to isogeny). Then $\mathbb{H}_{U,\mathbb{Q}}$ is generated by the image of $Z(x)_U$ in $\text{End}^0(J_U)$, and $\mathbb{T}_{U,\mathbb{Q}}$ is generated by the image of $Z(x)_U$ in $\text{End}^0(J_U)$ such that $x_v = 1$ at all places v such that U_v is not maximal. We can then define A_π to be $J_U \otimes_{\mathbb{T}_U} \text{End}(\pi)$ when π^U is one-dimensional. Clearly A_π has an action by $\text{End}(\pi)$ with the Betti cohomology $V(\pi)$ at place τ. This implies that $\text{End}(\pi) \subset \text{End}^0(A_\pi)$. Thus A is of GL(2)-type. By the Eichler–Shimura theory, the local L-series of A_π, given by L-series of forms appearing in the cotangent space of A_π, is exactly the L-series of π (up to a translation on s).

It remains to show that A_π is simple with $\mathrm{End}^0(A_\pi) = \mathrm{End}(\pi) =: M$. Write $A \simeq \oplus B_i^{n_i}$ up to isogeny, with B_i simple and non-isogenous to each other. Then we have an embedding

$$M \longrightarrow \oplus M_{n_i}(D_i), \qquad D_i := \mathrm{End}^0(B_i).$$

By Lemma 3.3, $\deg D_i \mid \dim B_i$. Thus $\deg M \mid n_i \dim B_i$. On the other hand

$$\deg M = \dim A_\pi = \sum_i n_i \dim B_i.$$

It follows that there is only one term in the decomposition. Thus we may write $A_\pi \simeq B^n$ with B simple and $D = \mathrm{End}^0(B)$. Then $\deg M = n \deg D$. It follows that D is commutative and included into M, and that for almost all places v,

$$L_v(s, A_\pi, M) = L_v(s, B, D)^n.$$

Thus almost all the local L-series of A_π have coefficients in D. It follows that the spherical eigenvalues on π take values in D. So we must have $M = D$, $A_\pi = B$. This shows that A_π is simple with $\mathrm{End}^0(A_\pi) = \mathrm{End}(\pi)$. $\qquad\square$

Let F be a totally real field. Let A be a simple abelian variety of GL(2)-type over F with $M = \mathrm{End}^0(A)$. We say that A is *automorphic* if there is a rational automorphic representation $\sigma \in \mathcal{A}(\mathbb{B}_0^\times)$ with $\mathrm{End}(\sigma) = M$ such that

$$L_v(s, A, M) = L(s - 1/2, \sigma_v)$$

for all finite places v of F. By the work of Carayol, it suffices to have equality for unramified primes. By Theorem 3.4, it is equivalent to the existence of a conjugacy class $\{\sigma^\iota : \iota\}$ in $\mathcal{A}(\mathbb{B}_0^\times)$ such that

$$L_v(s, A, \iota) = L(s - 1/2, \sigma_v^\iota).$$

The automorphy of an abelian variety of GL(2)-type depends only on its isogeny class, by Faltings's isogeny theorem in [Fa1].

We say that an automorphic representation $\sigma \in \mathcal{A}(\mathbb{B}_0^\times, \mathbb{Q})$ is *geometric* if it corresponds to an automorphic abelian variety of GL(2)-type over F. Denote by $\mathcal{A}^{geom}(\mathbb{B}_0, \mathbb{Q})$ the subset of geometric representation. Then we have defined a bijection between

$$\mathcal{A}^{geom}(\mathbb{B}_0, \mathbb{Q})$$

and

{Automorphic abelian varieties of GL(2)-type over F up to isogeny}.

CONJECTURE 3.10. *The following are true:*

(1) Every abelian varieties of GL(2)-type over F is automorphic.

(2) Every rational cuspidal representation in $\mathcal{A}(\mathbb{B}_0^\times, \mathbb{Q})$ is geometric.

3.2.4 Duality

Let A be an abelian variety over F parametrized by X, and denote $M = \mathrm{End}^0(A)$. Then the dual $A^\vee$ of A is also parametrized by X, and denote $M^\vee := \mathrm{End}(A^\vee)$. For any endomorphism $M : A \to A$, the pull-back map $m^* : \mathrm{Pic}^0(A) \to \mathrm{Pic}^0(A)$ gives a homomorphism $m^\vee : A^\vee \to A^\vee$. Thus we get a canonical isomophism

$$M \longrightarrow M^\vee, \quad m \longmapsto m^\vee.$$

Identify $M^\vee$ with M by this isomorphism.

The goal of this subsection is to consider the duality between

$$\pi_A = \mathrm{Hom}^0(J, A) = \varinjlim_U \mathrm{Hom}^0(J_U, A)$$

and

$$\pi_{A^\vee} = \mathrm{Hom}^0(J, A^\vee) = \varinjlim_U \mathrm{Hom}^0(J_U, A^\vee)$$

as representations of $\mathbb{B}^\times$ over M.

LEMMA 3.11. *The map*

$$(\cdot, \cdot)_U : \ \mathrm{Hom}^0(J_U, A) \times \mathrm{Hom}^0(J_U, A^\vee) \longrightarrow M$$

defined by

$$(f_1, f_2) \longmapsto f_1 \circ f_2^\vee$$

is a perfect pairing of vector spaces over M. Here for any $f_2 : J_U \to A^\vee$, the homomorphism $f_2^\vee : A \to J_U$ represents the homomorphism $f_2^* : \mathrm{Pic}^0(A^\vee) \to \mathrm{Pic}^0(J_U)$ under the canonical isomorphisms $(A^\vee)^\vee = A$ and $J_U^\vee = J_U$.

Moreover, the pairing is Hermitian in the sense that

$$(tx, y)_U = (x, t^\vee y)_U, \quad t \in \mathbb{H}_{U, \mathbb{Q}}.$$

Here the involution $t \mapsto t^\vee$ on $\mathbb{H}_{U, \mathbb{Q}}$ is induced by the transpose $Z(x)_U \mapsto Z(x^{-1})_U$.

PROOF. It is easy to see that the map

$$\mathrm{Hom}^0(J_U, A^\vee) \longrightarrow \mathrm{Hom}^0(A, J_U), \quad f_2 \longmapsto f_2^\vee$$

is an isomorphism. Via this isomorphism, the pairing $(\cdot, \cdot)_U$ becomes the composition map

$$\mathrm{Hom}^0(A, J_U) \times \mathrm{Hom}^0(J_U, A) \longrightarrow M.$$

It is verified to be perfect by writing J_U as a product of simple abelian varieties up to isogeny. The second result follows from the fact that the dual of the endomorphism $Z(x)_U : J_U \to J_U$ is exactly $Z(x^{-1})_U : J_U \to J_U$. It can be checked by definition. $\qquad\qquad\square$

Now we extend the pairing in the lemma to the direct limit. Define a pairing

$$(\cdot,\cdot): \ \pi_A \times \pi_{A^\vee} \longrightarrow M$$

by setting

$$(f_1, f_2) := \frac{1}{\mathrm{vol}(X_U)}(f_{1,U}, f_{2,U})_U = \frac{1}{\mathrm{vol}(X_U)}f_{1,U} \circ f_{2,U}^\vee.$$

Here $f_1 = \{f_{1,U}\}_U \in \pi_A$ and $f_2 = \{f_{2,U}\}_U \in \pi_{A^\vee}$, and U is any compact open subgroup of $\mathbb{B}_f^\times$ such that $f_{1,U}$ and $f_{2,U}$ are defined.

THEOREM 3.12. *The above definition does not depend on the choice of U and gives a perfect M-bilinear pairing*

$$(\cdot,\cdot): \ \pi_A \times \pi_{A^\vee} \longrightarrow M.$$

It is $\mathbb{B}^\times$-invariant in the sense that

$$(\pi_A(h)f_1, \pi_{A^\vee}(h)f_2) = (f_1, f_2), \quad \forall\, h \in \mathbb{B}^\times, \ f_1 \in \pi_A, \ f_2 \in \pi_{A^\vee}.$$

PROOF. We need only check the independence of U. Everything else follows from the lemma. For two compact open subgroups $U_1 \subset U_2$ of $\mathbb{B}_f^\times$, the projection $\phi: X_{U_1} \to X_{U_2}$ induces two morphisms

$$\phi^*: J_{U_2} \longrightarrow J_{U_1}, \quad \phi_*: J_{U_1} \longrightarrow J_{U_2}.$$

The composition is

$$\phi_* \circ \phi^* = \deg \phi = \mathrm{vol}(X_{U_1})/\mathrm{vol}(X_{U_2}).$$

Then the definition is independent of U. $\qquad\square$

Remark. In the limit level, the pairing is just the composition

$$A \xrightarrow{f_2^\vee} J^\vee \xrightarrow{\mathrm{vol}(X)^{-1}} J \xrightarrow{f_2} A.$$

Here the canonical element

$$\mathrm{vol}(X)^{-1} = \{\mathrm{vol}(X_U)^{-1}\}_U \in \mathrm{Hom}^0(J^\vee, J)$$

is as in §3.1.6.

Let $\lambda: A \to A^\vee$ be a polarization, and $\tau: M \to M$ be the Rosati involution induced by λ. Then we have an isomorphism $\pi_A \simeq \pi_{A^\vee}$ which is τ-linear under the action of M. In this way, we get a τ-Hermitian pairing

$$(\cdot,\cdot)_\lambda: \ \pi_A \times \pi_A \longrightarrow M.$$

For any nonzero $j \in \mathrm{Hom}^0(J_U, A)$, we can define a polarization by

$$\lambda := (j \circ j^\vee)^{-1}.$$

It can be shown that the induced Hermitian pairing on π_A is positive definite in the sense that, for nonzero element $f \in \pi_A$, the value $(f, f)_\lambda$ lies in the subgroup $M_+^\times$ of totally positive elements of $M^\times$. Especially, $(j, j)_\lambda = 1$.

Remark. If A is an elliptic curve, then $M = \mathbb{Q}$ and the canonical polarization gives an isomorphism $A \simeq A^\vee$. The resulting pairing $\pi_A \times \pi_A \to \mathbb{Q}$ on $\pi_A = \mathrm{Hom}_\xi^0(X, A)$ is defined by

$$(f_1, f_2) = \frac{1}{\mathrm{vol}(X_U)} f_{1,U} \circ f_{2,U}^\vee.$$

In particular,

$$(f, f) = \frac{1}{\mathrm{vol}(X_U)} \deg(f_U : X_U \to A).$$

3.2.5   The main theorem in complex coefficients

In the rest of the book, we will write the main theorem (Theorem 1.2) in complex coefficients. In fact,

$$L \otimes_{\mathbb{Q}} \mathbb{C} = \bigoplus_{\iota : L \hookrightarrow \mathbb{C}} \mathbb{C}.$$

THEOREM 3.13. *For any embedding $\iota : L \hookrightarrow \mathbb{C}$, we have*

$$\langle P_\chi(f_1)^\iota, \; P_{\chi^{-1}}(f_2)^\iota \rangle_{\mathrm{NT}} = \frac{\zeta_F(2) L'(1/2, \pi_A^\iota, \chi^\iota)}{4 L(1, \eta)^2 L(1, \pi_A^\iota, \mathrm{ad})} \alpha(f_1^\iota, f_2^\iota).$$

Here the basic setting is as in this section. Namely, X is a Shimura curve over a totally real field F, and A is an abelian variety parametrized by X. Some extra notations are as follows:

- L is a finite extension of $M = \mathrm{End}^0(A)$.

- $\chi : \mathrm{Gal}(E^{\mathrm{ab}}/E) \to L^\times$ be a character of finite order, also viewed as a character of $E^\times \backslash E_{\mathbb{A}}^\times$ by the reciprocity law.

- P be the CM point on X given by a CM extension E of F.

- The point

$$P_\chi(f_1) = \int_{\mathrm{Gal}(\overline{E}/E)} f_1(P^\tau) \otimes_M \chi(\tau) d\tau$$

lies in $A(E^{\mathrm{ab}})_{\mathbb{Q}} \otimes_M L$, and is χ-invariant under the action of $\mathrm{Gal}(E^{\mathrm{ab}}/E)$.

- The height

$$\langle P_\chi(f_1)^\iota, \ P_{\chi^{-1}}(f_2)^\iota \rangle_{\mathrm{NT}} \in \mathbb{C}$$

  is the projection of the L-linear height

$$\langle P_\chi(f_1), \ P_{\chi^{-1}}(f_2) \rangle_L \in L \otimes_{\mathbb{Q}} \mathbb{C}$$

  to the component $\mathbb{C}$ indexed by ι.

There is a more direct interpretation of the height pairing. The usual theory of Néron–Tate height gives a $\mathbb{Q}$-bilinear non-degenerate pairing

$$\langle \cdot, \cdot \rangle_{\mathrm{NT}} : A(\overline{F})_{\mathbb{Q}} \times A^\vee(\overline{F})_{\mathbb{Q}} \longrightarrow \mathbb{R}.$$

We refer to §7.1.1 for a quick review.

Recall that the field $M = \mathrm{End}^0(A)$ acts on $A(\overline{F})_{\mathbb{Q}}$ by definition, and acts on $A^\vee(\overline{F})_{\mathbb{Q}}$ through the duality. By the adjoint property of the height pairing in Proposition 7.3, the pairing $\langle \cdot, \cdot \rangle_{\mathrm{NT}}$ descends to a $\mathbb{Q}$-linear map

$$\langle \cdot, \cdot \rangle_{\mathrm{NT}} : A(\overline{F})_{\mathbb{Q}} \otimes_M A^\vee(\overline{F})_{\mathbb{Q}} \longrightarrow \mathbb{R}.$$

For simplicity, denote

$$V = A(\overline{F})_{\mathbb{Q}} \otimes_M A^\vee(\overline{F})_{\mathbb{Q}}.$$

It is an M-module. The $\mathbb{Q}$-linear map $\langle \cdot, \cdot \rangle_{\mathrm{NT}} : V \longrightarrow \mathbb{R}$ induces a $\mathbb{C}$-linear map

$$\langle \cdot, \cdot \rangle_{\mathrm{NT}} : V \otimes_{\mathbb{Q}} \mathbb{C} \longrightarrow \mathbb{C}.$$

By linear algebra,

$$V \otimes_{\mathbb{Q}} \mathbb{C} = V \otimes_M (M \otimes_{\mathbb{Q}} \mathbb{C}) = V \otimes_M (\oplus_{\iota: M \hookrightarrow \mathbb{C}} \mathbb{C}) = \oplus_{\iota: M \hookrightarrow \mathbb{C}} V^\iota.$$

Here we denote $W^\iota = W \otimes_{(M,\iota)} \mathbb{C}$ for any vector space W over M, and denote by w^ι the image of w in W^ι for any $w \in W$.

The $\mathbb{C}$-linear map induces, for each $\iota : M \hookrightarrow \mathbb{C}$, a $\mathbb{C}$-linear map

$$\langle \cdot, \cdot \rangle_{\mathrm{NT}}^\iota : V^\iota \longrightarrow \mathbb{C}.$$

In other words, we obtain a $\mathbb{C}$-bilinear pairing

$$\langle \cdot, \cdot \rangle_{\mathrm{NT}} : A(\overline{F})_{\mathbb{Q}}^\iota \times A^\vee(\overline{F})_{\mathbb{Q}}^\iota \longrightarrow \mathbb{C}.$$

3.3 MAIN THEOREM IN TERMS OF PROJECTORS

By the basic definitions and basic properties in §3.1.5 and §3.1.6, we define the projector $\mathrm{T}(f_1 \otimes f_2)$, and state an equivalent form of the main theorem (Theorem 1.2) in Theorem 3.15. In §3.3.4 we prove the equivalence between these two theorems, based on an expression of the projector in terms of parametrizations of modular abelian varieties in §3.3.3.

3.3.1   Projector I: Cohomological definition

The goal of this subsection is, for any $\pi \in \mathcal{A}(\mathbb{B}^\times)$, to introduce a homomorphism

$$\mathrm{T}: \ \pi \otimes \widetilde{\pi} \longrightarrow \mathrm{Hom}^0(J, J^\vee)_\mathbb{C}.$$

It will be given by Hecke correspondences.

Fix a decomposition

$$H^{1,0}(X_\tau) = \bigoplus_{\pi \in \mathcal{A}(\mathbb{B}^\times)} \pi.$$

For any open compact subgroup U of $\mathbb{B}_f^\times$, the decomposition induces a decomposition

$$H^{1,0}(X_{U,\tau}) = \bigoplus_{\pi \in \mathcal{A}(\mathbb{B}^\times)} \pi^U,$$

and a decomposition of the dual space

$$H^{1,0}(X_{U,\tau})^\vee = \bigoplus_{\pi \in \mathcal{A}(\mathbb{B}^\times)} \widetilde{\pi}^U.$$

It follows that

$$\mathrm{Hom}(H^{1,0}(X_{U,\tau}), \ H^{1,0}(X_{U,\tau})) = \bigoplus_{\pi_1, \pi_2 \in \mathcal{A}(\mathbb{B}^\times)} \pi_1^U \otimes \widetilde{\pi}_2^U.$$

It gives an injection

$$i_U : \Pi^{\Delta,U} \longrightarrow \mathrm{Hom}(H^{1,0}(X_{U,\tau}), \ H^{1,0}(X_{U,\tau})),$$

where the "diagonal"

$$\Pi^{\Delta,U} := \bigoplus_{\pi \in \mathcal{A}(\mathbb{B}^\times)} \pi^U \otimes \widetilde{\pi}^U.$$

Note that the choice of the isomorphism

$$H^{1,0}(X_\tau) \longrightarrow \bigoplus_{\pi \in \mathcal{A}(\mathbb{B}^\times)} \pi$$

as an $\mathcal{H}$-modules is not unique. One has the freedom of multiplying every component on the right-hand side by a constant. Once such an isomorphism is chosen, the corresponding isomorphisms of $H^{1,0}(X_{U,\tau})^\vee$ and $H^{1,0}(X_{U,\tau})^\vee$ are uniquely determined. In particular, the map i_U does not depend on the choice of the isomorphism for $H^{1,0}(X_\tau)$.

We will prove in Proposition 3.14 that the image of i_U is actually contained in the image of the inclusion

$$\mathrm{Hom}^0(J_U, J_U^\vee)_\mathbb{C} \longrightarrow \mathrm{Hom}(H^{1,0}(X_{U,\tau}), \ H^{1,0}(X_{U,\tau})).$$

Hence, it induces a well-defined map

$$T_U : \Pi^{\Delta, U} \hookrightarrow \mathrm{Hom}^0(J_U, J_U^\vee)_{\mathbb{C}}.$$

Now vary U. Denote

$$\Pi^\Delta := \bigoplus_{\pi \in \mathcal{A}(\mathbb{B}^\times)} \pi \otimes \widetilde{\pi}.$$

We will prove that the system

$$T = \{\mathrm{vol}(X_U) T_U\}_U$$

gives a well-defined map

$$T : \Pi^\Delta \longrightarrow \mathrm{Hom}^0(J, J^\vee)_{\mathbb{C}}.$$

It is the definition of the projector we propose in this section.

Next, we prove the algebraicity of the image of i_U, which is crucial for the definition of T_U and T. We also sketch the reason for the compatibility of the system $T = \{\mathrm{vol}(X_U) T_U\}_U$.

PROPOSITION 3.14. *(1) Let U be an open compact subgroup of $\mathbb{B}_f^\times$. For any $\alpha \in \Pi^{\Delta, U}$, there is a function $\phi \in \mathcal{H}_U$ such that $i_U(\alpha) = T(\phi)_U^*$ in $\mathrm{Hom}(H^{1,0}(X_{U,\tau}), H^{1,0}(X_{U,\tau}))$. Hence, the map*

$$T_U : \Pi^{\Delta, U} \hookrightarrow \mathrm{Hom}^0(J_U, J_U^\vee)_{\mathbb{C}}$$

is well-defined and independent of the choice of the embedding $\tau : F \hookrightarrow \mathbb{C}$.

(2) The system $T = \{\mathrm{vol}(X_U) T_U\}_U$ is a direct system and defines a map

$$T : \Pi^\Delta \longrightarrow \mathrm{Hom}^0(J, J^\vee)_{\mathbb{C}}.$$

PROOF. We first prove (1), which is the essential part of the proposition. Consider the natural map

$$R : \mathcal{H}_U \longrightarrow \bigoplus_{\pi \in \mathcal{A}(\mathbb{B}^\times)} \mathrm{End}_{\mathbb{C}}(\pi^U) = \Pi^{\Delta, U}.$$

Here the direct sums are actually finite sums of finite-dimensional representations of $\mathcal{H}_U$ since π^U is nonzero for only finitely many π.

The map R is surjective, and takes ϕ to be a preimage of α in $\mathcal{H}_U$. It satisfies the requirement. The surjectivity is also used in the proof of Theorem 3.4. It follows from the density theorem of Jacobson and Chevalley for semisimple modules (cf. Theorem 11.16 of [La]), and the property that $\{\pi^U \neq 0\}$ is a finite set of distinct finite-dimensional irreducible $\mathcal{H}_U$-modules.

Now we consider (2). It suffices to show that the system

$$i := \{\mathrm{vol}(X_U) i_U\}_U$$

gives a well-defined map

$$i : \Pi^\Delta \longrightarrow \mathrm{Hom}_{\mathrm{cont}}(H^{1,0}(X_\tau)', \ H^{1,0}(X_\tau)).$$

For any $U' \subset U$, the pull-back map $\pi^*_{U',U} : H^{1,0}(X_{U,\tau}) \to H^{1,0}(X_{U',\tau})$ induces a map $\pi^*_{U',U} : \pi^U \to \pi^{U'}$ exactly equal to the natural inclusion $\pi^U \hookrightarrow \pi^{U'}$. The push-forward map $(\pi_{U',U})_* : H^{1,0}(X_{U',\tau}) \to H^{1,0}(X_{U,\tau})$ induces a map $(\pi_{U',U})_* : \pi^{U'} \to \pi^U$ which is more complicated than the pull-back. However, we always have $(\pi_{U',U})_* \pi^*_{U',U} = \deg \pi_{U',U}$.

It suffices to verify that, for any $f_1 \otimes f_2 \in \pi^U \otimes \widetilde{\pi}^U$ and $f' \in \pi^{U'}$,

$$\mathrm{vol}(X_{U'}) \cdot \mathrm{T}'_{U'}(f_1 \otimes f_2)(f') = \mathrm{vol}(X_U) \cdot \pi^*_{U',U} \left(\mathrm{T}'_U(f_1 \otimes f_2)((\pi_{U',U})_* f') \right).$$

By definition,

$$\mathrm{T}'_{U'}(f_1 \otimes f_2)(f') = (f', f_2) \ f_1.$$

and

$$\mathrm{T}'_U(f_1 \otimes f_2)((\pi_{U',U})_* f') = ((\pi_{U',U})_* f', f_2) \ f_1.$$

It is reduced to check

$$\mathrm{vol}(X_{U'}) \cdot (f', f_2) = \mathrm{vol}(X_U) \cdot ((\pi_{U',U})_* f', f_2).$$

Notice that $f_2 \in \widetilde{\pi}^U$ is invariant under U, so

$$(f', f_2) = (f'', f_2), \quad ((\pi_{U',U})_* f', f_2) = ((\pi_{U',U})_* f'', f_2),$$

where $f'' \in \pi^U$ is the average

$$f'' = \fint_U \pi(h) f' dh.$$

Then we have

$$(\pi_{U',U})_* f'' = (\pi_{U',U})_* \pi^*_{U',U} f'' = \deg(\pi_{U',U}) f''.$$

It gives

$$\mathrm{vol}(X_{U'}) \cdot (f'', f_2) = \mathrm{vol}(X_U) \cdot ((\pi_{U',U})_* f'', f_2).$$

The result follows.   □

3.3.2 Formulation in terms of projectors

Assume the following notations and assumptions:

- F is a totally real field with adele ring $\mathbb{A} = \mathbb{A}_F$.

- $\mathbb{B}$ is a totally definite incoherent quaternion algebra over $\mathbb{A}$.

- E is a totally imaginary quadratic extension of F, with a fixed embedding $E_{\mathbb{A}} \hookrightarrow \mathbb{B}$ over $\mathbb{A}$.

- $\pi \in \mathcal{A}(\mathbb{B}^{\times})$, i.e., π is an irreducible admissible complex representation of $\mathbb{B}^{\times}$ such that its Jacquet–Langlands correspondence σ is a cuspidal automorphic representation of $\mathrm{GL}_2(\mathbb{A})$, discrete of weight two at all infinite places.

- $\chi : E^{\times}\backslash E_{\mathbb{A}}^{\times} \to \mathbb{C}^{\times}$ is a character of finite order with $\omega_{\pi} \cdot \chi|_{\mathbb{A}^{\times}} = 1$.

- Denote by $\eta : F^{\times}\backslash \mathbb{A}^{\times} \to \mathbb{C}^{\times}$ the quadratic character associated to the extension E/F.

Recall that P is an element in $X^{E^{\times}}(E^{\mathrm{ab}})$ which we fix once for all. Introduce the χ-eigencomponent

$$P_{\chi} = \fint_{T(F)\backslash T(\mathbb{A})/Z(\mathbb{A})} \mathrm{T}_t(P - \xi_P)\, \chi(t)dt \in J(\overline{F})_{\mathbb{C}}.$$

Here T_x denotes the Hecke action given by right-multiplication, acting as pushforward on the divisor. For the regularized integral, we refer to §1.6. The following is an equivalent form of Theorem 1.2.

THEOREM 3.15. *For any $f_1 \otimes f_2 \in \pi \otimes \widetilde{\pi}$,*

$$\langle \mathrm{T}(f_1 \otimes f_2)P_{\chi}, P_{\chi^{-1}}\rangle_{\mathrm{NT}} = \frac{\zeta_F(2)L'(1/2, \pi, \chi)}{4L(1, \eta)^2 L(1, \pi, \mathrm{ad})}\alpha(f_1 \otimes f_2).$$

We explain the height pairing in the theorem. By definition,

$$\mathrm{T}(f_1 \otimes f_2) \in \mathrm{Hom}^0(J, J^{\vee})_{\mathbb{C}} \subset \mathrm{Hom}(J(\overline{F})_{\mathbb{C}}, J^{\vee}(\overline{F})_{\mathbb{C}}).$$

Thus we have

$$\mathrm{T}(f_1 \otimes f_2)P_{\chi} \in J^{\vee}(\overline{F})_{\mathbb{C}}.$$

The height pairing in the theorem is just the natural height pairing

$$\langle \cdot, \cdot \rangle_{\mathrm{NT}} : J(\overline{F})_{\mathbb{C}} \times J^{\vee}(\overline{F})_{\mathbb{C}} \longrightarrow \mathbb{C}.$$

It is obtained as the limit of the usual Néron–Tate height pairing. See §3.1.6 for more details.

In the following, we prove the equivalence between Theorem 3.15 and Theorem 1.2.

3.3.3 Projector II: Algebraic interpretation

Assume the notation of Theorem 1.2. Recall that A is an abelian variety parametrized by X, and

$$\pi_A = \mathrm{Hom}^0(J, A) = \varinjlim_{U} \mathrm{Hom}^0(J_U, A).$$

It is an irreducible admissible representation of $\mathbb{B}^\times$ over $M = \operatorname{End}^0(A)$. By the Eichler–Shimura construction, there is a canonical action of M on J_U, compatible with both push-forward and pull-back when varying U. In this sense, M commutes with π_A.

Recall that we have introduced the duality

$$(\cdot, \cdot): \ \pi_A \times \pi_{A^\vee} \longrightarrow M$$

by

$$(f_1, f_2) = \frac{1}{\operatorname{vol}(X_U)}(f_{1,U}, f_{2,U})_U = \frac{1}{\operatorname{vol}(X_U)} f_{1,U} \circ f_{2,U}^\vee.$$

Now we consider the composition in the opposite order.

Denote

$$\mathrm{T}_{\mathrm{alg}}(f_1, f_2)_U := f_{2,U}^\vee \circ f_{1,U} \in \operatorname{Hom}^0(J_U, J_U).$$

It is easy to see that $\mathrm{T}_{\mathrm{alg}}(f_1, f_2) := \{\mathrm{T}_{\mathrm{alg}}(f_1, f_2)_U\}_U$ is a direct system. It defines an element

$$\mathrm{T}_{\mathrm{alg}}(f_1, f_2) \in \operatorname{Hom}^0(J, J^\vee).$$

It follows that we have obtained a map

$$\mathrm{T}_{\mathrm{alg}}: \pi_A \times \pi_{A^\vee} \longrightarrow \operatorname{Hom}^0(J, J^\vee).$$

Since M commutes with all the related maps here, the map descends to a map

$$\mathrm{T}_{\mathrm{alg}}: \pi_A \otimes_M \pi_{A^\vee} \longrightarrow \operatorname{Hom}^0(J, J^\vee).$$

For any embedding $\iota: M \hookrightarrow \mathbb{C}$, the base change by ι gives

$$\mathrm{T}_{\mathrm{alg}}^\iota: \pi_A^\iota \otimes_\mathbb{C} \pi_{A^\vee}^\iota \longrightarrow \operatorname{Hom}^0(J, J^\vee)^\iota.$$

Note that $\operatorname{Hom}^0(J, J^\vee)^\iota$ embeds naturally into $\operatorname{Hom}^0(J, J^\vee) \otimes_\mathbb{Q} \mathbb{C}$. See §1.6.8 for example. Therefore, we can also write the map as

$$\mathrm{T}_{\mathrm{alg}}^\iota: \pi_A^\iota \otimes_\mathbb{C} \pi_{A^\vee}^\iota \longrightarrow \operatorname{Hom}^0(J, J^\vee) \otimes_\mathbb{Q} \mathbb{C}.$$

It actually gives a decomposition

$$\mathrm{T}_{\mathrm{alg}} = \sum_{\iota \in \operatorname{Hom}(M, \mathbb{C})} \mathrm{T}_{\mathrm{alg}}^\iota$$

in $\operatorname{Hom}^0(J, J^\vee) \otimes_\mathbb{Q} \mathbb{C}$. It is just the decomposition induced by the spectral decomposition under the action of M.

Recall that, by the cohomological method, we have also defined a projector

$$\mathrm{T}: \pi_A^\iota \otimes_\mathbb{C} \pi_{A^\vee}^\iota \longrightarrow \operatorname{Hom}^0(J, J^\vee) \otimes_\mathbb{Q} \mathbb{C}.$$

The definition employs the duality map

$$(\cdot, \cdot): \ \pi_A^\iota \times \pi_{A^\vee}^\iota \longrightarrow \mathbb{C}$$

obtained from the duality between π_A and $\pi_{A^\vee}$ over M. It is reasonable to expect that these two definitions agree.

PROPOSITION 3.16. *For any $f_1 \in \pi_A$ and $f_2 \in \pi_{A^\vee}$, and embedding $\iota : M \hookrightarrow \mathbb{C}$, one has*

$$\mathrm{T}(f_1^\iota \otimes f_2^\iota) = \mathrm{T}_{\mathrm{alg}}(f_1 \otimes f_2)^\iota$$

in $\mathrm{Hom}^0(J, J^\vee) \otimes_{\mathbb{Q}} \mathbb{C}$.

PROOF. It suffices to prove the identity on the level of any open compact subgroup U of $\mathbb{B}_f^\times$. Let

$$f_1 \in \pi_A^U = \mathrm{Hom}^0(J_U, A), \quad f_2 \in \pi_{A^\vee}^U = \mathrm{Hom}^0(J_U, A^\vee).$$

We need to prove

$$\mathrm{vol}(X_U) \cdot \mathrm{T}(f_1^\iota \otimes f_2^\iota)_U = \mathrm{T}_{\mathrm{alg}}(f_1 \otimes f_2)_U^\iota.$$

Here the right-hand side is defined as an element of $\mathrm{Hom}^0(J_U, J_U^\vee) \otimes_{\mathbb{Q}} \mathbb{C}$ by a similar method. Since we are in the case of characteristic zero, it suffices to prove that they induce the same action on the tangent space, namely $H^{1,0}(J_{U,\tau})$.

By the property of GL(2)-type, $H^{1,0}(A_\tau)$ is a free module of rank one under the pull-back action of

$$M \otimes_{\mathbb{Q}} \mathbb{C} = \bigoplus_{\iota \in \mathrm{Hom}(M,\mathbb{C})} \mathbb{C}.$$

Thus we can decompose

$$H^{1,0}(A_\tau) = \bigoplus_{\iota \in \mathrm{Hom}(M,\mathbb{C})} H^{1,0}(A_\tau)(\iota)$$

by the idempotents. Each piece $H^{1,0}(A_\tau)(\iota)$ is just a $\mathbb{C}$-vector space of dimension one, on which M acts by ι.

On the other hand, the action of M on $H^{1,0}(J_{U,\tau})$ via pull-back induces a decomposition

$$H^{1,0}(J_{U,\tau}) = H^{1,0}(J_{U,\tau})(0) \oplus \bigoplus_{\iota \in \mathrm{Hom}(M,\mathbb{C})} H^{1,0}(J_{U,\tau})(\iota).$$

Here M kills $H^{1,0}(J_{U,\tau})(0)$, and acts on $H^{1,0}(J_{U,\tau})(\iota)$ by ι. Fix a nonzero element $\omega^\iota \in H^{1,0}(A_\tau)(\iota)$. Then we obtain an isomorphism

$$\pi_A^{U,\iota} \longrightarrow H^{1,0}(J_{U,\tau})(\iota), \quad g \longmapsto g^* \omega^\iota.$$

It is equivariant under the action of $\mathcal{H}_U$.

The pull-back action of $\mathrm{T}_{\mathrm{alg}}(f_1^\iota \otimes f_2^\iota)_U^\iota$ on $H^{1,0}(X_{U,\tau})$ is given by

$$(f_2^\vee \circ f_1)^* : H^{1,0}(X_{U,\tau})(\iota) \longrightarrow H^{1,0}(X_{U,\tau})(\iota).$$

We need to prove that

$$(f_2^\vee \circ f_1)^* \alpha = \mathrm{vol}(X_U) \cdot i(f_1^\iota \otimes f_2^\iota)_U \alpha$$

for any $\alpha \in H^{1,0}(X_{U,\tau})(\iota)$. Here $i(f_1^\iota \otimes f_2^\iota)_U$ is the induced action of $\mathrm{T}(f_1^\iota \otimes f_2^\iota)_U$ on $H^{1,0}(X_{U,\tau})$.

We can always write $\alpha = c \cdot f_1'^* \omega^\iota$ for some $f_1' \in \pi_A^U$ and $c \in \mathbb{C}$. Note that the pairing $\pi_A^{U,\iota} \times \pi_{A^\vee}^{U,\iota} \to \mathbb{C}$ is induced by the pairing

$$\pi_A^U \times \pi_{A^\vee}^U \longrightarrow M, \quad (g_1, g_2) \longmapsto \mathrm{vol}(X_U)^{-1} \cdot g_1 \circ g_2^\vee.$$

It follows that

$$i(f_1^\iota \otimes f_2^\iota)_U \; f_1'^* \omega^\iota = \mathrm{vol}(X_U)^{-1} \cdot (f_1' \circ f_2^\vee)^\iota \cdot f_1^* \omega^\iota.$$

Hence, the desired equality becomes

$$(f_2^\vee \circ f_1)^* f_1'^* \omega^\iota = (f_1' \circ f_2^\vee)^\iota \cdot f_1^* \omega^\iota.$$

Note that $(f_1' \circ f_2^\vee) \in M$ behaves as a scalar. We have

$$(f_2^\vee \circ f_1)^* f_1'^* = (f_1' \circ f_2^\vee \circ f_1)^* = (f_1' \circ f_2^\vee)^\iota \cdot f_1^*.$$

The equality follows.   □

3.3.4   The equivalence

Now we consider the equivalence between Theorem 3.15 and Theorem 1.2. Resume the notation of Theorem 1.2. It suffices to prove the height identity

$$\langle \mathrm{T}(f_1^\iota \otimes f_2^\iota) P_{\chi^\iota}, P_{(\chi^\iota)^{-1}} \rangle_{\mathrm{NT}} = \langle P_\chi(f_1)^\iota, \; P_{\chi^{-1}}(f_2)^\iota \rangle_{\mathrm{NT}}.$$

Under the reciprocity law

$$\mathrm{rec} : E^\times \backslash E_\mathbb{A}^\times \longrightarrow \mathrm{Gal}(E^{\mathrm{ab}}/E),$$

both sides depend on (χ, χ^{-1}) in the same manner. So it suffices to prove that, for any $P_1, P_2 \in J(\overline{F})_\mathbb{Q}$,

$$\langle \mathrm{T}(f_1^\iota \otimes f_2^\iota) P_1, P_2 \rangle_{\mathrm{NT}} = \langle P_1(f_1)^\iota, \; P_2(f_2)^\iota \rangle_{\mathrm{NT}}.$$

Here

$$P_1(f_1) = f_{1*} P_1 \in A(\overline{F})_\mathbb{Q}, \quad P_2(f_2) = f_{2*} P_2 \in A^\vee(\overline{F})_\mathbb{Q}.$$

By Proposition 3.16,

$$\mathrm{T}(f_1^\iota \otimes f_2^\iota) = \mathrm{T}_{\mathrm{alg}}(f_1 \otimes f_2)^\iota.$$

It follows that, in $J^\vee(\overline{F})_\mathbb{Q} \otimes_\mathbb{Q} \mathbb{C}$,

$$\mathrm{T}(f_1^\iota \otimes f_2^\iota) P_1 = \mathrm{T}_{\mathrm{alg}}(f_1 \otimes f_2)^\iota P_1 = \mathrm{T}_{\mathrm{alg}}(f_1 \otimes f_2)^\iota P_1^\iota = \mathrm{T}_{\mathrm{alg}}(f_1 \otimes f_2) P_1^\iota.$$

It is reduced to check

$$\langle \mathrm{T}_{\mathrm{alg}}(f_1 \otimes f_2) P_1^\iota, P_2^\iota \rangle_{\mathrm{NT}} = \langle P_1(f_1)^\iota, \; P_2(f_2)^\iota \rangle_{\mathrm{NT}}.$$

This follows from the projection formula. Assume that $f_1 \in \pi_A^U$ and $f_1 \in \pi_{A^\vee}^U$ for some U. Then they are realized as $f_1 : J_U \to A$ and $f_2 : J_U \to A^\vee$. By definition, we can view $\mathrm{T}_{\mathrm{alg}}(f_1 \otimes f_2) = f_2^\vee \circ f_1$ as an endomorphism of J_U. Realize P_1 and P_2 as points in $J_U(\overline{F})$. It follows that

$$\langle \mathrm{T}_{\mathrm{alg}}(f_1 \otimes f_2) P_1^\iota, P_2^\iota \rangle_{\mathrm{NT}} = \langle f_2^\vee(f_1(P_1^\iota)), P_2^\iota \rangle_{\mathrm{NT}} = \langle f_1(P_1^\iota), f_2(P_2^\iota) \rangle_{\mathrm{NT}}.$$

Here the last identity follows from Proposition 7.3. This finishes the proof.

Both Theorem 3.15 and Theorem 1.2 are not in the forms for which we can perform computations. By expressing the projector $\mathrm{T}(f_1 \otimes f_2)$ as an arithmetic theta lifting, we will obtain another equivalent form in Theorem 3.21. The statement of Theorem 3.21 is the major goal for the remaining sections of this chapter.

3.4 THE GENERATING SERIES

Let $\mathbb{V}$ be the orthogonal space $\mathbb{B}$ with reduced norm q. We consider the space $\mathcal{S}(\mathbb{V} \times \mathbb{A}^\times)$ with an action of $\mathbb{B}^\times \times \mathbb{B}^\times \times \mathrm{GL}_2(\mathbb{A})$ given by the Weil representation, and the space

$$\mathrm{Pic}(X \times X) := \varinjlim_U \mathrm{Pic}(X_U \times X_U)$$

with an action of $\mathbb{B}^\times \times \mathbb{B}^\times$ by *right multiplications*. In this section we want to construct an element

$$\widetilde{Z} \in \mathrm{Hom}_{\mathbb{B}^\times \times \mathbb{B}^\times \times \mathrm{GL}_2(\mathbb{A})}(\mathcal{S}(\mathbb{V} \times \mathbb{A}^\times), \quad C^\infty(\mathrm{GL}_2(F)\backslash\mathrm{GL}_2(\mathbb{A})) \otimes \mathrm{Pic}(X \times X))$$

using Kudla's generating series and modularity proved in [YZZ].

3.4.1 New class of Schwartz functions

Our first observation is that $\mathrm{Pic}(X \times X)$ is invariant under $\mathbb{B}_\infty^\times \times \mathbb{B}_\infty^\times$. Thus the element Z must factor through the maximal $\mathbb{B}_\infty^\times \times \mathbb{B}_\infty^\times$ quotient of $\mathcal{S}(\mathbb{V} \times \mathbb{A}^\times)$. Such a quotient is identified with the space $\overline{\mathcal{S}}(\mathbb{V} \times \mathbb{A}^\times)$ of functions on $\mathbb{V} \times \mathbb{A}^\times$ obtained as integrals

$$\int_{F_\infty^\times \backslash (\mathbb{B}_\infty^\times \times \mathbb{B}_\infty^\times)} r(h_\infty) \Phi \, dh_\infty, \qquad \Phi \in \mathcal{S}(\mathbb{V} \times \mathbb{A}^\times).$$

Here $F_\infty^\times$ is embedded diagonally into $\mathbb{B}_\infty^\times \times \mathbb{B}_\infty^\times$ and dh_∞ is any fixed Haar measure of the quotient.

Since

$$F_\infty^\times \backslash (\mathbb{B}_\infty^\times \times \mathbb{B}_\infty^\times) = F_\infty^\times \mathbb{B}_\infty^1 \times \mathbb{B}_\infty^1 / \{\pm 1\},$$

the integral is just averages on $F_\infty^\times$ and $\mathbb{B}_\infty^1 \times \mathbb{B}_\infty^1$. It is easy to see that the average on $\mathbb{B}_\infty^1 \times \mathbb{B}_\infty^1$ is the same as the average on one component of $\mathbb{B}_\infty^1 \times \mathbb{B}_\infty^1$. Thus we introduce the simplified integration

$$\overline{\Phi} := \int_{F_\infty^\times} \int_{\mathbb{B}_\infty^1} r(ch) \Phi \, dh dc, \qquad \Phi \in \mathcal{S}(\mathbb{V} \times \mathbb{A}^\times).$$

Here the integral on $\mathbb{B}^1_\infty$ uses the Haar measure of total volume one, and the integral on $F^\times_\infty$ uses the usual Haar measure introduced in §1.6. We normalize the quotient map by

$$\mathcal{S}(\mathbb{V} \times \mathbb{A}^\times) \longrightarrow \overline{\mathcal{S}}(\mathbb{V} \times \mathbb{A}^\times), \quad \Phi \longmapsto \overline{\Phi}.$$

It is easy to see that $\overline{\mathcal{S}}(\mathbb{V} \times \mathbb{A}^\times)$ has a decomposition

$$\overline{\mathcal{S}}(\mathbb{V} \times \mathbb{A}^\times) = \otimes_v \overline{\mathcal{S}}(\mathbb{V}_v \times F^\times_v).$$

Here $\overline{\mathcal{S}}(\mathbb{V}_v \times F^\times_v) = \mathcal{S}(\mathbb{V}_v \times F^\times_v)$ if v is non-archimedean.

Assume that v is archimedean. We will describe $\overline{\mathcal{S}}(\mathbb{V}_v \times F^\times_v)$ in more details. It consists of all functions of the form

$$\overline{\Phi}_v := \int_{F^\times_v} \int_{\mathbb{B}^1_v} r(ch)\Phi_v \, dh dc, \quad \Phi_v \in \mathcal{S}(\mathbb{V}_v \times F^\times_v).$$

Recall that $\mathcal{S}(\mathbb{V}_v \times F^\times_v)$ is the space of finite linear combinations of functions of the form

$$H(u)P(x)e^{-2\pi|u|q(x)}$$

where P is any polynomial function on $\mathbb{V}_v$, and H is any smooth and compactly supported function on $F^\times_v$. Then it is easy to verify that $\overline{\mathcal{S}}(\mathbb{V}_v \times F^\times_v)$ is the space of functions on $\mathbb{V}_v \times F^\times_v$ of the form

$$(P_1(uq(x)) + \mathrm{sgn}(u)P_2(uq(x))) \, e^{-2\pi|u|q(x)}$$

where P_1 and P_2 are polynomials with complex coefficients. Here $\mathrm{sgn}(u) = u/|u|$ denotes the sign of $u \in \mathbb{R}^\times$.

The Weil representation descends to an action of $\mathrm{GL}_2(F_v) \times \mathbb{B}^\times_v \times \mathbb{B}^\times_v$ on $\overline{\mathcal{S}}(\mathbb{V}_v \times F^\times_v)$. Here $\mathbb{B}^\times_v \times \mathbb{B}^\times_v$ acts trivially, and $\mathrm{GL}_2(F_v)$ acts by the same formula as $\mathcal{S}(\mathbb{V}_v \times F^\times_v)$. By the tensor product, we have the Weil representation of $\mathrm{GL}_2(\mathbb{A}) \times \mathbb{B}^\times \times \mathbb{B}^\times$ on $\overline{\mathcal{S}}(\mathbb{V} \times \mathbb{A}^\times)$.

3.4.2   Constant term of Eisenstein series

The constant term of the generating series is defined in terms of the constant term of the corresponding Siegel Eisenstein series. So we recall some results on the Eisenstein series in §2.5.2 in slightly different notations.

Fix $\phi \in \overline{\mathcal{S}}(\mathbb{V} \times \mathbb{A}^\times)$ and $u \in F^\times$. We have a Siegel Eisenstein series

$$E(s, g, u, \phi) = \sum_{\gamma \in P^1(F) \backslash \mathrm{SL}_2(F)} \delta(\gamma g)^s r(\gamma g)\phi(0, u), \quad g \in \mathrm{GL}_2(\mathbb{A}).$$

In the case $g \in \mathrm{SL}_2(\mathbb{A})$, in terms of the notation in §2.5.2, it is just $E(s, g, \phi(\cdot, u))$ defined by the quadratic space $(\mathbb{B}, uq)$.

It is standard to have the Fourier expansion

$$E(s, g, u, \phi) = \delta(g)^s r(g)\phi(0, u) + W_0(s, g, u, \phi) + \sum_{a \in F^\times} W_a(s, g, u, \phi)$$

where

$$W_a(s, g, u, \phi) := \int_{\mathbb{A}} \delta(wn(b)g)^s \ r(wn(b)g)\phi(0, u) \ \psi(-ab)db, \quad a \in F.$$

We are particularly interested in the value at $s = 0$ of the constant term

$$E_0(s, g, u, \phi) := \delta(g)^s r(g)\phi(0, u) + W_0(s, g, u, \phi).$$

The definition is naturally extended to all $u \in \mathbb{A}^\times$.

Analytic properties of the intertwining part $W_0(s, g, u, \phi)$ are discussed in §2.5.2. For the definition of the generating series, we need the following three basic properties:

- $W_0(s, g, u, \phi)$ has an analytic continuation to $s = 0$, and thus $W_0(0, g, u, \phi)$ is well-defined.

- $W_0(s, g, u, \phi) \neq 0$ only if $F = \mathbb{Q}$ and $\Sigma = \{\infty\}$.

- $W_0(0, g, u, \phi) = W_0(0, 1, u, r(g)\phi)$ and thus $E_0(0, g, u, \phi) = E_0(0, 1, u, r(g)\phi)$.

At $s = 0$, we get

$$E_0(0, g, u, \phi) = r(g)\phi(0, u) + W_0(0, g, u, \phi).$$

Here the $W_0(0, g, u, \phi) \neq 0$ only if $F = \mathbb{Q}$ and $\Sigma = \{\infty\}$.

For convenience, we sometimes write

$$E_0(g, u, \phi) := E_0(0, g, u, \phi), \quad E_0(u, \phi) := E_0(0, 1, u, \phi),$$

and

$$W_0(g, u, \phi) := W_0(0, g, u, \phi), \quad W_0(u, \phi) := W_0(0, 1, u, \phi).$$

3.4.3 Hecke correspondences

Fix an open and compact group U of $\mathbb{B}_f^\times$. Recall that we have defined $Z(x)_U$ to be the image of the morphism

$$(\pi_{U_x, U}, \pi_{U_x, U} \circ T_x) : \quad X_{U_x} \longrightarrow X_U \times X_U.$$

Here $U_x = U \cap xUx^{-1}$ is an open and compact subgroup of $\mathbb{B}_f^\times$.

In terms of complex uniformization above, the push-forward map by the Hecke correspondence $Z(x)_U$ gives

$$Z(x)_U : [z, \beta]_U \longmapsto \sum_{y \in UxU/U} [z, \beta y]_U.$$

Here $[z, \beta]_U$ represents the image of $(z, \beta) \in \mathcal{H}^\pm \times \mathbb{B}_f^\times$ in $X_{U,\tau}(\mathbb{C})$.

On the other hand, we can view $Z(x)_U$ as an element of $\text{Pic}(X_U \times X_U)$. It is the view point we will take here to define the generating series.

For convenience, for any $x = x_f x_\infty \in \mathbb{B}^\times$, by $Z(x)_U$ we mean $Z(x_f)_U$.

3.4.4 Hodge classes

On $M_K := X_U \times X_U$, one has a *Hodge class* $L_K \in \mathrm{Pic}(M_K) \otimes \mathbb{Q}$ defined as

$$L_K := \frac{1}{2}(p_1^* L_U + p_2^* L_U).$$

Here the Hodge class L_U of X_U is introduced in §3.1.3. Next, we introduce some notations for components of L_K.

After fixing a base point of X, the geometrically connected components of X_U are indexed by $F_+^\times \backslash \mathbb{A}_f^\times / q(U)$, and we use $X_{U,\alpha}$ to denote the corresponding component for $\alpha \in F_+^\times \backslash \mathbb{A}_f^\times / q(U)$. Then the geometrically connected components of $M_K = X_U \times X_U$ are naturally indexed by $(\alpha_1, \alpha_2) \in (F_+^\times \backslash \mathbb{A}_f^\times / q(U))^2$. For any $\alpha \in F_+^\times \backslash \mathbb{A}_f^\times / q(U)$, denote

$$M_{K,\alpha} = \coprod_{\beta \in F_+^\times \backslash \mathbb{A}_f^\times / q(U)} X_{U,\beta} \times X_{U,\alpha\beta}$$

as a subvariety of M_K. It is still defined over F by the reciprocity law which describes the Galois action on the components. Then

$$M_K = \coprod_{\alpha \in F_+^\times \backslash \mathbb{A}_f^\times / q(U)} M_{K,\alpha}.$$

View the Hodge bundle

$$L_{K,\alpha} := L_K|_{M_{K,\alpha}}$$

of $M_{K,\alpha}$ as a line bundle of M_K by trivial extension outside $M_{K,\alpha}$.

3.4.5 Generating series

For any $\phi \in \overline{\mathcal{S}}(\mathbb{V} \times \mathbb{A}^\times)$ invariant under $K = U \times U$, form a generating series

$$Z(\phi)_U := Z_0(\phi)_U + Z_*(\phi)_U. \tag{3.4.1}$$

Here the constant term and the non-constant part are respectively

$$Z_0(\phi)_U := -\sum_{\alpha \in F_+^\times \backslash \mathbb{A}_f^\times / q(U)} \sum_{u \in \mu_U^2 \backslash F^\times} E_0(\alpha^{-1} u, \phi) \, L_{K,\alpha}, \tag{3.4.2}$$

$$Z_*(\phi)_U := w_U \sum_{a \in F^\times} \sum_{x \in K \backslash \mathbb{B}_f^\times} \phi(x, aq(x)^{-1}) \, Z(x)_U. \tag{3.4.3}$$

Here $\mu_U = F^\times \cap U$, $\mu_U^2 = \{\alpha^2 : \alpha \in \mu_U\}$ and $w_U = |\{1, -1\} \cap U|$ is equal to 1 or 2. It is easy to see that both μ_U and μ_U^2 are subgroups of the unit group $O_F^\times$ of finite indexes, and that w_U is 1 for U small enough. Furthermore, $\mu_U^2 = \{1\}$ if $F = \mathbb{Q}$.

The constant term

$$E_0(u, \phi) = \phi(0, u) + W_0(u, \phi)$$

is introduced in §3.4.2. The intertwining part $W_0(u, \phi)$ is "the only nonholomorphic part" of the generating series. It is nonzero only when $F = \mathbb{Q}$ and $\Sigma = \{\infty\}$, which happens exactly when the Shimura curve has cusps. We will see the reason for $W_0(u, \phi)$ to appear in the generating series in §4.2 and §4.3.

For any $g \in \mathrm{GL}_2(\mathbb{A})$, we write

$$Z(g, \phi)_U = Z(r(g)\phi)_U, \quad Z_0(g, \phi)_U = Z_0(r(g)\phi)_U, \quad Z_*(g, \phi)_U = Z_*(r(g)\phi)_U.$$

They are considered as functions on $\mathrm{GL}_2(\mathbb{A})$ with vector values in $\mathrm{Pic}(X_U \times X_U)_{\mathbb{C}}$.

For any $\Phi \in \mathcal{S}(\mathbb{V} \times \mathbb{A}^{\times})$ invariant under $K = U \times U$, it is routine to define

$$Z(\Phi)_U = Z(\overline{\Phi})_U, \quad Z_0(\Phi)_U = Z_0(\overline{\Phi})_U, \quad Z_*(\Phi)_U = Z_*(\overline{\Phi})_U.$$

Here the averaging $\overline{\Phi} \in \overline{\mathcal{S}}(\mathbb{V} \times \mathbb{A}^{\times})$ of Φ is defined in §3.4.1.

THEOREM 3.17. *The series $Z(r(g)\Phi)_U$ is absolutely convergent and defines an automorphic form on $g \in \mathrm{GL}_2(\mathbb{A})$ with coefficients in $\mathrm{Pic}(X_U \times X_U)_{\mathbb{C}}$.*

By the modularity, we mean that $\ell(Z(r(g)\Phi)_U)$ is absolutely convergent and defines an automorphic form for any linear functional $\ell : \mathrm{Pic}(X_U \times X_U)_{\mathbb{C}} \to \mathbb{C}$. The theorem will be proved in §4.2. It is essentially the modularity result proved in [YZZ]. But here let us consider some of its functorial properties.

3.4.6 Normalization of the generating series

We normalize the generating series as follows:

$$\widetilde{Z}(\Phi)_U := \frac{2^{[F:\mathbb{Q}]-1} h_F |D_F|^{-\frac{1}{2}}}{[O_F^{\times} : \mu_U^2]} Z(\Phi)_U. \tag{3.4.4}$$

Here h_F denotes the class number of F and D_F denotes the discriminant of F. The whole normalizing factor is always 2 if $F = \mathbb{Q}$. The denominator is finite since μ_U^2 is a subgroup of $O_F^{\times}$ with finite index, and it makes the right-hand side compatible under pull-back from different levels. Only the numerator is independent of U, and we choose it for reasons we will see later. The normalizing factor can also be written as

$$\frac{2^{[F:\mathbb{Q}]-1} h_F |D_F|^{-\frac{1}{2}}}{[O_F^{\times} : \mu_U^2]} = \frac{1}{2R_{\mu_U^2}} \mathrm{Res}_{s=1} \zeta_F(s). \tag{3.4.5}$$

Here $R_{\mu_U^2}$ is the "regulator" of μ_U^2 in $\mathbb{R}^{[F:\mathbb{Q}]-1}$, whose definition is similar to the regulator of F. The equality follows from the class number formula.

This normalization makes the system

$$\widetilde{Z}(\Phi) := \{\widetilde{Z}(\Phi)_U\}_U$$

a well-defined element of $\mathrm{Pic}(X \times X)_{\mathbb{C}}$.

LEMMA 3.18. *The system $\widetilde{Z}(\Phi)$ is compatible with pull-back and thus defines an element of* $\mathrm{Pic}(X \times X)_{\mathbb{C}}$. *The map*

$$(\Phi, g) \mapsto \widetilde{Z}(r(g)\Phi)$$

defines an element

$$\widetilde{Z} \in \mathrm{Hom}_{\mathbb{B}^\times \times \mathbb{B}^\times \times \mathrm{GL}_2(\mathbb{A})}(\overline{\mathcal{S}}(\mathbb{V} \times \mathbb{A}^\times), \quad C^\infty(\mathrm{GL}_2(F) \backslash \mathrm{GL}_2(\mathbb{A})) \otimes \mathrm{Pic}(X \times X))_{\mathbb{C}}.$$

PROOF. By the modularity in Theorem 3.17, it suffices to prove the compatibility and the equivariance under $\mathbb{B}^\times \times \mathbb{B}^\times$. It suffices to check that, for any open compact subgroups U' and U of $\mathbb{B}_f^\times$ and any elements h_1 and h_2 of $\mathbb{B}_f^\times$, such that $h_1 U' h_1^{-1} \subset U$ and $h_2 U' h_2^{-1} \subset U$ and that U acts trivially on Φ, one has

$$\Pi^* Z(\Phi)_U = \frac{1}{[\mu_U^2 : \mu_{U'}^2]} Z(r(h_1, h_2)\Phi)_{U'},$$

where

$$\Pi : X_{U'} \times X_{U'} \longrightarrow X_U \times X_U$$

is the morphism given by the right multiplication by (h_1, h_2). The case $h_1 = h_2 = 1$ gives the independence of $\widetilde{Z}$ on U, and the general case gives the equivariance.

We check the corresponding identity for the constant term

$$Z_0(\Phi)_U = - \sum_{\alpha \in F_+^\times \backslash \mathbb{A}_f^\times / q(U)} \sum_{u \in \mu_U^2 \backslash F^\times} E_0(\alpha^{-1}u, \overline{\Phi}) \, L_{K,\alpha}$$

and the a-th Fourier coefficient

$$Z_a(\Phi)_U = w_U \sum_{x \in U \backslash \mathbb{B}_f^\times / U} \overline{\Phi}(x, aq(x)^{-1}) \, Z(x)_U$$

for all $a \in F^\times$.

Since the pull-backs of the Hodge bundles are still the Hodge bundles and Π maps $M_{K',\alpha'}$ to $M_{K',\alpha' q(h_1^{-1}h_2)}$, one has

$$\Pi^* L_{K,\alpha} = \sum_{\alpha' \in F_+^\times \backslash (F_+^\times \alpha q(h_1 h_2^{-1}) q(U)) / q(U')} L_{K',\alpha'}.$$

It follows that

$$
\begin{aligned}
\Pi^* Z_0(\Phi)_U &= - \sum_{\alpha' \in F_+^\times \backslash \mathbb{A}_f^\times / q(U')} \sum_{u \in \mu_U^2 \backslash F^\times} E_0(\alpha'^{-1} q(h_1 h_2^{-1})u, \overline{\Phi}) \, L_{K',\alpha'} \\
&= - \sum_{\alpha' \in F_+^\times \backslash \mathbb{A}_f^\times / q(U')} \sum_{u \in \mu_U^2 \backslash F^\times} E_0(\alpha'^{-1}u, r(h_1, h_2)\overline{\Phi}) \, L_{K',\alpha'}.
\end{aligned}
$$

Replace $\mu_U^2 \backslash F^\times$ by $\mu_{U'}^2 \backslash F^\times$ in the inner summation. We get

$$\Pi^* Z_0(\Phi)_U = \frac{1}{[\mu_U^2 : \mu_{U'}^2]} Z_0(r(h_1, h_2)\Phi)_{U'}.$$

To compute $\Pi^* Z_a(\Phi)_U$, the key is to express $\Pi^* Z(x)_U$ as a linear combination of $Z(x')_{U'}$. Note that two subvarieties $Z(x_1')_{U'} = Z(x_2')_{U'}$ if and only if the cosets $F^\times U' x_1' U' = F^\times U' x_2' U'$. And $Z(x')_{U'}$ is a component of $\Pi^* Z(x)_U$ if and only if $h_1^{-1} x' h_2 \subset F^\times U x U$. It follows that

$$\Pi^* Z(x)_U = \sum_{h_1^{-1} x' h_2 \in (F^\times U') \backslash F^\times U x U / U'} Z(x')_{U'}.$$

Arrange $\Pi^* Z_a(\Phi)_U$ in terms of $Z(x')_{U'}$. We have

$$\Pi^* Z_a(\Phi)_U = w_U \sum_{x' \in (F^\times U') \backslash \mathbb{B}_f^\times / U'} \sum_{x \in U \backslash F^\times U h_1^{-1} x' h_2 U / U} \overline{\Phi}(x, aq(x)^{-1}) \, Z(x')_{U'}.$$

It is easy to see tha $U \backslash F^\times U h_1^{-1} x' h_2 U / U = \mu_U \backslash F^\times h_1^{-1} x' h_2$. Thus

$$\Pi^* Z_a(\Phi)_U = w_U \sum_{x' \in (F^\times U') \backslash \mathbb{B}_f^\times / U'} \sum_{b \in \mu_U \backslash F^\times} r(h_1, h_2)\overline{\Phi}(bx', aq(bx')^{-1}) \, Z(x')_{U'}.$$

Replace $\mu_U \backslash F^\times$ by $\mu_{U'} \backslash F^\times$ in the inner summation. We get

$$\Pi^* Z_a(\Phi)_U = \frac{w_U}{[\mu_U : \mu_{U'}]} \sum_{x' \in (F^\times U') \backslash \mathbb{B}_f^\times / U'} \sum_{b \in \mu_{U'} \backslash F^\times}$$
$$r(h_1, h_2)\overline{\Phi}(bx', aq(bx')^{-1}) \, Z(x')_{U'}.$$

Note that $Z(x')_{U'} = Z(bx')_{U'}$ and $[\mu_U : \mu_U^2] = 2^{[F:\mathbb{Q}]-1} w_U$. The above becomes

$$\Pi^* Z_a(\Phi)_U = \frac{w_{U'}}{[\mu_U^2 : \mu_{U'}^2]} \sum_{x' \in U' \backslash \mathbb{B}_f^\times / U'} r(h_1, h_2)\overline{\Phi}(x', aq(x')^{-1}) \, Z(x')_{U'}.$$

It is just

$$\Pi^* Z_a(\Phi)_U = \frac{1}{[\mu_U^2 : \mu_{U'}^2]} Z_a(r(h_1, h_2)\Phi)_{U'}.$$

$\square$

3.5 GEOMETRIC KERNEL

Let χ be a finite character of $E^\times \backslash E_\mathbb{A}^\times$. In this section we define our geometric kernel function $Z(g, \chi, \Phi)$ by means of height pairing of CM points for each

$\Phi \in \mathcal{S}(\mathbb{V} \times \mathbb{A}^\times)$. More precisely we will construct an element $\widetilde{Z}(\cdot, \chi, \cdot)$ in the space

$$\mathrm{Hom}_{\mathrm{GL}_2(\mathbb{A}) \times E_\mathbb{A}^\times \times E_\mathbb{A}^\times}(\mathcal{S}(\mathbb{V} \times \mathbb{A}^\times) \boxtimes \chi \boxtimes \chi^{-1}, \ C_0^\infty(\mathrm{GL}_2(F) \backslash \mathrm{GL}_2(\mathbb{A}), \chi|_{\mathbb{A}^\times})).$$

Here $\mathcal{S}(\mathbb{V} \times \mathbb{A}^\times) \boxtimes \chi \boxtimes \chi^{-1}$ means the space $\mathcal{S}(\mathbb{V} \times \mathbb{A}^\times)$ with the action of $\mathrm{GL}_2(\mathbb{A}) \times E_\mathbb{A}^\times \times E_\mathbb{A}^\times$ by the Weil representation twisted by the character (χ, χ^{-1}) on $E_\mathbb{A}^\times \times E_\mathbb{A}^\times$; and $C_0^\infty(\mathrm{GL}_2(F) \backslash \mathrm{GL}_2(\mathbb{A})), \chi|_{\mathbb{A}^\times})$ means the space of cuspidal and smooth functions on $\mathrm{GL}_2(F) \backslash \mathrm{GL}_2(\mathbb{A})$ with the character $\chi|_{\mathbb{A}^\times}$ under translation by $Z(\mathbb{A})$ and with trivial action by $E_\mathbb{A}^\times \times E_\mathbb{A}^\times$.

The element is automatically cuspidal though the original generating series $Z(\Phi)$ does not need to be without some assumption (cf. Assumption 5.4).

3.5.1 Height series

Let $\Phi \in \overline{\mathcal{S}}(\mathbb{V} \times \mathbb{A}^\times)$ be a Schwartz function as above. Recall that we have a generating series $\widetilde{Z}(\Phi)$ whose coefficients lie in $\mathrm{Pic}(X \times X)_\mathbb{C}$. By 3.1.6, we see that the push-forward action of $\widetilde{Z}(\Phi)_U$ on X_U gives a natural map

$$\widetilde{Z}(\Phi): \quad J(\overline{F})_\mathbb{C} \longrightarrow J^\vee(\overline{F})_\mathbb{C}.$$

As in the introduction, fix an element $P \in X^{E^\times}(E^{\mathrm{ab}})$. For any $h \in \mathbb{B}^\times$, denote $[h] = T(h)P$ and $[h]^\circ = [h] - \xi_{q(h)}$. Here we have identified $\pi_0(X_{U, \overline{F}})$ with $F_+^\times \backslash \mathbb{A}_f^\times / q(U)$ so that P_U is indexed by 1 for each U. By definition, $[h]^\circ \in J(\overline{F})_\mathbb{C}$.

For any $h_1, h_2 \in \mathbb{B}^\times$, define the height series

$$\widetilde{Z}(g, (h_1, h_2), \Phi) := \langle \widetilde{Z}(g, \Phi) \, [h_1]^\circ, \ [h_2]^\circ \rangle_{\mathrm{NT}}, \quad g \in \mathrm{GL}_2(\mathbb{A}). \tag{3.5.1}$$

Here $\widetilde{Z}(g, \Phi) \, [h_1]^\circ \in J^\vee(\overline{F})_\mathbb{C}$ and $[h_2]^\circ \in J(\overline{F})_\mathbb{C}$, and the Néron–Tate height pairing

$$\langle \cdot, \cdot \rangle_{\mathrm{NT}} : J(\overline{F})_\mathbb{C} \times J^\vee(\overline{F})_\mathbb{C} \longrightarrow \mathbb{C}$$

is obtained naturally from the Néron–Tate height pairing on each level U. See §3.1.6 for more details.

It is clear that the definition does not depend on the infinite parts of h_1, h_2. And below we have more basic properties.

LEMMA 3.19. *The definition of $\widetilde{Z}(g, (h_1, h_2), \Phi)$ is independent of the choice of P, and it is invariant under the left action of $T(F) \times T(F)$ on (h_1, h_2). Furthermore, it is always a cusp form on $g \in \mathrm{GL}_2(\mathbb{A})$.*

PROOF. Take any $t_1, t_2 \in T(\mathbb{A})$. By definition,

$$\widetilde{Z}(g, (t_1 h_1, t_2 h_2), \Phi) = \langle \widetilde{Z}(g, \Phi) \, T(h_1) T(t_1) P^\circ, \ T(h_2) T(t_2) P^\circ \rangle_{\mathrm{NT}}.$$

View $T(t_1)P^\circ$ and $T(t_2)P^\circ$ as Galois conjugates of P° by the reciprocity law. If $t_1 = t_2$, then the height pairing does not change since the height pairing is invariant under the Galois action. If $t_1, t_2 \in T(F)$, then by the definition of P we get $T(t_1)P^\circ = T(t_2)P^\circ$, and the pairing still does not change.

The cuspidality of $\widetilde{Z}(g, (h_1, h_2), \Phi)$ follows from the cuspidality of $\widetilde{Z}(g, \Phi) [h_1]^\circ$. In other words, the constant term $\widetilde{Z}_0(g, \Phi) [h_1]^\circ = 0$. It is equivalent to check $Z_0(g, \Phi) [h_1]^\circ_U = 0$ for any open compact subgroup U of $\mathbb{B}_f^\times$ acting trivially on Φ. We will postpone this to §4.3.1. Roughly speaking, as a correspondence, $Z_0(g, \Phi)_U$ is a linear combination of Hodge classes $L_{K, \alpha, \beta}$, the (α, β)-component of the total Hodge bundle

$$L_K = \frac{1}{2}(p_1^* L_U + p_2^* L_U).$$

It is very easy to see that they map $\mathrm{Div}^0(X_{U, \overline{F}})$ to 0.   □

3.5.2   Geometric kernel

Recall that χ is a character on $T(F)\backslash T(\mathbb{A})$ which is trivial at infinity. In terms of the regularized integrations introduced in §1.6, we define

$$\widetilde{Z}(g, \chi, \Phi) := \int_{T(F)\backslash T(\mathbb{A})/Z(\mathbb{A})}^* \widetilde{Z}(g, (t, 1), \Phi) \, \chi(t) dt. \qquad (3.5.2)$$

It is called *the geometric kernel*.

LEMMA 3.20. *The automorphic form $\widetilde{Z}(g, \chi, \Phi)$ is cuspidal and of central character $\chi|_{\mathbb{A}^\times}$ on $g \in \mathrm{GL}_2(\mathbb{A})$. Moreover, for any $t_1, t_2 \in E_\mathbb{A}^\times$,*

$$\widetilde{Z}(g, \chi, r(t_1, t_2)\Phi) = \chi(t_1^{-1} t_2)\widetilde{Z}(g, \chi, \Phi).$$

Thus the map $\Phi \mapsto \widetilde{Z}(g, \chi, \Phi)$ defines an element

$$\widetilde{Z}(\cdot, \chi, \cdot) \in \mathrm{Hom}_{\mathrm{GL}_2(\mathbb{A}) \times E_\mathbb{A}^\times \times E_\mathbb{A}^\times}(\mathcal{S}(\mathbb{V} \times \mathbb{A}^\times) \boxtimes \chi \boxtimes \chi^{-1},$$
$$C_0^\infty(\mathrm{GL}_2(F)Z(\mathbb{A})\backslash \mathrm{GL}_2(\mathbb{A}), \chi|_{\mathbb{A}^\times})).$$

PROOF. Only the central character needs justification. Let $c \in \mathbb{A}^\times$ be in the center of $\mathrm{GL}_2(\mathbb{A})$. We need to show $\widetilde{Z}(cg, \chi, \Phi) = \chi(c)\,\widetilde{Z}(g, \chi, \Phi)$. By definition,

$$\widetilde{Z}(cg, \chi, \Phi) = \int_{T(F)\backslash T(\mathbb{A})/Z(\mathbb{A})}^* \widetilde{Z}(cg, (t, 1), \Phi) \, \chi(t) dt.$$

We claim that $\widetilde{Z}(cg, (t, 1), \Phi) = \widetilde{Z}(g, (c^{-1}t, 1), \Phi)$. It is easy to see that the claim implies the desired result.

It suffices to check

$$\widetilde{Z}(cg, \Phi) [t]^\circ = \widetilde{Z}(g, \Phi) [c^{-1}t]^\circ.$$

And we only need to check

$$Z(cg, \Phi)_U \; [t]_U^\circ = Z(g, \Phi)_U \; [c^{-1}t]_U^\circ$$

on the level U. By the cuspidality, we only need to show

$$Z_a(cg, \Phi)_U \; [t]_U^\circ = Z_a(g, \Phi)_U \; [c^{-1}t]_U^\circ$$

for any $a \in F^\times$. By definition,

$$Z_a(cg, \Phi)_U = \sum_{x \in U \backslash \mathbb{B}_f^\times / U} r(cg)\Phi(x, aq(x)^{-1}) \; Z(x)_U.$$

Note that $r(cg)\Phi(x, aq(x)^{-1}) = r(g)\Phi(cx, aq(cx)^{-1})$. The above becomes

$$Z_a(cg, \Phi)_U = \sum_{x \in U \backslash \mathbb{B}_f^\times / U} r(g)\Phi(x, aq(x)^{-1}) \; Z(c^{-1}x)_U.$$

Since c^{-1} is in the center, right multiplication by c^{-1} gives an automorphism of X_U which switches the geometrically connected components of X_U by c^{-2}. This automorphism is exactly the Hecke correspondence $Z(c^{-1})_U$. Then we have $Z(c^{-1}x)_U = Z(x)_U \circ Z(c^{-1})_U$ as operators, which gives

$$Z_a(cg, \Phi)_U = Z_a(g, \Phi)_U \circ Z(c^{-1})_U.$$

It finishes the proof. □

3.6 ANALYTIC KERNEL AND KERNEL IDENTITY

We recall the analytic kernel, and state the main result on a relation between the analytic kernel and the geometric kernel.

3.6.1 The analytic kernel

Fix a Schwartz function $\Phi \in \mathcal{S}(\mathbb{V} \times \mathbb{A}^\times)$. Recall the associated series

$$I(s, g, \Phi) = \sum_{\gamma \in P^1(F) \backslash \mathrm{SL}_2(F)} \delta(\gamma g)^s \sum_{(x_1, u) \in E \times F^\times} r(\gamma g)\Phi(x_1, u)$$

and the twisted average

$$I(s, g, \chi, \Phi) = \int_{T(F) \backslash T(\mathbb{A})} I(s, g, r(t, 1)\Phi) \; \chi(t)dt.$$

Up to some simple factors, we have

$$L(\frac{s+1}{2}, \pi, \chi) \approx (I(s, g, \chi, \Phi), \; \varphi(g))_{\mathrm{Pet}}$$

for $\varphi \in \sigma$. Taking derivative, we get

$$L'(\frac{1}{2}, \pi, \chi) \approx (I'(0, g, \chi, \Phi), \ \varphi(g))_{\mathrm{Pet}}.$$

It follows that $I'(0, g, \chi, \Phi)$ is the kernel function representing $L'(\frac{1}{2}, \pi, \chi)$. We call it the *analytic kernel*.

3.6.2   Kernel identity

THEOREM 3.21. *Let* $\Phi \in \mathcal{S}(\mathbb{V} \times \mathbb{A}^\times)$ *be any Schwartz function. Then*

$$(I'(0, \cdot, \chi, \Phi), \ \varphi)_{\mathrm{Pet}} = 2 \left(\widetilde{Z}(\cdot, \chi, \Phi), \ \varphi \right)_{\mathrm{Pet}}, \quad \forall \ \varphi \in \sigma.$$

Recall that the Petersson bilinear pairing

$$(\varphi_1, \varphi_2)_{\mathrm{Pet}} := \int_{Z(\mathbb{A})\mathrm{GL}_2(F)\backslash \mathrm{GL}_2(\mathbb{A})} \varphi_1(g)\varphi_2(g)dg.$$

It only makes sense if φ_1, φ_2 are automorphic with inverse central characters. Note that the pairing is not Hermitian, but bilinear in our definition. The integration uses the Tamagawa measure, though it does not matter in the current theorem.

The pairings in the theorem make sense. The central character of $\widetilde{Z}(g, \chi, \Phi)$ is $\chi|_{\mathbb{A}^\times} = \omega_\sigma^{-1}$ by Lemma 3.20, and it is easy to check that the same result is true for $I'(0, g, \chi, \Phi)$.

3.6.3   Projector III: Arithmetic theta lifting

Let $\Phi \in \mathcal{S}(\mathbb{V} \times \mathbb{A}^\times)$ be a Schwartz function. Recall that we have a generating series $\widetilde{Z}(g, \Phi)$, which is a modular form in g with coefficients in $\mathrm{Pic}(X \times X)_{\mathbb{C}}$. Consider the "arithmetic theta lifting"

$$\widetilde{Z}(\Phi \otimes \varphi) := \int_{\mathrm{GL}_2(F)\backslash \mathrm{GL}_2(\mathbb{A})/Z(\mathbb{A})}^* \varphi(g) \, \widetilde{Z}(g, \Phi) \, dg, \quad \varphi \in \sigma.$$

Note that $\widetilde{Z}(g, \Phi)$ has no central character, so the integral is a regularized integral introduced in §1.6. By definition, we have

$$\widetilde{Z}(\Phi \otimes \varphi) \in \mathrm{Pic}(X \times X)_{\mathbb{C}}.$$

By §3.1.6, via the push-forward action, we can view

$$\widetilde{Z}(\Phi \otimes \varphi) \in \mathrm{Hom}^0(J, J^\vee)_{\mathbb{C}}.$$

On the other hand, we have a theta lifting

$$\theta : \mathcal{S}(\mathbb{V} \times \mathbb{A}^\times) \otimes \sigma \longrightarrow \pi \otimes \widetilde{\pi}.$$

It is normalized in (2.2.3) place by place. By §3.3.1, one has a map

$$\mathrm{T} : \pi \otimes \widetilde{\pi} \longrightarrow \mathrm{Hom}^0(J, J^\vee)_\mathbb{C}.$$

The composition gives

$$\mathrm{T} \circ \theta : \mathcal{S}(\mathbb{V} \times \mathbb{A}^\times) \otimes \sigma \longrightarrow \mathrm{Hom}^0(J, J^\vee)_\mathbb{C}.$$

THEOREM 3.22. *As an identity in* $\mathrm{Hom}^0(J, J^\vee)_\mathbb{C}$, *one has*

$$\widetilde{Z}(\Phi \otimes \varphi) = \frac{L(1, \pi, \mathrm{ad})}{2\zeta_F(2)} \mathrm{T}(\theta(\Phi \otimes \varphi)), \quad \Phi \in \mathcal{S}(\mathbb{V} \times \mathbb{A}^\times), \; \varphi \in \sigma.$$

It is not surprising that the theorem is true up to a constant. Namely, there is a constant $c(\pi)$ independent of Φ and φ such that

$$\widetilde{Z}(\Phi \otimes \varphi) = c(\pi)\mathrm{T}(\theta(\Phi \otimes \varphi)).$$

To see the existence of $c(\pi)$, it suffices to check this in

$$\mathrm{Hom}_{\mathrm{cont}}(H^{1,0}(X_\tau)', \; H^{1,0}(X_\tau))$$

by the natural embedding

$$\mathrm{Hom}^0(J, J^\vee)_\mathbb{C} \longhookrightarrow \mathrm{Hom}_{\mathrm{cont}}(H^{1,0}(X_\tau)', \; H^{1,0}(X_\tau)).$$

Fix an embedding $\tau : F \to \mathbb{C}$, and fix a decomposition

$$H^{1,0}(X_\tau) = \bigoplus_{\pi_1 \in \mathcal{A}(\mathbb{B}^\times)} \pi_1.$$

It gives a decomposition

$$H^{1,0}(X_{U,\tau}) = \bigoplus_{\pi_1 \in \mathcal{A}(\mathbb{B}^\times)} \pi_1^U,$$

and a decomposition of the dual space

$$H^{1,0}(X_{U,\tau})'^\vee = \bigoplus_{\pi_2 \in \mathcal{A}(\mathbb{B}^\times)} \widetilde{\pi}_2^U.$$

Then

$$\mathrm{Hom}(H^{1,0}(X_{U,\tau})', \; H^{1,0}(X_{U,\tau})) = \bigoplus_{\pi_1, \pi_2 \in \mathcal{A}(\mathbb{B}^\times)} \pi_1^U \otimes \widetilde{\pi}_2^U.$$

Taking direct limit, we obtain

$$\mathrm{Hom}_{\mathrm{cont}}(H^{1,0}(X_\tau)', \; H^{1,0}(X_\tau)) \simeq \bigoplus_{\pi_1, \pi_2 \in \mathcal{A}(\mathbb{B}^\times)} \pi_1 \otimes \widetilde{\pi}_2.$$

It follows that both $\widetilde{Z}$ and $T \circ \theta$ can be viewed as elements of

$$\text{Hom}(\mathcal{S}(\mathbb{V} \times \mathbb{A}^\times) \otimes \sigma, \bigoplus_{\pi_1, \pi_2 \in \mathcal{A}(\mathbb{B}^\times)} \pi_1 \otimes \widetilde{\pi}_2).$$

The key point now is that both of them are invariant under the diagonal action of $\text{GL}_2(\mathbb{A})$ on $\Phi \otimes \varphi$, and the action of $\mathbb{B}^\times \times \mathbb{B}^\times$ on Φ. So both of them must be multiples of the Shimizu lifting. Namely, they belong to the one-dimensional space

$$\text{Hom}_{\text{GL}_2(\mathbb{A}) \times \mathbb{B}^\times \times \mathbb{B}^\times}(\mathcal{S}(\mathbb{V} \times \mathbb{A}^\times) \otimes \sigma, \ \pi \otimes \widetilde{\pi}).$$

We already know that $T \circ \theta$ is a nonzero element in this space. So there is a constant $c(\pi)$ such that

$$\widetilde{Z} = c(\pi) \cdot T \circ \theta. \tag{3.6.1}$$

In the next chapter, we will use the Lefschetz fixed point theorem to prove that the constant $c(\pi)$ is the same as in Theorem 3.22.

3.6.4 Proof of Theorem 3.15 by the kernel identity

Here we deduce Theorem 3.15 from Theorem 3.21 and Theorem 3.22.

Let f_1 and f_2 be as Theorem 3.15. Let

$$\theta = \otimes_v \theta_v : \mathcal{S}(\mathbb{B} \times \mathbb{A}^\times) \otimes \sigma \longrightarrow \pi \otimes \widetilde{\pi}$$

be the Shimizu lifting normalized in (2.2.3). The map is surjective since $\pi \otimes \widetilde{\pi}$ is an irreducible representation of $\mathbb{B}^\times \times \mathbb{B}^\times$. Then $f_1 \otimes f_2$ must be in the image of θ. By linearity, we can assume that

$$f_1 \otimes f_2 = \theta(\varphi \otimes \Phi), \quad \varphi \in \sigma, \ \Phi \in \mathcal{S}(\mathbb{B} \times \mathbb{A}^\times).$$

Apply Theorem 3.21 to (φ, Φ) above. We have

$$(I'(0, g, \chi, \Phi), \ \varphi(g))_{\text{Pet}} = 2 \left(\widetilde{Z}(g, \chi, \Phi), \ \varphi(g)\right)_{\text{Pet}}.$$

Recall in (2.3.1) that we defined that

$$P(s, \chi, \Phi, \varphi) = (I(s, g, \chi, \Phi), \ \varphi(g))_{\text{Pet}}.$$

By Proposition 2.6,

$$P'(0, \chi, \Phi, \varphi) = \frac{L'(1/2, \pi, \chi)}{2L(1, \eta)} \alpha(\theta(\Phi \otimes \varphi)).$$

It remains to convert the right-hand side. We claim that

$$\left(\widetilde{Z}(g, \chi, \Phi), \ \varphi(g)\right)_{\text{Pet}} = \frac{L(1, \pi, \text{ad})L(1, \eta)}{\zeta_F(2)} \langle T(f_1 \otimes f_2)P_\chi, \ P_{\chi^{-1}} \rangle_{\text{NT}}.$$

It obviously implies the theorem.

To prove the claim, recall that

$$
\begin{aligned}
\widetilde{Z}(g,\chi,\Phi) &= \int_{T(F)\backslash T(\mathbb{A})/Z(\mathbb{A})}^{*} \langle \widetilde{Z}(g,\Phi)[t]^{\circ},\ 1\rangle_{\mathrm{NT}}\ \chi(t)dt \\
&= 2L(1,\eta)\fint_{T(F)\backslash T(\mathbb{A})/Z(\mathbb{A})} \langle \widetilde{Z}(g,\Phi)[t]^{\circ},\ 1\rangle_{\mathrm{NT}}\ \chi(t)dt.
\end{aligned}
$$

Recall that for any $t' \in T(\mathbb{A})$,

$$
\langle \widetilde{Z}(g,\Phi)[t't]^{\circ},\ [t']^{\circ}\rangle_{\mathrm{NT}} = \langle \widetilde{Z}(g,\Phi)[t]^{\circ},\ [1]^{\circ}\rangle_{\mathrm{NT}}.
$$

The identity is true by viewing the action of t' on $([t]^{\circ},[1]^{\circ})$ as a Galois action which does not change the height pairing. Therefore, we obtain

$$
\widetilde{Z}(g,\chi,\Phi) = 2L(1,\eta)\fint_{T(F)\backslash T(\mathbb{A})/Z(\mathbb{A})}\fint_{T(F)\backslash T(\mathbb{A})/Z(\mathbb{A})}
$$
$$
\langle \widetilde{Z}(g,\Phi)[t't]^{\circ},\ [t']^{\circ}\rangle_{\mathrm{NT}}\ \chi(t)dt'dt.
$$

By a change of variable, it becomes

$$
\widetilde{Z}(g,\chi,\Phi) = 2L(1,\eta)\fint_{T(F)\backslash T(\mathbb{A})/Z(\mathbb{A})}\fint_{T(F)\backslash T(\mathbb{A})/Z(\mathbb{A})}
$$
$$
\langle \widetilde{Z}(g,\Phi)[t_1]^{\circ},\ [t_2]^{\circ}\rangle_{\mathrm{NT}}\ \chi(t_1 t_2^{-1})dt_1 dt_2.
$$

It is just

$$
\widetilde{Z}(g,\chi,\Phi) = 2L(1,\eta)\ \langle \widetilde{Z}(g,\Phi)P_\chi,\ P_{\chi^{-1}}\rangle_{\mathrm{NT}}.
$$

Here

$$
P_\chi = \fint_{T(F)\backslash T(\mathbb{A})/Z(\mathbb{A})} \chi(t)[t]^{\circ}dt.
$$

It follows that

$$
\left(\widetilde{Z}(g,\chi,\Phi),\ \varphi(g)\right)_{\mathrm{Pet}} = 2L(1,\eta)\ \langle \widetilde{Z}(\Phi\otimes\varphi)P_\chi,\ P_{\chi^{-1}}\rangle_{\mathrm{NT}}.
$$

Here the arithmetic theta lifting

$$
\widetilde{Z}(\Phi\otimes\varphi) = \int_{\mathrm{GL}_2(F)\backslash \mathrm{GL}_2(\mathbb{A})/Z(\mathbb{A})}^{*} \varphi(g)\ \widetilde{Z}(g,\Phi)\ dg
$$

is defined above. Notice that the Petersson pairing is the usual integration, but it is equal to our regularized integration. By Theorem 3.22,

$$
\widetilde{Z}(\Phi\otimes\varphi) = \frac{L(1,\pi,\mathrm{ad})}{2\zeta_F(2)}\mathrm{T}(\theta(\Phi\otimes\varphi)).
$$

Hence,

$$
\left(\widetilde{Z}(g,\chi,\Phi),\ \varphi(g)\right)_{\mathrm{Pet}} = \frac{L(1,\pi,\mathrm{ad})L(1,\eta)}{\zeta_F(2)}\langle \mathrm{T}(f_1\otimes f_2)P_\chi,\ P_{\chi^{-1}}\rangle_{\mathrm{NT}}.
$$

It finishes the proof.

3.6.5 Interpretation by linear functionals

Similar to the interpretation of Theorem 1.2, Theorem 3.15 is an identity of two vectors in the complex vector space $\mathcal{P}(\pi, \chi) \otimes \mathcal{P}(\tilde{\pi}, \chi^{-1})$. Here we recall that

$$\mathcal{P}(\pi, \chi) = \mathrm{Hom}_{T(\mathbb{A})}(\pi \otimes \chi, \mathbb{C}).$$

It is at most one-dimensional by Theorem 1.3. Note that α is a generator of $\mathcal{P}(\pi, \chi) \otimes \mathcal{P}(\tilde{\pi}, \chi^{-1})$, and $\langle T(\cdot)P_\chi, P_{\chi^{-1}} \rangle_{\mathrm{NT}}$ must be a multiple of α.

To get an essential case of the theorems, we need to assume that $\mathbb{B}$ is determined by the local root numbers as in Theorem 1.3. To be precise, we list all the notations and assumptions in the following:

(1) F is a totally real number field and E is a totally imaginary quadratic extension of F.

(2) σ is a cuspidal automorphic representation of $\mathrm{GL}_2(\mathbb{A})$, discrete of weight 2 at all infinite places of F.

(3) χ is a character on $E^\times \backslash E_\mathbb{A}^\times$ of finite order.

(4) $\Sigma = \left\{ v : \epsilon(\frac{1}{2}, \pi_v, \chi_v) \neq \chi_v \eta_v(-1) \right\}$ has an odd cardinality.

(5) $\mathbb{B}$ is an incoherent quaternion algebra over $\mathbb{A}$ with ramification set Σ, endowed with a fixed embedding $E_\mathbb{A} \hookrightarrow \mathbb{B}$ over $\mathbb{A}$.

(6) π is the admissible representation of $\mathbb{B}^\times$ whose Jacquel–Langlands correspondence is σ.

By (2) and (3), the set Σ contains all the archimedean places of F. It follows that $\mathbb{B}$ is totally definite quaternion algebra over $\mathbb{A}$. The embedding $E_\mathbb{A} \hookrightarrow \mathbb{B}$ and the representation π exist by the construction of $\mathbb{B}$, and π is trivial under the action of the infinite part $\mathbb{B}_\infty^\times$. We refer to the assumptions as *"geometric assumptions"* since they are the basic assumptions to obtain a non-trivial formula for the height pairing of CM cycles on Shimura curves.

As a consequence of the multiplicity one property, we immediately obtain the following result under the geometric assumptions.

LEMMA 3.23. *The following are true:*

(1) Theorem 3.15 holds for all (f_1, f_2) if and only if it holds for some (f_1, f_2) with $\alpha(f_1, f_2) \neq 0$.

(2) Theorem 3.21 holds for all (Φ, φ) if and only if it holds for some (Φ, φ) with $\alpha(\theta(\Phi \otimes \varphi)) \neq 0$.

Note that (2) is just a consequence of (1) by the relation of these two theorems described above. Under assumption (5), we can also interpret Theorem 3.21 as an identity in

$$\mathrm{Hom}_{\mathrm{GL}_2(\mathbb{A}) \times E_\mathbb{A}^\times \times E_\mathbb{A}^\times}(\sigma \boxtimes \mathcal{S}(\mathbb{B} \times \mathbb{A}^\times) \boxtimes (\chi, \chi^{-1}), \ \mathbb{C}).$$

It is one-dimensional by its relation to $\mathcal{P}(\pi, \chi) \otimes \mathcal{P}(\tilde{\pi}, \chi^{-1})$ via the theta lifting.

Chapter Four

Trace of the Generating Series

The goal of this chapter is to prove Theorem 3.17 and Theorem 3.22 in the last chapter.

Before going to the proofs, in §4.1 we give more details on the new space $\overline{\mathcal{S}}(V \times \mathbb{R}^\times)$ of Schwartz functions including the formation of theta series and Eisenstein series by them.

Theorem 3.17 asserts the modularity of the generating series. The major part of its proof is in §4.2, where we reduce the problem to the results in [YZZ]. In this way, the modularity is proved on the open Shimura variety. To extend to the compactification, it suffices to prove the degree (as a correspondence) is modular. The degree is proved to be given by Eisenstein series in §4.3, and thus the modularity follows.

Theorem 3.22 is an identity between $T(f_1 \otimes f_2)$ and $(Z(\phi), \varphi)$. We know the identity is true up to constant. To determine the constant, it suffices to compare the traces of both operators. By the Lefschetz trace formula, the trace of $(Z(\phi), \varphi)$ is reduced to the degree of the pull-back of $(Z(\phi), \varphi)$ by the diagonal map $X \to X \times X$ in §4.4. The computation of the degree takes up §4.5, §4.6 and §4.7. The compact case is treated in §4.5, while the non-compact case takes up the last two sections.

4.1 DISCRETE SERIES AT INFINITE PLACES

In §2.1, we reviewed Waldspurger's extension of the Weil representation to the Schwartz function space $\mathcal{S}(V \times k^\times)$. Following §3.4.1, we will consider a new class $\overline{\mathcal{S}}(V \times k^\times)$ of Schwartz functions. It is the same as the original one in the non-archimedean case, but different in the archimedean case. We will also construct theta series and Eisenstein series from such functions.

4.1.1 Discrete series at infinity

Let V be a positive definite quadratic space over $\mathbb{R}$. Let $\overline{\mathcal{S}}(V \times \mathbb{R}^\times)$ denote the space of functions on $V \times \mathbb{R}^\times$ of the form

$$\phi(x, u) = (P_1(uq(x)) + \mathrm{sgn}(u)P_2(uq(x))) \, e^{-2\pi|u|q(x)}$$

with polynomials P_i of complex coefficients. Here $\text{sgn}(u) = u/|u|$ denotes the sign of $u \in \mathbb{R}^\times$. As in §3.4.1, we have a (surjective) quotient map

$$\mathcal{S}(V \times \mathbb{R}^\times) \longrightarrow \overline{\mathcal{S}}(V \times \mathbb{R}^\times), \quad \Phi \longmapsto \overline{\Phi} := \int_{\mathbb{R}^\times} \int_{O(V)} r(ch)\Phi \, dh dc.$$

By the formulae for the Weil representation on $\mathcal{S}(V \times \mathbb{R}^\times)$ in §2.1, one has a smooth function $r(g,h)\phi$ on $V \times \mathbb{R}^\times$ for any $\phi \in \overline{\mathcal{S}}(V \times \mathbb{R}^\times)$ and $(g,h) \in \text{GL}_2(\mathbb{R}) \times \text{GO}(\mathbb{R})$. It induces an action r of $(\mathfrak{gl}_2(\mathbb{R}), O_2(\mathbb{R})) \times \text{GO}(V)$ on $\overline{\mathcal{S}}(V \times \mathbb{R}^\times)$. The action of $\text{GO}(V)$ is trivial by definition.

The *standard Schwartz function* $\phi \in \overline{\mathcal{S}}(V \times \mathbb{R}^\times)$ is the Gaussian

$$\phi(x, u) = e^{-2\pi u q(x)} \, 1_{\mathbb{R}_+}(u).$$

Here $1_{\mathbb{R}_+}$ is the characteristic function of the set $\mathbb{R}_+$ of positive real numbers. Assume that $\dim V = 2d$ is even for simplicity. Then one verifies that, for any $g \in \text{GL}_2(\mathbb{R})$,

$$r(g)\phi(x, u) = \begin{cases} W^{(d)}_{uq(x)}(g) & \text{if } x \neq 0, \\ W^{(d)}_0(g, u) & \text{if } x = 0. \end{cases}$$

Here the *standard holomorphic Whittaker function* $W^{(d)}$ is as follows:

$$W^{(d)}_a(g) = |y_0|^{\frac{d}{2}} e^{2\pi i a(x_0 + i y_0)} e^{di\theta} \, 1_{\mathbb{R}_+}(a \det(g)), \quad a \in \mathbb{R}^\times;$$
$$W^{(d)}_0(g, u) = |y_0|^{\frac{d}{2}} e^{di\theta} \, 1_{\mathbb{R}_+}(u \det(g)), \quad u \in \mathbb{R}^\times.$$

Here we express $g \in \text{GL}_2(\mathbb{R})$ by the Iwasawa decomposition as follows:

$$g = \begin{pmatrix} z_0 & \\ & z_0 \end{pmatrix} \begin{pmatrix} y_0 & x_0 \\ & 1 \end{pmatrix} \begin{pmatrix} \cos\theta & \sin\theta \\ -\sin\theta & \cos\theta \end{pmatrix}.$$

The space $\overline{\mathcal{S}}(V \times \mathbb{R}^\times)$ actually gives the discrete series of $\text{GL}_2(\mathbb{R})$ of weight d. The standard Schwartz function gives an element of the lowest weight. Furthermore, the action of $\mathfrak{gl}_2(\mathbb{R})$ on the standard function generates the whole space $\overline{\mathcal{S}}(V \times \mathbb{R}^\times)$. By this fact, all our major computations will be reduced to the standard function.

4.1.2   Some notations

For convenience, we introduce some global notations. Let F be a totally real field with adele ring $\mathbb{A}$. For any (coherent or incoherent) quadratic space $\mathbb{V}$ over $\mathbb{A}$ which is positive definite at infinity, denote restricted tensor product

$$\overline{\mathcal{S}}(\mathbb{V} \times \mathbb{A}^\times) := \otimes_v \overline{\mathcal{S}}(\mathbb{V}_v \times F_v^\times).$$

Here $\overline{\mathcal{S}}(\mathbb{V}_v \times F_v^\times) = \mathcal{S}(\mathbb{V}_v \times F_v^\times)$ if v is non-archimedean. The same integration on $F_\infty^\times \times O(F_\infty)$ gives a surjection $\mathcal{S}(\mathbb{V} \times \mathbb{A}^\times) \to \overline{\mathcal{S}}(\mathbb{V} \times \mathbb{A}^\times)$.

For any $a \in \mathbb{A}^\times$ and $x \in \mathbb{V}$ with $q(x) \neq 0$, denote

$$r(g, h)\phi(x)_a := r(g, h)\phi(x, aq(x)^{-1}). \tag{4.1.1}$$

Use similar notations in the local case.

Note that the infinite part $r(g_\infty, h_\infty)\phi_\infty(x_\infty, a_\infty q(x_\infty)^{-1})$ is independent of x_∞, h_∞. Thus we use the notation

$$r(g, h)\phi(x)_a := r(g_f, h_f)\phi_f(x, aq(x)^{-1}) \, r(g_\infty)\phi_\infty(x_\infty, a_\infty q(x_\infty)^{-1}) \tag{4.1.2}$$

for $(g, h) \in \mathrm{GL}_2(\mathbb{A}) \times \mathrm{GO}(\mathbb{V}_f)$, $x \in \mathbb{V}_f$ and $a \in \mathbb{A}^\times$. It is independent of the choice of $x_\infty \in \mathbb{V}_\infty$.

Similarly, for $(g, h) \in \mathrm{GL}_2(\mathbb{A}) \times \mathrm{GO}(\mathbb{V}_f)$, $(x, u) \in \mathbb{V}_f \times \mathbb{A}_f^\times$ with $uq(x) \in F^\times$, it makes sense to define

$$r(g, h)\phi(x, u) := r(g_f, h_f)\phi_f(x, u) \, r(g_\infty)\phi_\infty(x_\infty, u_\infty) \tag{4.1.3}$$

for any $(x_\infty, u_\infty) \in \mathbb{V}_\infty \times F_\infty^\times$ with $u_\infty q(x_\infty) \in F^\times$ and equal to $uq(x) \in F^\times$.

4.1.3 Theta series

First let us consider the definition of theta series. Let V be a positive definite even-dimensional quadratic space over a totally real field F. Let $\phi \in \overline{\mathcal{S}}(V(\mathbb{A}) \times \mathbb{A}^\times)$.

Similar to the case for $\mathcal{S}(V(\mathbb{A}) \times \mathbb{A}^\times)$, we have the theta series

$$\theta(g, u, \phi) := \sum_{x \in V} r(g)\phi(x, u), \quad g \in \mathrm{GL}_2(\mathbb{A}), \ u \in \mathbb{A}^\times.$$

If $u \in F^\times$, it is invariant under the left action of $\mathrm{SL}_2(F)$ on g.

To get an automorphic form on $\mathrm{GL}_2(\mathbb{A})$, we need to sum over $u \in F^\times$. This time the definition is slightly different due to a convergence issue here. There is an open compact subgroup $K \subset \mathrm{GO}(\mathbb{A}_f)$ such that ϕ_f is invariant under the action of K by Weil representation. Denote $\mu_K = F^\times \cap K$. Then μ_K is a subgroup of the unit group $O_F^\times$, and thus is a finitely generated abelian group. Our theta series is of the following form:

$$\theta(g, \phi)_K := \sum_{u \in \mu_K^2 \backslash F^\times} \sum_{x \in V} r(g)\phi(x, u). \tag{4.1.4}$$

If $F = \mathbb{Q}$, then $\mu_K^2 = 1$ and thus the definition has the same form as the original definition. In terms of $\theta(g, u, \phi)$, the theta series is just

$$\theta(g, \phi)_K = \sum_{u \in \mu_K^2 \backslash F^\times} \theta(g, u, \phi).$$

The summation is well-defined since for any $\alpha \in \mu_K$,

$$\theta(g, \alpha^2 u, \phi) = \sum_{x \in V} r(g)\phi(x, \alpha^2 u) = \sum_{x \in V} r(g)\phi(\alpha x, u)$$

$$= \sum_{x \in V} r(g)\phi(x, u) = \theta(g, u, \phi).$$

We will show that the summations are absolutely convergent. Then $\theta(g, \phi)_K$ is an automorphic form on $g \in \mathrm{GL}_2(\mathbb{A})$, and $\theta(g, r(h)\phi)_K$ is an automorphic form on $(g, h) \in \mathrm{GL}_2(\mathbb{A}) \times \mathrm{GO}(\mathbb{A})$. Furthermore, if ϕ_∞ is standard, then $\theta(g, \phi)_K$ is holomorphic of weight $\frac{1}{2} \dim V$. It is the major reason for introducing this new class of functions.

By choosing fundamental domains, we can rewrite the sum as

$$\theta(g, \phi)_K = \sum_{u \in \mu_K^2 \backslash F^\times} r(g)\phi(0, u) + w_K \sum_{(x,u) \in \mu_K \backslash ((V - \{0\}) \times F^\times)} r(g)\phi(x, u).$$

Here the natural action of μ_K on $V \times F^\times$ is just $\alpha \circ (x, u) \mapsto (\alpha x, \alpha^{-2} u)$. The summation over u is well-defined since $\phi(\alpha x, \alpha^{-2} u) = r(\alpha^{-1})\phi(x, u) = \phi(x, u)$ for any $\alpha \in \mu_K$. The factor $w_K = |\{1, -1\} \cap K| \in \{1, 2\}$, and it is 1 for K small enough.

Now we check the convergence of (4.1.4). We claim that the summation over u is actually a finite sum depending on (g, h). For fixed (g, h), there is a compact subset $A \subset \mathbb{A}_f^\times$ such that $r(g, h)\phi_f(x, u) \neq 0$ only if $u \in A$. Thus the summation is taken over $u \in \mu_K^2 \backslash (F^\times \cap A)$, which is a finite set since μ_K^2 is a finite-index subgroup of the unit group $O_F^\times$, and $F^\times \cap A$ is included in a finite union of cosets of $O_F^\times$.

The definition depends on the choice of K. In fact, if $K' \subset K$ is another open compact subgroup, then

$$\theta(g, \phi)_{K'} = [\mu_K^2 : \mu_{K'}^2]\, \theta(g, \phi)_K.$$

4.1.4 Eisenstein series

We can also define Eisenstein series for such Schwartz functions. The definition is valid for both coherent or incoherent quadratic space. Let $\phi \in \overline{\mathcal{S}}(\mathbb{V} \times \mathbb{A}^\times)$, where $\mathbb{V}$ is a quadratic space over $\mathbb{A}$ which is positive definite of even dimension at infinity.

First, as in the case of $\mathcal{S}(\mathbb{V} \times \mathbb{A}^\times)$, we introduce

$$E(s, g, u, \phi) := \sum_{\gamma \in P^1(F) \backslash \mathrm{SL}_2(F)} \delta(\gamma g)^s r(\gamma g)\phi(0, u), \quad g \in \mathrm{GL}_2(\mathbb{A}), \ u \in F^\times.$$

It is invariant under the left action of $\mathrm{SL}_2(F)$ on g.

Second, take an open compact subgroup K of $\mathrm{GO}(\mathbb{A})$ such that ϕ is invariant under K as above. Define

$$E(s, g, \phi)_K := \sum_{u \in \mu_K^2 \backslash F^\times} E(s, g, u, \phi).$$

Similar to the argument above, the summation over u is well-defined, and it is essentially a finite sum. It follows that $E(s, g, \phi)_K$ is an automorphic form on $g \in \mathrm{GL}_2(\mathbb{A})$.

4.1.5   Relation with the old definitions

In the following we want to describe a connection from the definitions of theta series and Eisenstein here to those in §2.1.

Take theta series as an example. Let V be a positive definite even dimensional quadratic space over a totally real field F. Let $\Phi \in \mathcal{S}(V(\mathbb{A}) \times \mathbb{A}^\times)$. The theta series in §2.1 is simply

$$\theta(g, \Phi) = \sum_{(x,u) \in V \times F^\times} r(g)\Phi(x, u).$$

By the quotient map, we set

$$\phi = \overline{\Phi} = \int_{F_\infty^\times} \fint_{O(V_\infty)} r(ch)\Phi \; dhdc \; \in \overline{\mathcal{S}}(V \times \mathbb{R}^\times).$$

Assume that $K \subset \mathrm{GO}(\mathbb{A}_f)$ is an open compact subgroup acting trivially on ϕ. Then we have

$$\theta(g, \phi)_K = \sum_{u \in \mu_K^2 \backslash F^\times} \sum_{x \in V} r(g)\phi(x, u).$$

The relation is that

$$\int_{\mu_K \backslash F_\infty^\times} \fint_{O(V_\infty)} \theta(zg, r(h)\Phi) \; dhdz = \frac{1}{w_K}\theta(g, \phi)_K. \qquad (4.1.5)$$

By this identity, we can switch between these two kinds of Schwartz functions. The same relation holds for the Eisenstein series.

To check the identity, it suffices to assume that Φ is invariant under the action of $O(V_\infty)$. In fact, since $O(V_\infty)$ is compact, we can replace Φ by its average on $O(V_\infty)$. Then the identity comes from a coset identity.

4.2   MODULARITY OF THE GENERATING SERIES

In this section, we prove Theorem 3.17 on the modularity of the generating series. It is essentially an example of the modularity theorem in [YZZ].

Let $\phi \in \overline{S}(\mathbb{V} \times \mathbb{A}^\times)$ be a Schwartz function invariant under the action of $K = U \times U$ for some open compact subgroup U of $\mathbb{B}_f^\times$. We need to prove the modularity of

$$Z(g, \phi)_U = Z_0(g, \phi)_U + Z_*(g, \phi)_U, \quad g \in \mathrm{GL}_2(\mathbb{A}).$$

Here the constant term

$$Z_0(g, \phi)_U = - \sum_{\alpha \in F_+^\times \backslash \mathbb{A}_f^\times / q(U)} \sum_{u \in \mu_U^2 \backslash F^\times} E_0(\alpha^{-1} u, r(g)\phi) \, L_{K,\alpha}, \quad (4.2.1)$$

and the non-constant part

$$Z_*(g, \phi)_U = w_U \sum_{a \in F^\times} \sum_{x \in K \backslash \mathbb{B}_f^\times} r(g)\phi(x)_a \, Z(x)_U. \quad (4.2.2)$$

Here the intertwining part $W_0(u, \phi)$ is introduced in §3.4.2, and the notation $r(g)\phi(x)_a$ is introduced in (4.1.2). We also write $Z(\phi)_U = Z(1, \phi)_U$ and thus $Z(g, \phi)_U = Z(r(g)\phi)_U$. Use similar notations for $Z_0(1, \phi)_U$, $Z_*(1, \phi)_U$, $Z_a(1, \phi)_U$.

4.2.1 Reduction to the standard Gaussian

By linearity we can assume that $\phi = \phi_f \otimes \phi_\infty$. It suffices to prove the modularity of $Z(g, \phi)_U$ in the case that ϕ_∞ is the standard Schwartz function. In fact, denote by $\phi_\infty^{(2)}$ the standard Schwartz function at infinity, and denote $\phi' = \phi_f \otimes \phi_\infty^{(2)}$. Since $\phi_\infty^{(2)}$ generates $\overline{S}(\mathbb{V}_\infty \times F_\infty^\times)$ under the action of $\mathrm{gl}_2(F_\infty)$, we can assume that $\phi_\infty = r(\partial)\phi_\infty^{(2)}$ from some element $\partial \in \mathrm{gl}_2(F_\infty)$. The modularity of $Z(g, \phi')_U = Z(r(g)\phi')_U$ implies the modularity of $Z(r(ge^{t\partial})\phi')_U$ for any $t \in \mathbb{R}$. Taking the derivative at $t = 0$, we get the modularity of $Z(r(g)\phi)_U$ as desired.

In the following, we assume that $\phi = \phi_f \otimes \phi_\infty$ with ϕ_∞ standard.

4.2.2 Result for the general spin group

Now we state the main result of [YZZ], which can be seen as a continuation of the previous works [HZ, KM1, KM2, KM3, GKZ, Bor, Zha].

Before going to the details, we point out that the result in [YZZ] is modularity on open Shimura varieties. To get a modularity on the compactified Shimura varieties, we also need to consider the boundary behavior. It is reduced to check the modularity of the degree of the generating series at the end of this section. The degree is computed in the next section.

The result is stated in terms of the general spin group of $(\mathbb{V}, q)$. Recall that

$$\mathrm{GSpin}(\mathbb{V}) = \{(h_1, h_2) \in \mathbb{B}^\times \times \mathbb{B}^\times : q(h_1) = q(h_2)\}$$

is naturally a subgroup of $\mathbb{B}^\times \times \mathbb{B}^\times$ which is compatible with the action on $\mathbb{V}$. Let $M_{K'}'$ be the compactified Shimura surface associated to $\mathrm{GSpin}(\mathbb{V}_f)$ for any open compact subgroup $K' \subset \mathrm{GSpin}(\mathbb{V}_f)$. For any Schwartz function $\phi' \in \mathcal{S}(\mathbb{V})$ invariant under $K' \times O(\mathbb{V}_\infty)$, and any $g \in \mathrm{SL}_2(\mathbb{A})$, one has a generating series

$$Z(g, \phi')_{K'} = -r(g)\phi'(0) \ L_{K'} + w_{K'} \sum_{a \in F_+^\times} \sum_{y \in K' \backslash \mathbb{V}_f(a)} r(g)\phi'(y) \ Z(y)_{K'}.$$

Here $w_{K'} = |\{1, -1\} \cap K'|$, $L_{K'}$ is the Hodge bundle on $M'_{K'}$, and $\mathbb{V}_f(a)$ is the set of elements of $\mathbb{V}_f$ with norm a. The special curve $Z(y)_{K'}$ is defined for $y \in \mathbb{B}_f^\times$ if $q(y) \in F_+^\times$. In the case $K' = K \cap \mathrm{GSpin}(\mathbb{V}_f)$, the image of $Z(y)_{K'}$ under the natural map $M'_{K'} \to M_K$ is just $Z(y)_U$.

Denote by $M'^o_{K'}$ the open part of $M'_{K'}$ before compactification. They are different unless $F = \mathbb{Q}$ and $\mathbb{B}_f = M_2(\mathbb{A}_f)$. In [YZZ], we have proved that if $\phi' = \phi'_f \otimes \phi'_\infty$ with ϕ'_∞ equal to the standard Gaussian, then the restriction $Z(g, \phi')_{K'}|_{M'^o_{K'}}$ to the open part is absolutely convergent and defines an automorphic form on $\mathrm{SL}_2(\mathbb{A})$ with coefficients in $\mathrm{Pic}(M'^o_{K'})$. We claim that the modularity can easily be extended to any $\phi' \in \mathcal{S}(\mathbb{V})$.

In fact, we can assume $\phi' = \phi'_f \otimes \phi'_\infty$ by linearity. Denote by ϕ''_∞ the standard Gaussian in $\mathcal{S}(\mathbb{V})$, and denote $\phi'' = \phi'_f \otimes \phi''_\infty$. The key is that ϕ''_∞ generates $\mathcal{S}(\mathbb{V}_\infty)$ under the action of the Lie algebra $\mathrm{sl}_2(F_\infty)$. Thus we can assume that $\phi'_\infty = r(\partial)\phi''_\infty$ for some element $\partial \in \mathrm{sl}_2(F_\infty)$. The modularity of $Z(g, \phi'')_{K'} = Z(r(g)\phi'')_{K'}$ on $M'^o_{K'}$ implies the modularity of $Z(r(ge^{t\partial})\phi'')_{K'}$ for any $t \in \mathbb{R}$. Taking the derivative at $t = 0$, we get the modularity of $Z(r(g)\phi')_U$ on $M'^o_{K'}$ as desired.

More generally, we can change the quadratic form by a fixed constant $u \in F^\times$ and get a new series

$$Z(g, \phi', uq)_{K'} = -r_u(g)\phi'(0) \ L_{K'} + w_{K'} \sum_{a \in F_+^\times} \sum_{y \in K' \backslash \mathbb{V}_f(a)} r_u(g)\phi'(y) \ Z(y)_{K'}$$

for any $g \in \mathrm{SL}_2(\mathbb{A})$. Here r_u is the Weil representation with respect to the quadratic space $(\mathbb{V}_f, uq)$. Then the result of [YZZ] also gives the modularity of $Z(g, \phi', uq)|_{M'^o_{K'}}$. We will come back to discuss the boundary later.

When K' is sufficiently small, one has $w_{K'} = 1$. In general one needs the factor $w_{K'}$ in the definition of the generating series. The modularity result for big level K' can be obtained by push-forward from small levels.

4.2.3 Map between the Shimura varieties

For any $h \in \mathbb{B}_f^\times \times \mathbb{B}_f^\times$ and any open compact subgroup $K = U \times U$ of $\mathbb{B}_f^\times \times \mathbb{B}_f^\times$ as above, the group $K^h := \mathrm{GSpin}(\mathbb{V}_f) \cap hKh^{-1}$ is an open compact subgroup of $\mathrm{GSpin}(\mathbb{V}_f)$. Let $i_h : M'_{K^h} \to M_K$ be the finite map given by right multiplication by h.

LEMMA 4.1. *The image of $i_h : M'_{K^h} \to M_K$ is exactly $M_{K, \nu(h)^{-1}}$, and the degree of $i_h : M'_{K^h} \to M_{K, \nu(h)}$ is equal to $[\mu'_U : \mu_U^2]$ everywhere. Here $\mu_U = F^\times \cap U$ and $\mu'_U = F_+^\times \cap q(U)$.*

PROOF. Fix an archimedean place τ, denote by $B = B(\tau)$ the nearby quaternion algebra, and by $G = \mathrm{GSpin}(B, q)$. Then we have uniformizations

$$M'_{K^h,\tau}(\mathbb{C}) = G(F)_+ \backslash (\mathcal{H} \times \mathcal{H}) \times G(\mathbb{A}_f)/K^h,$$
$$M_{K,\tau}(\mathbb{C}) = (B_+^\times \times B_+^\times) \backslash (\mathcal{H} \times \mathcal{H}) \times \mathbb{B}_f^\times \times \mathbb{B}_f^\times /K.$$

Here $G(F)_+$ is the intersection of $G(F)$ with $B_+^\times \times B_+^\times$. If $F = \mathbb{Q}$ and $B = M_2(\mathbb{Q})$ we need to add two boundary divisors $\mathcal{H} \times \{\text{cusps}\}$ and $\{\text{cusps}\} \times \mathcal{H}$.

The image of $i_h : M'_{K^h} \to M_K$ is represented by

$$(B_+^\times \times B_+^\times)\left((\mathcal{H} \times \mathcal{H}) \times G(\mathbb{A}_f)hK\right).$$

Note that $G(\mathbb{A}_f)$ is a normal subgroup of $\mathbb{B}_f^\times \times \mathbb{B}_f^\times$. It is easy the see the image is exactly $M_{K,\nu(h)^{-1}}$. Both the geometrically connected components of M'_{K^h} and $M_{K,\nu(h)^{-1}}$ are parametrized by $F_+^\times \backslash \mathbb{A}_f^\times / q(U)$. So i_h is one-to-one on the geometrically connected components.

Now we figure out the degree. Any geometrically connected component of $M'_{K^h,\tau}(\mathbb{C})$ is of the form

$$\Gamma' \backslash \mathcal{H} \times \mathcal{H}, \quad \Gamma' = G(F)_+ \cap (gK^h g^{-1})$$

for some $g \in G(\mathbb{A}_f)$. Its image in $M_{K,\tau}(\mathbb{C})$ is just

$$\Gamma \backslash \mathcal{H} \times \mathcal{H}, \quad \Gamma = (B_+^\times \times B_+^\times) \cap (ghKh^{-1}g^{-1}).$$

Note that the stabilizer of a general point (i.e., not a CM point) of $\mathcal{H} \times \mathcal{H}$ in $B_+^\times \times B_+^\times$ is just the center $F^\times \times F^\times$. It follows that the degree of i_h on that component is equal to $[\Gamma : Z_\Gamma \Gamma']$, where

$$Z_\Gamma = (F^\times \times F^\times) \cap \Gamma = \mu_U \times \mu_U.$$

Then the degree is just $[\Gamma : (\mu_U \times \mu_U)(\Gamma \cap G(F))]$.

Write $\Gamma = (\Gamma_1, \Gamma_2)$ in terms of components. The action of $\Gamma \cap G(F)$ on Γ can always transfer the first component in Γ_1 into 1. Then it is easy to see that

$$[\Gamma : (\mu_U \times \mu_U)(\Gamma \cap G(F))] = [\Gamma_2 : \mu_U \Gamma_2^1] = [q(\Gamma_2) : \mu_U^2].$$

Here Γ_2^1 is the subgroup of elements of Γ_2 with norm 1. The second equality is induced by the bijective norm map

$$q : \Gamma_2/(\mu_U \Gamma_2^1) \longrightarrow q(\Gamma_2)/\mu_U^2.$$

It remains to prove $q(\Gamma_2) = \mu'_U$, or equivalently

$$q(B_+^\times \cap U') = F_+^\times \cap q(U') \tag{4.2.3}$$

where $U' = g_2 h_2 U h_2^{-1} g_2^{-1}$ if we write $h = (h_1, h_2)$ and $g = (g_1, g_2)$. Only the implication $F_+^\times \cap q(U') \subset q(B_+^\times \cap U')$ needs justification. Let $a \in F_+^\times \cap q(U')$ be any element. Then we can find $b \in B_+^{\times}$ and $u \in U'$ with $q(b) = q(u) = a$. The existence of b follows from the Hasse principle. To show that $a \in q(B_+^{\times} \cap U')$, it is equivalent to showing that $bB^1 \cap uU'^1 = b(B^1 \cap b^{-1}uU'^1)$ is non-empty. Here B^1 and U'^1 still denote the subgroups of elements with norm 1. By the strong approximation theorem, B^1 is dense in $\mathbb{B}_f^1$. Thus $B^1 \cap b^{-1}uU'^1$ is non-empty since $b^{-1}uU'^1$ is open in $\mathbb{B}_f^1$. □

4.2.4 Components of the generating series

Let $\phi \in \overline{\mathcal{S}}(\mathbb{V} \times \mathbb{A}^{\times})$ be a Schwartz function invariant under $K = U \times U$ for some open compact subgroup U of $\mathbb{B}_f^{\times}$. We need to prove the modularity of

$$Z(g, \phi)_U = Z_0(g, \phi)_U + Z_*(g, \phi)_U, \quad g \in \mathrm{GL}_2(\mathbb{A}).$$

It suffices to prove the modularity of the component

$$Z(g, \phi)_{U,\alpha} := Z(g, \phi)_U|_{M_{K,\alpha}}$$

for all $\alpha \in F_+^{\times} \backslash \mathbb{A}_f^{\times}/q(U)$. We further decompose

$$Z(g, \phi)_{U,\alpha} = Z_0(g, \phi)_{U,\alpha} + Z_*(g, \phi)_{U,\alpha}$$

into the sum of the constant term and the non-constant part.

Fix an $h \in \mathbb{B}_f^{\times} \times \mathbb{B}_f^{\times}$ such that $\nu(h) \in \alpha^{-1} F_+^{\times} q(U)$. It is easy to have

$$Z_0(g, \phi)_{U,\alpha} = - \sum_{u \in \mu_U^2 \backslash F^{\times}} (r(g, h)\phi(0, u) + W_0(u, r(g, h)\phi)) \, L_{K,\alpha}.$$

Now we treat $Z_*(g, \phi)_{U,\alpha}$. For any $x \in \mathbb{B}_f^{\times}$, the Hecke operator $Z(x)_U = UxU$ is completely contained in $M_{K,q(x)}$. It has contribution to $Z(g, \phi)_{U,\alpha}$ if and only if $q(x) \in \alpha F_+^{\times} q(U)$. In that case we have $x \in Kh^{-1}y$ for some $y \in \mathbb{B}_f^{\times}$ with norm in $q(y) \in F_+^{\times}$. A different y' yields $Kh^{-1}y' = Kh^{-1}y$ if and only if $y' = hKh^{-1}y$. It follows that $q(y') \in q(y)\mu_U'$ with $\mu_U' = F_+^{\times} \cap q(U)$ as above. Taking these into account, we have

$$Z_*(g, \phi)_{U,\alpha} = w_U \sum_{a \in F^{\times}} \sum_{b \in \mu_U' \backslash F_+^{\times}} \sum_{y \in K^h \backslash \mathbb{V}_f(b)} r(g, h)\phi(y, aq(y)^{-1}) \, Z(h^{-1}y)_U.$$

Here $\mathbb{V}_f(b) = \{x \in \mathbb{V} : q(x) = b\}$. Therefore,

$$Z_*(g, \phi)_{U,\alpha} = w_U \sum_{u \in \mu_U' \backslash F^{\times}} \sum_{a \in F_+^{\times}} \sum_{y \in K^h \backslash \mathbb{V}_f(a)} r(g, h)\phi(y, u) \, Z(h^{-1}y)_U. \quad (4.2.4)$$

4.2.5 Modularity on open Shimura varieties

Fix one component $\alpha \in F_+^\times \backslash \mathbb{A}_f^\times / q(U)$. We need to prove the modularity of $Z(g, \phi)_{U,\alpha}$ above. Let $h \in \mathbb{B}_f^\times \times \mathbb{B}_f^\times$ such that $\nu(h) \in \alpha^{-1} F_+^\times q(U)$ as above. Consider the finite map $i_h : M'_{K^h} \to M_{K,\alpha}$. We want to express $Z(g, \phi)_{U,\alpha}$ as the push-forward of some generating series on M'_{K^h}.

Note that $i_{h*} L_{K^h} = \deg(i_h) L_{K,\alpha}$ with $\deg(i_h) = [\mu'_U : \mu_U^2]$ by the above lemma. Then

$$Z_0(g, \phi)_{U,\alpha} = -i_{h*} \sum_{u \in \mu'_U \backslash F^\times} r(g,h)\phi(0,u)\, L_{K^h} + Z_{0,\text{int}}(g,\phi)_{U,\alpha}.$$

Here $Z_{0,\text{int}}(g, \phi)_{U,\alpha} = Z_{0,\text{int}}(g,\phi)_U|_{M_{K,\alpha}}$ is the α-component of the intertwining part

$$Z_{0,\text{int}}(g,\phi)_U = - \sum_{\beta \in F_+^\times \backslash \mathbb{A}_f^\times / q(U)} \sum_{u \in \mu_U^2 \backslash F^\times} W_0(\beta^{-1}u, r(g)\phi)\, L_{K,\beta}.$$

By $i_{h*} Z(y)_{K^h} = Z(h^{-1}y)_U$, we see that

$$Z_*(g,\phi)_{U,\alpha} = w_U \sum_{u \in \mu'_U \backslash F^\times} i_{h*} \sum_{a \in F_+^\times} \sum_{y \in K^h \backslash \mathbb{V}(a)} r(g,h)\phi(y,u)\, Z(y)_{K^h}.$$

In summary, we have

$$Z(g,\phi)_{U,\alpha} = Z_{0,\text{int}}(g,\phi)_{U,\alpha} + \sum_{u \in \mu'_U \backslash F^\times} i_{h*} Z(1, r(g,h)\phi(\cdot, u), uq)_{K^h}.$$

Here we view $r(g,h)\phi(\cdot, u)$ as a Schwartz function on $\mathcal{S}(\mathbb{V})$. The summation on u is a finite sum depending on (g, h), by the reason similar to the convergence of (4.1.4).

Denote by

$$M'^\circ_{K'}, \quad M^\circ_K, \quad M^\circ_{K,\alpha}$$

respectively the open parts of

$$M'_{K'}, \quad M_K, \quad M_{K,\alpha}$$

before completion.

We claim that $Z_{0,\text{int}}(g,\phi)_{U,\alpha}|_{M^\circ_{K,\alpha}} = 0$ as a divisor class on the open part $M^\circ_{K,\alpha}$ for all α. In fact, $Z_{0,\text{int}}(g,\phi)_{U,\alpha}$ is a linear combination of

$$L_{\beta,\beta'} = \frac{1}{2}(p_1^* L_{U,\beta} + p_2^* L_{U,\beta'})$$

where p_1, p_2 denote the projections of $X_{U,\beta} \times X_{U,\beta'}$ to the two components, and $L_{U,\beta}, L_{U,\beta'}$ denote the Hodge bundles on these two components. A well-known result asserts that $L_{U,\beta}$ is linearly equivalent to a divisor supported on cusps on $X_{U,\beta}$. Thus it is linearly equivalent to zero on the open part $X^\circ_{U,\beta}$. The same is true for $L_{U,\beta'}$. It proves the claimed result.

By the modularity of the generating series on the open part $M'^\circ_{K'}$, we see that $Z(g,\phi)_{U,\alpha}|_{M^\circ_{K,\alpha}}$ is absolutely convergent and invariant under the left action by $\mathrm{SL}_2(F)$. It is easy to check that it is also invariant under the left action by the diagonal group $d(F^\times)$. Thus it is invariant under the left action of $\mathrm{GL}_2(F)$. It proves the modularity of $Z(g,\phi)_{U,\alpha}|_{M^\circ_{K,\alpha}}$, and thus the modularity of $Z(g,\phi)_U|_{M^\circ_K}$.

It gives Theorem 3.17 on the open part M°_K. Note that $M_K = M^\circ_K$ unless $F = \mathbb{Q}$ and $\mathbb{B}_f = M_2(\mathbb{A}_f)$.

4.2.6 Extending to the boundary

Here we consider the case $M_K \neq M^\circ_K$. It happens if and only if $F = \mathbb{Q}$ and $\mathbb{B}_f = M_2(\mathbb{A}_f)$. We have already proved the modularity of $Z(g,\phi)_U|_{M^\circ_K}$, and we will extend it to the modularity of $Z(g,\phi)_U$.

Recall that the modularity of $Z(g,\phi)_U|_{M^\circ_K}$ means that $\ell(Z(g,\Phi)_U|_{M^\circ_K})$ is an automorphic form for any linear functional $\ell : \mathrm{SC} \to \mathbb{C}$. Here SC denotes the subspace of $\mathrm{Pic}(M^\circ_K)_\mathbb{C}$ generated by the coefficients of $Z(g,\phi)_U|_{M^\circ_K}$ over $\mathbb{C}$.

We first claim that SC is finite-dimensional. For that it suffices to assume that $\phi = \phi_f \times \phi_\infty$ with ϕ_∞ equal to the standard Schwartz function. It is easy to see from the definition that $Z(g,\phi)_U$ is invariant under the right action of some open compact subgroup W of $\mathrm{GL}_2(\mathbb{A}_f)$ on g. Then we have a injection

$$\mathrm{SC}^\vee \hookrightarrow \mathrm{Hilb}_2(\mathrm{GL}_2(F), W), \quad \ell \longmapsto \ell(Z(g,\Phi)_U|_{M^\circ_K}).$$

Here $\mathrm{Hilb}_2(\mathrm{GL}_2(F), W)$ is the space of holomorphic Hilbert modular forms on $\mathrm{GL}_2(\mathbb{A})$ with parallel weight two and level W. The space is finite-dimensional, and thus SC is finite-dimensional.

Go back to the modularity of $Z(g,\phi)_U$ for a general ϕ. Pick a basis $\{D_1, \cdots, D_r\}$ of SC, and write

$$Z(g,\phi)_U|_{M^\circ_K} = \sum_{i=1}^r f_i(g) D_i.$$

Here every f_i is a complex-valued function of g. By the modularity, every f_i is a usual holomorphic Hilbert modular forms of parallel weight two.

It suffices to prove the modularity of the difference

$$A(g) := Z(g,\phi)_U - \sum_{i=1}^r f_i(g)\overline{D}_i.$$

Here $\overline{D}_i$ denotes the Zariski closure of D_i in M_K. Since the restriction $A(g)|_{M^\circ_K}$ is zero, the coefficients of $A(g)$ (after possibly moving by linear equivalence) must be supported on the boundary

$$M_K - M_K^\circ = (X_U \times \{\text{cusps}\}) \cup (X_U \times \{\text{cusps}\}).$$

By a result of Manin [Ma] and Drinfel'd [Dr], the difference of any two cusps on the same geometrically connected component of X_U is a torsion divisor in $\text{Pic}(X_U)$. Fix a cusp P_α on $X_{U,\alpha}$ for any component $\alpha \in F_+^\times \backslash \mathbb{A}_f^\times / q(U)$. Then any divisor supported on $M_K - M_K^\circ$ is a linear combination of $X_{U,\alpha} \times P_\beta$ and $P_\alpha \times X_{U,\beta}$. Therefore, we can always write

$$A(g) = \sum_{\alpha,\beta \in F_+^\times \backslash \mathbb{A}_f^\times / q(U)} (B_{\alpha,\beta}(g)(X_{U,\alpha} \times P_\beta) + C_{\alpha,\beta}(g)(P_\alpha \times X_{U,\beta})).$$

Here $B_{\alpha,\beta}$ and $C_{\alpha,\beta}$ are complex-valued functions of g. It suffices to prove that $B_{\alpha,\beta}$ and $C_{\alpha,\beta}$ are automorphic.

Fix (α, β) and consider $B_{\alpha,\beta}(g)$. By intersection on M_K,

$$B_{\alpha,\beta}(g) = A(g) \cdot (P_\alpha \times X_{U,\beta}) = Z(g,\phi)_U \cdot (P_\alpha \times X_{U,\beta}) - \sum_{i=1}^r f_i(g) \overline{D}_i \cdot (P_\alpha \times X_{U,\beta}).$$

Therefore, it suffices to prove the modularity of

$$Z(g,\phi)_U \cdot (P_\alpha \times X_{U,\beta}) = Z(g,\phi)_{U,\alpha^{-1}\beta} \cdot (P_\alpha \times X_{U,\beta}) = \deg Z(g,\phi)_{U,\alpha^{-1}\beta}.$$

Here $\deg Z(g,\phi)_{U,\alpha^{-1}\beta}$ denotes the degree as a correspondence. Namely, it is the degree of $Z(g,\phi)_{U,\alpha^{-1}\beta}D$ for any divisor D on X_U of degree one.

A similar result is true for $C_{\alpha,\beta}(g)$. Indeed, it is reduced to the modularity of the degree of the transpose correspondence $Z(g,\phi)_{U,\alpha^{-1}\beta}^{\text{t}}$. The coefficients of $Z(g,\phi)_{U,\alpha^{-1}\beta}$ are the Hodge bundles and divisors of the form $Z(x)_U$. By this, it is easy to check that

$$\deg Z(g,\phi)_{U,\alpha^{-1}\beta}^{\text{t}} = \deg Z(g,\phi)_{U,\alpha^{-1}\beta}.$$

Therefore, Theorem 3.17 is reduced to the modularity of $\deg Z(g,\phi)_{U,\alpha}$ for any $\alpha \in F_+^\times \backslash \mathbb{A}_f^\times / q(U)$. In Proposition 4.2, we will prove that the degree is an Eisenstein series. It finishes the proof of Theorem 3.17.

4.3   DEGREE OF THE GENERATING SERIES

Here we describe the actions of the generating series $Z(g,\phi)_U$ and its components on a divisor, and compute the degrees of the components as correspondences.

4.3.1   Action of the generating series

We consider the action of

$$Z(g,\phi)_U = Z_0(g,\phi)_U + Z_*(g,\phi)_U$$

on a point $[z, \beta]_U$ of X_U represented by $(z, \beta) \in \mathcal{H} \times \mathbb{B}_f^\times$.

The result is simple for the constant term

$$Z_0(g, \phi)_U = - \sum_{\alpha \in F_+^\times \backslash \mathbb{A}_f^\times / q(U)} \sum_{u \in \mu_U^2 \backslash F^\times} E_0(\alpha^{-1}u, r(g)\phi) \, L_{K,\alpha}.$$

By definition

$$L_{K,\alpha} = \sum_{\alpha' \in F_+^\times \backslash \mathbb{A}_f^\times / q(U)} L_{K,\alpha',\alpha\alpha'}, \quad L_{K,\alpha',\alpha\alpha'} = \frac{1}{2}(p_1^* L_{U,\alpha'} + p_2^* L_{U,\alpha\alpha'}).$$

Here p_1, p_2 are the projections of $X_{U,\alpha} \times X_{U,\alpha\alpha'}$ to its components. It follows that

$$L_{K,\alpha} \circ [z, \beta]_U = \frac{1}{2} L_{U,\alpha q(\beta)}.$$

Therefore,

$$Z_0(g, \phi)_U[z, \beta]_U = -\frac{1}{2} \sum_{\alpha \in F_+^\times \backslash \mathbb{A}_f^\times / q(U)} \sum_{u \in \mu_U^2 \backslash F^\times} E_0(\alpha^{-1}u, r(g)\phi) \, L_{U,\alpha q(\beta)}.$$

Now we write down the action of the non-constant part

$$Z_*(g, \phi)_U = w_U \sum_{a \in F^\times} \sum_{x \in U \backslash \mathbb{B}_f^\times / U} r(g)\phi(x)_a Z(x)_U.$$

By definition,

$$Z(x)_U[z, \beta] = \sum_j [z, \beta\alpha_j], \quad \text{if } UxU = \coprod_j \alpha_j U.$$

It follows from (4.2.2) that

$$Z_*(g, \phi)_U[z, \beta] = w_U \sum_{a \in F^\times} \sum_{x \in \mathbb{B}_f^\times / U} r(g)\phi(x)_a [z, \beta x].$$

For any $\alpha \in F_+^\times \backslash \mathbb{A}_f^\times / q(U)$, we also consider the action of the generating series

$$Z(g, \phi)_{U,\alpha} = Z(g, \phi)|_{M_{K,\alpha}}.$$

Note that the constant term

$$Z_0(g, \phi)_{U,\alpha} = - \sum_{u \in \mu_U^2 \backslash F^\times} E_0(\alpha^{-1}u, r(g)\phi) \, L_{K,\alpha}.$$

Then

$$Z_0(g, \phi)_{U,\alpha}[z, \beta]_U = -\frac{1}{2} \sum_{u \in \mu_U^2 \backslash F^\times} E_0(\alpha^{-1}u, r(g)\phi) \, L_{U,\alpha q(\beta)}.$$

By (4.2.4), the non-constant part is given by

$$Z_*(g, \phi)_{U,\alpha} = w_U \sum_{u \in \mu'_U \backslash F^\times} \sum_{y \in K^h \backslash \mathbb{B}_f^{ad}} r(g, (h, 1)) \phi(y, u) Z(h^{-1}y)_U.$$

Here h is any element of $\mathbb{B}_f^\times$ such that $q(h) \in \alpha^{-1} F_+^\times q(U)$, and $K^h = \mathrm{GSpin}(\mathbb{V}_f) \cap hKh^{-1}$. Here $\mathbb{B}_f^{ad}$ is defined by:

$$\mathbb{B}_f^{ad} = \{x \in \mathbb{B}_f : q(x) \in F_+^\times\} = \bigcup_{a \in F_+^\times} \mathbb{B}_f(a),$$

$$\mathbb{B}_f^a = \mathbb{B}_f(a) = \{x \in \mathbb{B}_f : q(x) = a\}, \quad a \in \mathbb{A}_f^\times.$$

Note that the second notation is also valid in the local case. And the infinite component of $r(g, (h, 1)) \phi(y, u)$ is understood to be $W_{uq(y)}^{(2)}(g_\infty)$, which makes sense for $q(y) \in F_+^\times$.

We are going to write down the action of $Z_*(g, \phi)_{U,\alpha}$ on X_U. Assume that h is an element of $\mathbb{B}_f^\times$ for simplicity. That is, the second component is trivial. The action of $Z(h^{-1}y)_U$ is given by the coset $Uh^{-1}yU/U$. We have identities

$$Uh^{-1}yU/U = Kh^{-1}y/U = h^{-1}(hKh^{-1}y/U) = h^{-1}(K^h y/U^1).$$

Here $U^1 = U \cap \mathbb{B}_f^1 = \{b \in U : q(b) = 1\}$. By this it is easy to see that

$$Z_*(g, \phi)_{U,\alpha} [z, \beta] = w_U \sum_{u \in \mu'_U \backslash F^\times} \sum_{y \in \mathbb{B}_f^{ad}/U^1} r(g, (h, 1)) \phi(y, u) [z, \beta h^{-1}y].$$

Equivalently,

$$Z_*(g, \phi)_{U,\alpha} [z, \beta] = w_U \sum_{u \in \mu'_U \backslash F^\times} \sum_{a \in F_+^\times} \sum_{y \in \mathbb{B}_f(a)/U^1} r(g, (h, 1)) \phi(y, u) [z, \beta h^{-1}y].$$

$$(4.3.1)$$

Here h is any element of $\mathbb{B}_f^\times$ such that $q(h) \in \alpha^{-1} F_+^\times q(U)$.

4.3.2 Degree of the generating series

Now we compute the degree of $Z(g, \phi)_{U,\alpha}$ for any $\alpha \in F_+^\times \backslash \mathbb{A}_f^\times / q(U)$. It is a correspondence from $X_{U,\beta}$ to $X_{U,\alpha\beta}$ for any $\beta \in F_+^\times \backslash \mathbb{A}_f^\times / q(U)$. The degree of this correspondence is just the degree of $Z(g, \phi)_{U,\alpha} D$ for any degree-one divisor D on X_U. It is independent of D. By definition, we see that $\deg Z(x)_U = [UxU : U]$.

Recall that in §4.1 the weight two Eisenstein series are defined as follows:

$$E(s, g, u, \phi) = \sum_{\gamma \in P^1(F) \backslash \mathrm{SL}_2(F)} \delta(\gamma g)^s r(\gamma g) \phi(0, u),$$

$$E(s, g, \phi)_U = \sum_{u \in \mu_U^2 \backslash F^\times} E(s, g, u, \phi).$$

The result can be viewed as a geometric variant of the Siegel–Weil formula.

PROPOSITION 4.2. *For any* $\alpha \in F_+^\times \backslash \mathbb{A}_f^\times / q(U)$,

$$\deg Z(g, \phi)_{U,\alpha} = -\frac{1}{2}(\deg L_{U,\alpha}) \, E(0, g, r(h)\phi)_U.$$

Here h is any element of $\mathbb{B}_f^\times \times \mathbb{B}_f^\times$ such that $\nu(h) \in \alpha^{-1} F_+^\times q(U)$, and $\deg L_{U,\alpha}$ is independent of α.

PROOF. Note that X_U is connected, but not geometrically connected. Then the Galois group acts transitively on the set of geometrically connected components, and the action switches the Hodge bundles $L_{U,\alpha}$ between components. Thus the degree $\kappa_U^\circ = \deg L_{U,\alpha}$ is independent of α.

By the action described above, we immediately see that

$$\deg Z_0(g, \phi)_{U,\alpha} = -\frac{1}{2}\kappa_U^\circ \sum_{u \in \mu_U^2 \backslash F^\times} E_0(\alpha^{-1}u, r(g)\phi),$$

$$\deg Z_*(g, \phi)_{U,\alpha} = w_U \sum_{u \in \mu_U' \backslash F^\times} \sum_{a \in F_+^\times} \sum_{x \in \mathbb{B}_f(a)/U^1} r(g, h)\phi(x, u).$$

Write

$$E(0, g, \phi)_U = E_0(0, g, \phi)_U + E_*(0, g, \phi)_U.$$

Here $E_0(0, g, \phi)_U$ is the constant term, and

$$E_*(0, g, \phi)_U = \sum_{a \in F^\times} E_a(0, g, \phi)_U$$

is the non-constant part. We are going to compare the constant terms and the non-constant parts of these two series.

By definition,

$$E_0(u, r(g)\phi) = E_0(0, g, u, \phi).$$

Thus we get the identity for constant terms:

$$\deg Z_0(g, \phi)_{U,\alpha} = -\frac{1}{2}\kappa_U^\circ \, E_0(0, g, r(h)\phi)_U.$$

The corresponding identity for the non-constant parts follows from the local results in Proposition 2.9. Consider

$$\deg Z_*(g, \phi)_{U,\alpha} = w_U \sum_{u \in \mu_U' \backslash F^\times} \sum_{a \in F_+^\times} \sum_{x \in \mathbb{B}_f(a)/U^1} r(g, h)\phi(x, u).$$

Let x_a be any fixed element in $\mathbb{B}_f^\times$ with norm a. The last summation above equals

$$\frac{1}{\text{vol}(U^1)\text{vol}(\mathbb{B}_\infty^1)} \int_{\mathbb{B}^1} r(g, h)\phi(bx_a, u)db$$

$$= -\frac{1}{\text{vol}(U^1)\text{vol}(\mathbb{B}_\infty^1)} W_{au}(0, g, u, r(h)\phi).$$

The equality follows from Proposition 2.9. Both sides are zero if a lies in $F^\times - F_+^\times$.

It follows that

$$\deg Z_*(g, \phi)_{U,\alpha} = -\frac{w_U}{\text{vol}(U^1)\text{vol}(\mathbb{B}_\infty^1)} \sum_{u \in \mu_U' \backslash F^\times} \sum_{a \in F^\times} W_{au}(0, g, u, r(h)\phi).$$

Thus,

$$\deg Z_*(g, \phi)_{U,\alpha} = -\frac{w_U}{[\mu_U' : \mu_U^2]\text{vol}(U^1)\text{vol}(\mathbb{B}_\infty^1)} E_*(0, g, r(h)\phi).$$

It remains to check

$$\kappa_U^\circ = \frac{2\, w_U}{[\mu_U' : \mu_U^2]\text{vol}(U^1)\text{vol}(\mathbb{B}_\infty^1)}. \tag{4.3.2}$$

Here the Haar measure on $\mathbb{B}_v^1$ for each place v is as in §1.6.2. In particular, $\text{vol}(\mathbb{B}_v^1) = 4\pi^2$ for any archimedean place v.

Fix an archimedean place τ, and denote by $B = B(\tau)$ the nearby quaternion algebra. The factor $[\mu_U' : \mu_U^2]$ is exactly the degree of the natural map

$$(B^1 \cap U^1)\backslash\mathcal{H} \longrightarrow (B_+^\times \cap U)\backslash\mathcal{H}.$$

In fact, the degree is just

$$[(B_+^\times \cap U) : (B^1 \cap U^1)\mu_U] = [q(B_+^\times \cap U) : \mu_U^2] = [\mu_U' : \mu_U^2].$$

The last identity follows from (4.2.3). Denote by $\kappa_{U^1}^\circ$ the degree of the Hodge bundle of $(B^1 \cap U^1)\backslash\mathcal{H}$. It is also the integration of $\frac{dxdy}{2\pi y^2}$ on $(B^1 \cap U^1)\backslash\mathcal{H}$. Then we have the relation $\kappa_{U^1}^\circ = [\mu_U' : \mu_U^2]\kappa_U^\circ$. Now (4.3.2) is equivalent to

$$\kappa_{U^1}^\circ = \frac{2\, w_U}{\text{vol}(U^1)\text{vol}(\mathbb{B}_\infty^1)}. \tag{4.3.3}$$

We will interpret the equality as the fact that the Tamagawa number of B^1 is 1.

Endow the Haar measure on B_v^1 for every place v as in §1.6.2. It gives a product measure on $B_\mathbb{A}^1$, which is exactly the Tamagawa measure. Since B^1 is simply connected, we have $\text{vol}(B^1\backslash B_\mathbb{A}^1) = 1$. By the strong approximation theorem, $B_\mathbb{A}^1 = B^1 B_\infty^1 U^1$. It follows that

$$B^1\backslash B_\mathbb{A}^1 = B^1\backslash B^1 B_\infty^1 U^1 = (\Gamma\backslash B_\tau^1)B_\infty^{1;\tau}U^1.$$

Here we denote $\Gamma = B^1 \cap U^1$. It follows that

$$\text{vol}(\Gamma\backslash B_\tau^1)\text{vol}(B_\infty^{1;\tau})\text{vol}(U^1) = 1 \tag{4.3.4}$$

Consider the volume of $\Gamma\backslash B_\tau^1$. Note that $B_\tau^1 \simeq SL_2(\mathbb{R})$. By the Iwasawa decomposition, any element is uniquely of the form

$$\begin{pmatrix} 1 & x \\ & 1 \end{pmatrix} \begin{pmatrix} y^{\frac{1}{2}} & \\ & y^{-\frac{1}{2}} \end{pmatrix} \begin{pmatrix} \cos\theta & \sin\theta \\ -\sin\theta & \cos\theta \end{pmatrix}, \quad x \in \mathbb{R}, \ y \in \mathbb{R}_+, \ \theta \in [0, 2\pi).$$

The measure on B_τ^1 is just $\dfrac{dxdy}{2y^2} d\theta$. It follows that

$$\text{vol}(\Gamma\backslash B_\tau^1) = \frac{2\pi}{w_U} \int_{\Gamma\backslash\mathcal{H}} \frac{dxdy}{2y^2}.$$

Here $\mathcal{H}$ is the space of (x, y), and the factor w_U appears due to the action of $-1 \in \Gamma$ (in the case $w_U = 2$) on the matrix $\begin{pmatrix} \cos\theta & \sin\theta \\ -\sin\theta & \cos\theta \end{pmatrix}$.

On the other hand, the degree of the Hodge bundle

$$\kappa_{U^1}^\circ = \int_{\Gamma\backslash\mathcal{H}} \frac{dxdy}{2\pi y^2}.$$

It yields that

$$\text{vol}(\Gamma\backslash B_\tau^1) = \frac{2\pi^2}{w_U} \kappa_{U^1}^\circ.$$

Then (4.3.4) becomes

$$\frac{2\pi^2}{w_U} \kappa_{U^1}^\circ \text{vol}(B_\infty^{1,\tau}) \text{vol}(U^1) = 1.$$

It is equivalent to (4.3.3) because $\text{vol}(\mathbb{B}_\tau^1) = 4\pi^2$. It finishes the proof.   □

Remark. If U is maximal, by computing the volumes, we can obtain an explicit formula

$$\kappa_{U^1}^\circ = 4(4\pi)^{-[F:\mathbb{Q}]}|D_F|^{\frac{3}{2}}\zeta_F(2) \prod_{v \in \Sigma, \ v\nmid\infty} (N_v - 1).$$

See also Vignéras [Vi].

4.4   THE TRACE IDENTITY

The goal for the rest of this chapter is to prove Theorem 3.22. We first recall the content of the theorem.

Let $\Phi \in \mathcal{S}(\mathbb{V} \times \mathbb{A}^\times)$ be a Schwartz function. We have defined the "arithmetic theta lifting"

$$\widetilde{Z}(\Phi \otimes \varphi) = \int_{GL_2(F)\backslash GL_2(\mathbb{A})/Z(\mathbb{A})}^* \varphi(g)\, \widetilde{Z}(g, \Phi)\, dg, \quad \varphi \in \sigma.$$

As a correspondence, it gives a map

$$\widetilde{Z}(\Phi \otimes \varphi): \quad J(\overline{F})_{\mathbb{C}} \longrightarrow J^{\vee}(\overline{F})_{\mathbb{C}}.$$

On the other hand, we have a theta lifting

$$\theta: \quad \mathcal{S}(\mathbb{V} \times \mathbb{A}^{\times}) \otimes \sigma \longrightarrow \pi \otimes \widetilde{\pi}.$$

It is normalized in (2.2.3) place by place. By §3.3.1, one has a map

$$\mathrm{T}: \pi \otimes \widetilde{\pi} \longrightarrow \mathrm{Hom}^0(J, J^{\vee})_{\mathbb{C}}.$$

Theorem 3.22 asserts

$$\widetilde{Z}(\Phi \otimes \varphi) = \frac{L(1, \pi, \mathrm{ad})}{2\zeta_F(2)} \mathrm{T}(\theta(\Phi \otimes \varphi))$$

as an identity in $\mathrm{Hom}^0(J, J^{\vee})_{\mathbb{C}}$. By the argument right after the theorem, there is a constant $c(\pi)$ independent of Φ and φ such that

$$\widetilde{Z}(\Phi \otimes \varphi) = c(\pi)\mathrm{T}(\theta(\Phi \otimes \varphi)). \tag{4.4.1}$$

The task is to prove that $c(\pi)$ is the given one.

4.4.1 Lefschetz trace formula

Fix an open compact subgroup U of $\mathbb{B}_f^{\times}$ acting trivially on Φ. Fix an embedding $\tau: F \hookrightarrow \mathbb{C}$. Write $H^i(X_U) = H^i(X_{U,\tau}(\mathbb{C}), \mathbb{C})$ and $H^{p,q}(X_U) = H^{p,q}(X_{U,\tau}(\mathbb{C}), \mathbb{C})$ for simplicity.

View (4.4.1) as an identity of operators on $H^{1,0}(X_U)$. Taking traces, we obtain

$$\mathrm{tr}(\ \widetilde{Z}(\Phi \otimes \varphi) \ | H^{1,0}(X_U) \ ) = c(\pi) \ \mathrm{tr}(\ \mathrm{T}(\theta(\Phi \otimes \varphi)) \ | H^{1,0}(X_U) \ ).$$

To compute $c(\pi)$, it suffices to compute the traces in both sides.

Trace of projectors

Let $f_1 \otimes f_2 \in \pi^U \otimes \widetilde{\pi}^U$ be any vector. By definition,

$$\mathrm{T}(f_1 \otimes f_2) = \mathrm{vol}(X_U) \ \mathrm{T}(f_1 \otimes f_2)_U.$$

Fix a decomposition

$$H^{1,0}(X_U) = \bigoplus_{\pi' \in \mathcal{A}(\mathbb{B}^{\times})} \pi'^U.$$

The action of $\mathrm{T}(f_1 \otimes f_2)_U$ on $H^{1,0}(X_U)$ is simply

$$f \longmapsto (f, f_2) \ f_1.$$

Its trace is simply the pairing (f_1, f_2). It follows that

$$\text{tr}(\ T(\theta(\Phi \otimes \varphi))\ |H^{1,0}(X_U)\ ) = \text{vol}(X_U)\ \mathcal{F}\theta(\Phi \otimes \varphi).$$

Here $\mathcal{F} = (\cdot, \cdot)$ is the canonical form $\mathcal{F} : \pi \otimes \tilde{\pi} \to \mathbb{C}$. Recall that the Shimizu lifting is normalized in (2.2.3) so that

$$\mathcal{F}\theta(\Phi \otimes \varphi) = \prod_v \frac{\zeta_v(2)}{L(1, \pi_v, \text{ad})} \int_{N(F_v)\backslash \text{GL}_2(F_v)} W_{\varphi_v, -1}(g) r(g) \Phi_v(1, 1) dg.$$

Trace of the arithmetic theta lifting

Write $\tilde{Z}(\Phi \otimes \varphi)_U$ for the corresponding divisor of $\tilde{Z}(\Phi \otimes \varphi)$ on $X_U \times X_U$. The key to compute the trace is the following Lefschetz trace formula

$$\deg \Delta^* \tilde{Z}(\Phi \otimes \varphi)_U = \sum_{i=0,1,2} (-1)^i \text{tr}(\tilde{Z}(g, \Phi)_U | H^i(X_U)).$$

Here $\Delta : X_U \to X_U \times X_U$ is the diagonal embedding, and the left-hand side is just the number of fixed points of the correspondence $\tilde{Z}(\Phi \otimes \varphi)_U$.

It is easy to see that

$$\text{tr}(\tilde{Z}(\Phi \otimes \varphi)_U | H^0(X_U)) = \text{tr}(\tilde{Z}(\Phi \otimes \varphi)_U | H^2(X_U)) = \deg(\tilde{Z}(\Phi \otimes \varphi)_U).$$

Here $\deg(\tilde{Z}(\Phi \otimes \varphi)_U)$ is in the sense of correspondence. Namely, it is the degree of $\tilde{Z}(\Phi \otimes \varphi)_U D$ for any divisor D on X_U of degree one.

As for $H^1(X_U)$, we have the decomposition

$$H^1(X_U) = H^{1,0}(X_U) \oplus H^{0,1}(X_U) = H^{1,0}(X_U) \oplus \overline{H^{1,0}(X_U)}.$$

Here

$$H^{1,0}(X_U) = \Gamma(X_U, \Omega^1_{X_U}) \otimes_F \mathbb{C}.$$

The tensor product is through $\tau : F \hookrightarrow \mathbb{C}$. The complex conjugation is taken on the $\mathbb{C}$ part, which fixes $\Gamma(X_U, \Omega^1_{X_U})$ because $\tau(F) \subset \mathbb{R}$. Note that $\tilde{Z}(\Phi \otimes \varphi)_U$ is defined over F since so is $\tilde{Z}(g, \Phi)_U$. So their traces on $H^{1,0}(X_U)$ are the same as on $\overline{H^{1,0}(X_U)}$. It follows that

$$\text{tr}(\tilde{Z}(\Phi \otimes \varphi)_U | \overline{H^{1,0}(X_U)}) = \text{tr}(\tilde{Z}(\Phi \otimes \varphi)_U | H^{1,0}(X_U)).$$

In summary, we have

$$\text{tr}(\tilde{Z}(\Phi \otimes \varphi) | H^{1,0}(X_U)) = -\frac{1}{2} \deg \Delta^* \tilde{Z}(\Phi \otimes \varphi)_U + \deg \tilde{Z}(\Phi \otimes \varphi)_U.$$

Then Theorem 3.22 is implied by the following two results:

PROPOSITION 4.3.

$$\int_{\text{GL}_2(F)\backslash \text{GL}_2(\mathbb{A})/Z(\mathbb{A})}^* \varphi(g)\ \deg \tilde{Z}(g, \Phi)_U\ dg\ =\ 0.$$

PROPOSITION 4.4.

$$\int_{GL_2(F)\backslash GL_2(\mathbb{A})/Z(\mathbb{A})}^{*} \varphi(g) \ \deg \Delta^* \widetilde{Z}(g, \Phi)_U \ dg$$

$$= -\mathrm{vol}(X_U) \frac{L(1, \pi, \mathrm{ad})}{\zeta_F(2)} \mathcal{F}\theta(\Phi \otimes \varphi).$$

Both propositions are proved by the explicit expressions of the degrees. We will see that Proposition 4.3 is true because $\deg \widetilde{Z}(g, \Phi)_U$ is essentially an Eisenstein series, and Proposition 4.4 is the geometric case of Proposition 2.3.

4.4.2   Degrees of correspondences

Here we prove Proposition 4.3. Denote by $\phi = \overline{\Phi} \in \overline{\mathcal{S}}(\mathbb{V} \times \mathbb{A}^\times)$ the average of Φ at infinity. It suffices to prove the same result for $\deg Z(g, \phi)_U$.

By definition,

$$\deg Z(g, \phi)_U = \sum_{\alpha \in F_+^\times \backslash \mathbb{A}_f^\times / q(U)} \deg Z(g, \phi)_{U, \alpha}.$$

By Proposition 4.2, each component

$$\deg Z(g, \phi)_{U, \alpha} = -\frac{1}{2} \kappa_U^\circ E(0, g, r(h, 1)\phi)_U.$$

Here h is any element of $\mathbb{B}_f^\times$ such that $q(h) \in \alpha^{-1} F_+^\times q(U)$, and the Eisenstein series are given by

$$E(s, g, u, \phi) = \sum_{\gamma \in P^1(F)\backslash SL_2(F)} \delta(\gamma g)^s r(\gamma g)\phi(0, u),$$

$$E(s, g, \phi)_U = \sum_{u \in \mu_U^2 \backslash F^\times} E(s, g, u, \phi).$$

It suffices to show that

$$\int_{GL_2(F)\backslash GL_2(\mathbb{A})/Z(\mathbb{A})}^{*} \varphi(g) \ E(0, g, \phi)_U \ dg = 0.$$

The general case $E(0, g, r(h, 1)\phi)_U$ is obtained by replacing ϕ by $r(h, 1)\phi$. By definition, the integral equals

$$\int_{GL_2(F)\backslash GL_2(\mathbb{A})/Z(\mathbb{A})} \fint_{F^\times \backslash \mathbb{A}_f^\times / U_Z} \varphi(zg) \ E(0, zg, \phi)_U \ dz \ dg.$$

Here $U_Z = U \cap Z(\mathbb{A})$ acts trivially on φ and E. It further becomes

$$(\varphi(g), \ E(0, g, \omega_\sigma, \phi)_U)_{\mathrm{pet}}.$$

Here ω_σ is the central character of σ, and

$$E(s, g, \omega_\sigma, \phi)_U = \fint_{F^\times \backslash \mathbb{A}_f^\times / U_Z} E(s, zg, \phi)_U \, \omega_\sigma(z) dz.$$

By definition,

$$E(s, g, \omega_\sigma, \phi)_U$$

$$= \sum_{\gamma \in P(F) \backslash \mathrm{GL}_2(F)} \fint_{F^\times \backslash \mathbb{A}_f^\times / U_Z} \sum_{u \in \mu_U^2 \backslash F^\times} \delta(\gamma g)^s r(\gamma z g) \phi(0, u) \omega_\sigma(z) dz.$$

It is a P-series defined by

$$B(s, g) = \fint_{F^\times \backslash \mathbb{A}_f^\times / U_Z} \sum_{u \in \mu_U^2 \backslash F^\times} \delta(g)^s r(zg) \phi(0, u) \, \omega_\sigma(z) dz,$$

which transfers as

$$B(s, pnzg) = \omega_\sigma^{-1}(z) B(s, g), \quad p \in P(F), \ n \in N(\mathbb{A}), \ z \in Z(\mathbb{A}).$$

By the standard theory, we see that $E(s, g, \omega_\sigma, \phi)_U$ is orthogonal to σ.

4.4.3   Degree of pull-back

The proof of Proposition 4.4 will take up the rest of this chapter. It is implied by an expression of $\deg \Delta^* Z(g, \phi)_U$ in terms of the incoherent Eisenstein series of weight $3/2$ in §2.5.3. We call this formula the pull-back formula. The treatment is naturally divided into two cases:

(1) Compact case $\#\Sigma > 1$. In this case, the Shimura curve X_U° is compact and the related Eisenstein series of weight $3/2$ is holomorphic.

(2) Non-compact case $\#\Sigma = 1$. It happens if and only if $F = \mathbb{Q}$ and $\Sigma = \{\infty\}$. The Shimura curve X_U is just a classical modular curve (if choosing suitable U). In this case, the Shimura curve X_U° is non-compact and the related Eisenstein series of weight $3/2$ is non-holomorphic. Then our computation is complicated by extra terms coming from the cusp of X_U and the non-holomorphic terms of the Eisenstein series.

To illustrate the idea, we list the pull-back formula in the compact case in the following.

PROPOSITION 4.5. *Let $\phi \in \overline{\mathcal{S}}(\mathbb{A} \times \mathbb{A}^\times)$ be invariant under $U \times U$. Assume that $\#\Sigma > 1$. Then*

$$\deg \Delta^* Z(g, \phi)_U \quad = \quad -\mathrm{vol}(X_U) \, J(0, g, \phi)_U, \quad g \in \mathrm{GL}_2(\mathbb{A}).$$

Here the mixed Eisenstein–theta series

$$J(s, g, \phi)_U = \sum_{\gamma \in P(F) \backslash \mathrm{GL}_2(F)} \delta(\gamma g)^s \sum_{u \in \mu_U^2 \backslash F^\times} \sum_{x \in F} r(\gamma g) \phi(x, u).$$

We will prove this formula in the next section. Its counterpart in the non-compact case is Theorem 4.15 in §4.6, where the difference of two sides is nonzero and given by some extra terms. In §4.7, we obtain a description of the extra terms in terms of a Poincaré series in Theorem 4.20. It finishes Proposition 4.4 (in the non-compact case).

Now let us see how Proposition 4.5 deduces Proposition 4.4 (in the compact case). Assume that $\#\Sigma > 1$. By taking $\phi = \overline{\Phi}$, we need to show:

$$\int_{\mathrm{GL}_2(F)\backslash \mathrm{GL}_2(\mathbb{A})/Z(\mathbb{A})}^{*} \varphi(g)\ \widetilde{J}(0,g,\phi)_U\ dg = \frac{L(1,\pi,\mathrm{ad})}{\zeta_F(2)}\mathcal{F}\theta(\Phi \otimes \varphi). \qquad (4.4.2)$$

Here we normalize

$$\widetilde{J}(s,g,\phi)_U := \frac{2^{[F:\mathbb{Q}]-1}h_F|D_F|^{-\frac{1}{2}}}{[O_F^{\times} : \mu_U^2]} J(s,g,\phi)_U.$$

The normalizing factor is the same as in the definition of $\widetilde{Z}(\Phi)_U$ in (3.4.4). We also define

$$J(s,g,\Phi) = \sum_{\gamma \in P(F)\backslash \mathrm{GL}_2(F)} \delta(\gamma g)^s \sum_{u \in F^{\times}} \sum_{x \in F} r(\gamma g)\Phi(x,u).$$

We can assume that Φ is invariant under the action of $\mathbb{B}_{\infty}^1$ as usual.

We prove it by a few steps as follows:

$$J(s,g,\phi)_U = w_U \int_{\mu_U \backslash F_{\infty}^{\times}} J(s,zg,\Phi)dz, \qquad (4.4.3)$$

$$\int_{\mathrm{GL}_2(F)\backslash \mathrm{GL}_2(\mathbb{A})/Z(\mathbb{A})}^{*} \varphi(g)\ \widetilde{J}(s,g,\phi)_U\ dg = \int_{\mathrm{GL}_2(F)\backslash \mathrm{GL}_2(\mathbb{A})} \varphi(g)J(s,g,\Phi)dg, \qquad (4.4.4)$$

$$\int_{\mathrm{GL}_2(F)\backslash \mathrm{GL}_2(\mathbb{A})} \varphi(g)\ J(0,g,\Phi)dg = \frac{L(1,\pi,\mathrm{ad})}{\zeta_F(2)}\mathcal{F}\theta(\Phi \otimes \varphi). \qquad (4.4.5)$$

First verify (4.4.3). Similar to (4.1.5), it follows from the relation $\phi = \int_{F_{\infty}^{\times}} r(z)\Phi\ dz$. To check it, we can assume that $-1 \notin U$ since both sides change by the same multiple when varying U. Then it suffices to check

$$\int_{\mu_U \backslash F_{\infty}^{\times}} \sum_{u \in F^{\times}} \sum_{x \in F} r(z\gamma g)\Phi(x,u)dz = \sum_{u \in \mu_U^2 \backslash F^{\times}} \sum_{x \in F} r(\gamma g)\phi(x,u).$$

The key is that $r(\alpha)\Phi = \Phi$ for $\alpha \in \mu_U$. So the left-hand side becomes

$$\sum_{(x,u) \in \mu_U \backslash (F \times F^{\times})} \int_{F_{\infty}^{\times}} r(z\gamma g)\Phi(x,u)dz = \sum_{(x,u) \in \mu_U \backslash (F \times F^{\times})} r(\gamma g)\phi(x,u).$$

It gives the result.

Now we verify (4.4.4). The left-hand side is equal to

$$w_U \frac{2^{[F:\mathbb{Q}]-1} h_F |D_F|^{-\frac{1}{2}}}{[O_F^\times : \mu_U^2]} \int_{\mathrm{GL}_2(F)\backslash \mathrm{GL}_2(\mathbb{A})/Z(\mathbb{A})}^* \varphi(g) \int_{\mu_U \backslash F_\infty^\times} J(s, g, r(z)\Phi)dzdg.$$

By the definition in (1.6.1), the above double integral equals

$$\frac{2}{\mathrm{Res}_{s=1}\zeta_F(s)} \int_{\mathrm{GL}_2(F)\backslash \mathrm{GL}_2(\mathbb{A})/Z(\mathbb{A})} \int_{F^\times \backslash \mathbb{A}^\times / F_\tau^\times} \int_{\mu_U \backslash F_\infty^\times} \varphi(z'zg) J(s, z'zg, \Phi)dzdz'dg.$$

Here τ is any archimedean place of F. Use the identity

$$\mu_U \backslash F_\infty^\times \approx (\mu_U \backslash F_\infty^{\tau,\times}) \cdot F_\tau^\times,$$

and split the inner integration. Then the integrations on $F_\tau^\times$ and on $F^\times \backslash \mathbb{A}^\times / F_\tau^\times$ collapse to an integration on $F^\times \backslash \mathbb{A}^\times$. The above becomes

$$\frac{2 \,\mathrm{vol}(\mu_U \backslash F_\infty^{\tau,\times})}{\mathrm{Res}_{s=1}\zeta_F(s)} \int_{\mathrm{GL}_2(F)\backslash \mathrm{GL}_2(\mathbb{A})/Z(\mathbb{A})} \int_{F^\times \backslash \mathbb{A}^\times} \varphi(z'g) J(s, z'g, \Phi)dz'dg$$

$$= \frac{2 \,\mathrm{vol}(\mu_U \backslash F_\infty^{\tau,\times})}{\mathrm{Res}_{s=1}\zeta_F(s)} \int_{\mathrm{GL}_2(F)\backslash \mathrm{GL}_2(\mathbb{A})} \varphi(g) J(s, g, \Phi)dg.$$

Note that μ_U is a subgroup of rank $n-1$ in $F_\infty^{\tau,\times} \cong (\mathbb{R}^\times)^{n-1}$. Here $n = [F : \mathbb{Q}]$. The volume

$$\mathrm{vol}(\mu_U \backslash (\mathbb{R}^\times)^{n-1}) = \frac{1}{2^{n-1}w_U} \mathrm{vol}(\mu_U^2 \backslash (\mathbb{R}^\times)^{n-1})$$

$$= \frac{1}{w_U} \mathrm{vol}(\mu_U^2 \backslash (\mathbb{R}_+^\times)^{n-1}) = \frac{1}{w_U} R_{\mu_U^2}.$$

Here $R_{\mu_U^2}$ is the regulator of μ_U^2 defined similar to the regulator of $O_F^\times$. So we only need to check

$$w_U \frac{2^{[F:\mathbb{Q}]-1} h_F |D_F|^{-\frac{1}{2}}}{[O_F^\times : \mu_U^2]} \cdot \frac{2 \, R_{\mu_U^2}/w_U}{\mathrm{Res}_{s=1}\zeta_F(s)} = 1.$$

It is equivalent to (3.4.5), another description of the normalizing factor. It finishes (4.4.4).

The proof of (4.4.5) is just the standard folding-unfolding process. It is not different from the computation of $A(s)$ in Proposition 2.3. The proof there also applies to the incoherent case.

4.5 PULL-BACK FORMULA: COMPACT CASE

In this section, we prove the formula for the degree of $\Delta^* Z(g, \phi)_U$ in Proposition 4.5 when the Shimura curve X_U has no cusps (or equivalently the open part X_U° is compact). It happens if and only if $\#\Sigma > 1$. It includes "almost all" Σ, but is much simpler than the non-compact case. It has the same flavor as [YZZ, Proposition 3.1]. For the convenience of readers, we give a detailed proof here.

4.5.1   CM cycles on the Shimura curve

Denote by $\Delta : X_U \to X_U \times X_U$ the diagonal embedding. Then $\Delta^* Z(g, \phi)_U$ is an automorphic form on $\mathrm{GL}_2(\mathbb{A})$ with coefficients in $\mathrm{Pic}(X_U)_{\mathbb{C}}$. Our goal is to obtain a formula for it.

Let $\mathbb{B}_0$ be the set of elements in $\mathbb{B}$ with trace zero. It gives an orthogonal decomposition $\mathbb{B} = \mathbb{A} \oplus \mathbb{B}_0$ of quadratic spaces. Denote the conjugation action of $\mathbb{B}^\times$ on $\mathbb{B}_0$ by Ad, i.e., $\mathrm{Ad}(h) \circ x = hxh^{-1}$ for any $h \in \mathbb{B}^\times$ and $x \in \mathbb{B}_0$. The action keeps the norm.

Fix an archimedean place τ of F, denote by $B = B(\tau)$ the nearby quaternion algebra, and identify $B^\times_{\mathbb{A}_f} = \mathbb{B}^\times_f$ as usual. Similar to $\mathbb{B}$, we have an orthogonal decomposition $B = F \oplus B_0$, and a conjugation action Ad of $B^\times$ on B_0.

For any element $y \in B_0$ with $q(y) \neq 0$, denote by B_y (resp. $\mathbb{B}_{f,y}$) the centralizer of y in B (resp. $\mathbb{B}_f$). Then $B_y = F[y] = F \oplus Fy$ is a quadratic extension of F. In particular, B_y is a CM extension of F if and only if y lies in the subset $B_{0,+}$ of elements of B_0 with totally positive norms.

Assume $y \in B_{0,+}$ so that B_y is a CM extension. Let τ_y be the unique point in $\mathcal{H}$ fixed by the action of $B_y^\times$ through the embedding $B_y^\times \subset B_\tau(\mathbb{R})$. Then $\overline{\tau}_y$ is the unique fixed point of $B_y^\times$ in $\mathcal{H}^-$. Write $\tau_y^\pm = \{\tau_y, \overline{\tau}_y\} \subset \mathcal{H}^\pm$. The (zero-dimensional) Shimura variety associated to B_y is just

$$\mathrm{Sh}(B_y^\times, W) = \tau_y^\pm \times B_y^\times \backslash \mathbb{B}^\times_{f,y}/W,$$

where W is an open compact subgroup of $\mathbb{B}^\times_{f,y}$.

For any $h \in \mathbb{B}^\times_f$, denote by $C(y,h)_U$ the push-forward of the map $\mathrm{Sh}(B_y^\times, U_h) \to X_U$ given by

$$\tau_y^\pm \times B_y^\times \backslash \mathbb{B}^\times_{f,y}/U_h \longrightarrow B^\times \backslash \mathcal{H}^\pm \times \mathbb{B}^\times_f/U, \quad (\tau, b) \longmapsto (\tau, bh).$$

Here $U_h = \mathbb{B}^\times_{f,y} \cap hUh^{-1}$ and the map is always an embedding if U is sufficiently small. Then

$$C(y,h)_U = B^\times \backslash B^\times (\tau_y^\pm \times \mathbb{B}^\times_{f,y}hU/U).$$

It depends only on the coset of h in $\mathbb{B}^\times_{f,y} \backslash \mathbb{B}^\times_f/U$. By the reciprocity law, $C(y,h)_U$ is a divisor of X_U defined over F.

LEMMA 4.6. *Let* $y, y' \in B_{0,+}$, *and* $h, h' \in \mathbb{B}^\times_f$. *Then*

$$C(y,h)_U = C(y',h')_U \iff F^\times \cdot \mathrm{Ad}(U) \circ (h^{-1}yh) = F^\times \cdot \mathrm{Ad}(U) \circ (h'^{-1}y'h').$$

PROOF. We first prove "$\Longleftarrow$." Assume that the right-hand side is true. The subvariety $C(y,h)_U$ is represented by the set

$$B^\times (\tau_y^\pm \times \mathbb{B}^\times_{f,y}hU).$$

It is easy to see that it does not change if we multiply y by an element in $F^\times$ or multiply h by an element in U. Thus we can assume that $h^{-1}yh = h'^{-1}y'h'$.

Then $q(y) = q(y')$ and we can find $\gamma \in B^\times$ such that $y' = \gamma^{-1}y\gamma$. It follows that $\tau_{y'}^\pm = \gamma^{-1}\tau_y^\pm$, and $\mathbb{B}_{f,y'} = \gamma^{-1}\mathbb{B}_{f,y}\gamma$. Furthermore, $h^{-1}yh = h'^{-1}\gamma^{-1}y\gamma h'$ implies that $\gamma h'h^{-1}$ commutes with y, and thus lies in $\mathbb{B}_{f,y}^\times$. Hence,

$$
\begin{aligned}
B^\times(\tau_{y'}^\pm \times \mathbb{B}_{f,y'}^\times h'U) &= B^\times(\gamma^{-1}\tau_y^\pm \times \gamma^{-1}\mathbb{B}_{f,y}^\times \gamma h'U) \\
&= B^\times(\tau_y^\pm \times \mathbb{B}_{f,y}^\times \gamma h'U) = B^\times(\tau_y^\pm \times \mathbb{B}_{f,y}^\times hU).
\end{aligned}
$$

It proves that $C(y, h)_U = C(y', h')_U$.

Now we prove "$\Longrightarrow$." Assume that $C(y, h)_U = C(y', h')_U$. Then there is a $\gamma \in B^\times$ such that $\tau_{y'} = \gamma^{-1}\tau_y$ or $\tau_{y'} = \gamma^{-1}\overline{\tau}_y$. The stabilizer of $\tau_{y'}$ in $B^\times$ is $B_{y'}^\times$, and the stabilizer of $\gamma^{-1}\tau_y$ (or $\gamma^{-1}\overline{\tau}_y$) in $B^\times$ is exactly $B_{\gamma^{-1}y\gamma}^\times$. It follows that $B_{y'} = B_{\gamma^{-1}y\gamma}$ in both cases. Note that $B_{y'}$ is a CM extension over F, and both y' and $\gamma^{-1}y\gamma$ have trace zero. It follows that $y' = c\gamma^{-1}y\gamma$ for some $c \in F^\times$. By the direction we have proved,

$$
C(y, h)_U = C(y', h')_U = C(\gamma^{-1}y\gamma, h')_U = C(y, \gamma h')_U.
$$

It follows that $\mathbb{B}_{f,y}^\times hU = \mathbb{B}_{f,y}^\times \gamma h'U$. The result follows. □

With the lemma, we can define $C(y)_U$ for any y in

$$
\mathbb{B}_{f,0}^{\mathrm{ad}} := \{x \in \mathbb{B}_{f,0} : q(x) \in F_+^\times\}.
$$

In fact, write $y = h^{-1}y_0 h$ for any $y_0 \in B_{0,+}$ and $h \in \mathbb{B}_f^\times$. Then define

$$
C(y)_U := C(y_0, h)_U.
$$

It is independent of the choice of (y_0, h). We further know that $C(y)_U = C(y')_U$ if and only if $F^\times \cdot \mathrm{Ad}(U) \circ y = F^\times \cdot \mathrm{Ad}(U) \circ y'$.

4.5.2 Pull-back as cycles

LEMMA 4.7. *The following are true for $x \in \mathbb{B}_f^\times$:*

(1) If $x \in F^\times U$, then $\Delta^* Z(x)_U = -\omega_{X_U}$. Here ω_{X_U} denotes the canonical bundle of X_U.

(2) If $x \notin F^\times U$, then

$$
\Delta^* Z(x)_U = \sum_{y \in \mathrm{Ad}(U)\backslash \mathbb{B}_{f,0}^{\mathrm{ad}}/F^\times} [F[y]^\times U \cap UxU : U]\, C(y)_U.
$$

Here $F[y] = F + Fy$ is the totally imaginary quadratic field over F generated by y in $\mathbb{B}_f$, and $F[y]^\times$ denotes its multiplicative group.

PROOF. If $x \in F^\times U$, then $Z(x)_U = Z(1)_U = \Delta$. The result follows from the definition of the canonical bundle. Next, we assume that $x \notin F^\times U$. Then $\Delta \cdot Z(x)_U$ is a proper intersection.

Let τ be an archimedean place and $B = B(\tau)$ be the nearby quaternion algebra. Recall the uniformization

$$X_{U,\tau}(\mathbb{C}) \times X_{U,\tau}(\mathbb{C}) = (B^\times \backslash \mathcal{H}^\pm \times \mathbb{B}_f^\times / U) \times (B^\times \backslash \mathcal{H}^\pm \times \mathbb{B}_f^\times / U).$$

The divisor $Z(x)_U$ of $X_U \times X_U$ is represented by

$$\{(\tau, hU) \times (\tau, hUxU) : \tau \in \mathcal{H}^\pm, h \in \mathbb{B}_f^\times\}.$$

Assume that some point $(\tau, h)_U$ lies in the intersection $\Delta \cdot Z(x)_U$. Then there exist $z \in UxU$ and $\gamma \in B_+^\times$ such that

$$\gamma\tau = \tau, \quad \gamma hU = hzU.$$

We have $\gamma \notin F^\times$ since $x \notin F^\times U$. It follows that τ is a CM point, and $F[\gamma] = F + F\gamma = F + Fy$ is a CM extension of F. Here $y \in B_0$ is the trace-free part of γ, and $q(y)$ is totally positive. Note that y is determined by τ.

The condition on z is equivalent to $z \in h^{-1}\gamma hU$. Such a γ exists for z if and only if $z \in h^{-1}F[y]^\times hU = F[h^{-1}yh]^\times U$. It follows that the number of such $z \in UxU/U$, which is just the the multiplicity of $(\tau, h)_U$ in $\Delta \cdot Z(x)_U$, is exactly equal to

$$\left[F[h^{-1}yh]^\times U \cap UxU : U\right].$$

In particular, it depends only on $h^{-1}yh$. So this number is exactly the multiplicity of each point of $C(h^{-1}yh)_U$ in $\Delta \cdot Z(x)_U$. A geometric reason for these multiplicities to be equal is that all these points lie in a Galois orbit. $\square$

Now we can give a formula for the pull-back $\Delta^* Z(g, \phi)_U$. Recall from (4.2.1) and (4.2.2) that

$$Z(g, \phi)_U = Z_0(g, \phi)_U + Z_*(g, \phi)_U,$$

where

$$Z_0(g, \phi)_U = - \sum_{\alpha \in F_+^\times \backslash \mathbb{A}_f^\times / q(U)} \sum_{u \in \mu_U^2 \backslash F^\times} r(g)\phi(0, \alpha^{-1}u) \, L_{K,\alpha},$$

$$Z_*(g, \phi)_U = w_U \sum_{a \in F^\times} \sum_{x \in U \backslash \mathbb{B}_f^\times / U} r(g)\phi(x)_a Z(x)_U.$$

Here the intertwining part vanishes since we assume that X_U° is compact.

PROPOSITION 4.8 (Pull-back as cycles). *Assume that $\#\Sigma > 1$ and $\phi = \phi^0 \otimes \phi_0$ under the orthogonal decomposition $\mathbb{B} = \mathbb{A} \oplus \mathbb{B}_0$. Then*

$$\Delta^* Z(g, \phi)_U = \sum_{u \in \mu_U^2 \backslash F^\times} \theta(g, u, \phi^0) C(g, u, \phi_0)_U, \quad g \in \widetilde{\mathrm{SL}}_2(\mathbb{A}).$$

Here the generating series on X_U is defined by

$$
\begin{aligned}
C(g, u, \phi_0)_U \;=\; & -r(g)\phi_0(0, u)\, L_U \\
& + \sum_{y \in \mathrm{Ad}(U)\backslash \mathbb{B}_{f,0}^{\mathrm{ad}}} \frac{1}{[F[y]^\times \cap U : \mu_U]} r(g)\phi_0(y, u)\, C(y)_U.
\end{aligned}
$$

It is an automorphic form on $g \in \widetilde{\mathrm{SL}}_2(\mathbb{A})$ with coefficients in $\mathrm{Pic}(X_U)_\mathbb{C}$. Here $F[y] = F + Fy$ is the CM extension over F generated by y, and $r(g)\phi_0(y, u) = r(g)\phi_0(1, uq(y))$ is understood in the sense of (4.1.3).

PROOF. The computation below can be simplified in the case where U is small enough so that all the ramification indices $[F[y]^\times \cap U : \mu_U] = 1$. But we include all cases to see the matching of the ramifications.

Divide the summation in $Z_*(\phi)$ into $x \in F^\times U$ and $x \notin F^\times U$. We get

$$
\Delta^* Z(\phi)_U = P + Q + R
$$

with

$$
\begin{aligned}
P \;=\; & -\sum_{\alpha \in F_+^\times \backslash \mathbb{A}_f^\times / q(U)} \; \sum_{u \in \mu_U^2 \backslash F^\times} \phi(0, \alpha^{-1}u)\, \Delta^* L_{K,\alpha}, \\
Q \;=\; & w_U \sum_{a \in F^\times} \; \sum_{x \in U\backslash(F^\times U)/U} \phi(x)_a\, \Delta^* Z(x)_U, \\
R \;=\; & w_U \sum_{a \in F^\times} \; \sum_{x \in U\backslash(\mathbb{B}_f^\times - F^\times U)/U} \phi(x)_a\, \Delta^* Z(x)_U.
\end{aligned}
$$

We first consider P. The pull-back $\Delta^* L_{K,\alpha}$ is nontrivial if and only if $L_{K,\alpha}$ lies in the same component as Δ. Then $\alpha = 1$. It is easy to see that $\Delta^* L_{K,1}$ equals the Hodge bundle L_U of X_U. Then

$$
P = -\sum_{u \in \mu_U^2 \backslash F^\times} \phi(0, u)\, L_U.
$$

By Lemma 4.7,

$$
Q = -w_U \sum_{a \in F^\times} \; \sum_{x \in \mu_U \backslash F^\times} \phi(x)_a \omega_{X_U} = -\sum_{u \in \mu_U^2 \backslash F^\times} \; \sum_{x \in F^\times} \phi(x, u)\omega_{X_U}.
$$

It remains to treat R. Consider the a-th coefficient

$$
R_a := w_U \sum_{x \in U\backslash(\mathbb{B}_f^\times - F^\times U)/U} \phi(x)_a\, \Delta^* Z(x)_U.
$$

The lemma also implies

$$
R_a = w_U \sum_{y \in \mathrm{Ad}(U)\backslash \mathbb{B}_{f,0}^{\mathrm{ad}}/F^\times} \; \sum_{x \in U\backslash(\mathbb{B}_f^\times - F^\times U)/U} \left[(F[y]^\times)U \cap UxU : U \right] \phi(x)_a\, C(y)_U.
$$

Write

$$\left[(F[y]^\times)U \cap UxU : U\right] = \sum_{z \in (F[y]^\times U - F^\times U)/U} [zU \cap UxU : U].$$

In order to make $[zU \cap UxU : U]$ nonzero we need $x \in UzU$. It follows that

$$R_a = w_U \sum_{y \in \mathrm{Ad}(U)\backslash \mathbb{B}_{f,0}^{\mathrm{ad}}/F^\times} \sum_{z \in (F[y]^\times U - F^\times U)/U} \phi(z)_a \, C(y)_U.$$

The summation over z transforms as

$$\sum_{z \in F[y]^\times U/U} - \sum_{z \in F^\times U/U} = \sum_{z \in F[y]^\times/(F[y]^\times \cap U)} - \sum_{z \in F^\times/\mu_U}.$$

Note that $y^2 = -q(y) \in F$ is totally negative. The algebra $F[y] = F + Fy$ is a CM extension over F. By Dirichlet's unit theorem, the index

$$e_y := [F[y]^\times \cap U : \mu_U]$$

is finite. Then we can replace the sum over $F[y]^\times/(F[y]^\times \cap U)$ by that over $F[y]^\times/\mu_U$ divided by e_y. It follows that

$$R_a = w_U \sum_{y \in \mathrm{Ad}(U)\backslash \mathbb{B}_{f,0}^{\mathrm{ad}}/F^\times} \frac{1}{e_y} \sum_{z \in F[y]^\times/\mu_U} \phi(z)_a \, C(y)_U$$

$$- w_U \sum_{y \in \mathrm{Ad}(U)\backslash \mathbb{B}_{f,0}^{\mathrm{ad}}/F^\times} \sum_{z \in F^\times/\mu_U} \phi(z)_a \, C(y)_U.$$

Using $F[y]^\times = (F + F^\times y) \coprod F^\times$, it becomes

$$R_a = w_U \sum_{y \in \mathrm{Ad}(U)\backslash \mathbb{B}_{f,0}^{\mathrm{ad}}/F^\times} \frac{1}{e_y} \sum_{z \in (F + F^\times y)/\mu_U} \phi(z)_a \, C(y)_U$$

$$- w_U \sum_{y \in \mathrm{Ad}(U)\backslash \mathbb{B}_{f,0}^{\mathrm{ad}}/F^\times} \left(1 - \frac{1}{e_y}\right) \sum_{z \in F^\times/\mu_U} \phi(z)_a \, C(y)_U.$$

Go back to $R = \sum_{a \in F^\times} R_a$. It is easy to get

$$R = R^1 + R^2$$

with

$$R^1 = \sum_{u \in \mu_U^2 \backslash F^\times} \sum_{y \in \mathrm{Ad}(U)\backslash \mathbb{B}_{f,0}^{\mathrm{ad}}/F^\times} \frac{1}{e_y} \sum_{z \in F + F^\times y} \phi(z, u) \, C(y)_U,$$

$$R^2 = - \sum_{u \in \mu_U^2 \backslash F^\times} \sum_{z \in F^\times} \phi(z, u) \sum_{y \in \mathrm{Ad}(U)\backslash \mathbb{B}_{f,0}^{\mathrm{ad}}/F^\times} \left(1 - \frac{1}{e_y}\right) C(y)_U.$$

Using the splitting $\phi = \phi^0 \otimes \phi_0$, one has

$$
R^1 = \sum_{u \in \mu_U^2 \setminus F^\times} \sum_{y \in \mathrm{Ad}(U) \setminus \mathbb{B}_{f,0}^{\mathrm{ad}} / F^\times} \frac{1}{e_y} \sum_{z \in F} \phi^0(z, u) \sum_{x \in F^\times y} \phi_0(x, u) \, C(y)_U
$$

$$
= \sum_{u \in \mu_U^2 \setminus F^\times} \theta(g, u, \phi^0) \sum_{y \in \mathrm{Ad}(U) \setminus \mathbb{B}_{f,0}^{\mathrm{ad}}} \frac{1}{e_y} \phi_0(y, u) \, C(y)_U.
$$

It is the major part of the pull-back formula.

As for R^2, use the formula

$$
L_U = \omega_{X_U} + \sum_{y \in \mathrm{Ad}(U) \setminus \mathbb{B}_{f,0}^{\mathrm{ad}} / F^\times} \left(1 - \frac{1}{e_y} \right) C(y)_U.
$$

It is easy to get

$$
R^2 + Q + P = - \sum_{u \in \mu_U^2 \setminus F^\times} \sum_{x \in F} \phi(x, u) L_U = - \sum_{u \in \mu_U^2 \setminus F^\times} \theta(g, u, \phi^0) \, \phi_0(0, u) L_U.
$$

The pull-back formula follows. The formula between the Hodge bundle and the canonical bundle can be proved by the definition of the Hodge bundle.

The modularity of $C(g, u, \phi_0)_U$ can be derived from that of $Z(g, \phi)_U$ using the pull-back formula. For fixed ϕ_0 we can vary ϕ^0 to get information on a single $C(g, u, \phi_0)_U$. It is actually an example of the modularity in [YZZ], and the proof in [YZZ] uses this pull-back method. When U is small enough, the factor $[F[y]^\times \cap U : \mu_U]^{-1} = 1$ and the series is the same as that in [YZZ]. In general, the factor should be there and the modularity result for big level U can be obtained by push-forward from small levels. □

4.5.3 Degree of the pull-back

By Proposition 4.8, the proof of Proposition 4.5 (in the compact case) is reduced to the following result. Note that we can first obtain Proposition 4.5 for $g \in \mathrm{SL}_2(\mathbb{A})$, and extend it to $\mathrm{GL}_2(\mathbb{A})$ for the action of an element $d(a)$ with $a \in \mathbb{A}^\times$.

PROPOSITION 4.9. *Assume that* $\#\Sigma > 1$. *Let* $C(g, u, \phi_0)_U$ *be the generating function on* X_U *as in Proposition 4.8. Then*

$$
\deg C(g, u, \phi_0)_U = -\mathrm{vol}(X_U) \, E(0, g, u, \phi_0), \quad g \in \widetilde{\mathrm{SL}}_2(\mathbb{A}).
$$

Here the Eisenstein series

$$
E(s, g, u, \phi_0) = \sum_{\gamma \in P^1(F) \setminus \mathrm{SL}_2(F)} \delta(\gamma g)^s r(\gamma g) \phi_0(0, u), \quad g \in \widetilde{\mathrm{SL}}_2(\mathbb{A}).
$$

PROOF. It follows from the local Siegel–Weil formula in §2.5.3. The proof is very similar to Proposition 4.2, except that the computation is more complicated.

Recall that

$$
\begin{aligned}
C(g, u, \phi_0)_U &= -r(g)\phi_0(0, u) \, L_U \\
&+ \sum_{y \in \mathrm{Ad}(U) \backslash \mathbb{B}_{f,0}^{\mathrm{ad}}} \frac{1}{[F[y]^\times \cap U : \mu_U]} r(g)\phi_0(y, u) \, C(y)_U.
\end{aligned}
$$

Recall from Proposition 2.10 that

$$
E(0, g, u, \phi_0) = r(g)\phi_0(0, u) - \sum_{a \in F^\times} L(1, \eta_{-ua}) \int_{\mathbb{B}_{y_a}^\times \backslash \mathbb{B}^\times} r(g, (h, h))\phi_0(y_a, u) dh.
$$

Here $y_a \in \mathbb{B}$ is any element with $uq(y_a) = a$, and the integration is considered to be zero if such y_a does not exist.

By Lemma 3.1, it is immediate to have the identity on the constant terms:

$$
\deg C_0(g, u, \phi_0)_U = -\mathrm{vol}(X_U) \, E_0(0, g, u, \phi_0).
$$

It implies the result without much computation. In fact, we first verify that both $E(0, g, u, \phi_0)$ and $\mathrm{vol}(X_U)^{-1} \deg C(g, u, \phi_0)_U$ define elements in the one-dimensional space

$$
\mathrm{Hom}_{\mathrm{SO}(\mathbb{B}_0) \times \widetilde{\mathrm{SL}}_2(\mathbb{A})}(\mathcal{S}(\mathbb{B}_0), C^\infty(\mathrm{SL}_2(F) \backslash \widetilde{\mathrm{SL}}_2(\mathbb{A}))).
$$

Then they must be proportional, and their quotient is given by the identity between the constant terms.

The argument has used the modularity of $C_0(g, u, \phi_0)_U$, which is not valid if $\#\Sigma = 1$. In the following we propose a computation which can (and will) be used in the case $\#\Sigma = 1$.

Fix an $a \in F^\times$. In the following we verify

$$
\deg C_a(g, u, \phi_0)_U = -\mathrm{vol}(X_U) \, E_a(0, g, u, \phi_0).
$$

It is easy to have

$$
C_a(g, u, \phi_0)_U = \sum_{y \in \mathrm{Ad}(U) \backslash \mathbb{B}_{f,0}(a)} \frac{1}{[F[y]^\times \cap U : \mu_U]} r(g)\phi_0(y, u) \, C(y)_U.
$$

By the choice of y_a, we can rewrite it as

$$
\begin{aligned}
&C_a(g, u, \phi_0)_U \\
&= \sum_{h \in \mathbb{B}_{f, y_a}^\times \backslash \mathbb{B}_f^\times / U} \frac{1}{[B_{y_a}^\times \cap (hUh^{-1}) : \mu_U]} r(g, (h, h))\phi_0(y_a, u) \, C(y_a, h)_U.
\end{aligned}
$$

In the following we abbreviate y_a as y for simplicity.

Let $B = B(\tau)$ be the nearby quaternion algebra for some archimedean place τ. Identifying $B(\mathbb{A}^\tau) = \mathbb{B}^\tau$, the y is an element of $B_0(\mathbb{A}^\tau)$. We can assume that y actually lies in B_0 since the result depends only on the norm of y. The divisor $C(y, h)_U$ is represented by the set

$$B^\times \backslash (B^\times \tau_y^\pm \times \mathbb{B}_{f,y}^\times hU/U) = \tau_y^\pm \times B_y^\times \backslash \mathbb{B}_{f,y}^\times hU/U.$$

Its degree is given by twice of the order of

$$B_y^\times \backslash \mathbb{B}_{f,y}^\times hU/U = B_y^\times \backslash \mathbb{B}_{f,y}^\times / (\mathbb{B}_{f,y}^\times \cap hUh^{-1}).$$

It follows that

$$\begin{aligned}
&\deg C_a(g, u, \phi_0)_U \\
&= \; 2 \sum_{h \in \mathbb{B}_{f,y}^\times \backslash \mathbb{B}_f^\times /U} \frac{|B_y^\times \backslash \mathbb{B}_{f,y}^\times / (\mathbb{B}_{f,y}^\times \cap hUh^{-1})|}{[B_y^\times \cap (hUh^{-1}) : \mu_U]} r(g, (h, h)) \phi_0(y, u).
\end{aligned}$$

On the other hand,

$$\begin{aligned}
E_a(0, g, u, \phi_0) &= \; -L(1, \eta_{-ua}) \int_{\mathbb{B}_y^\times \backslash \mathbb{B}^\times} r(g, (h, h)) \phi_0(y, u) dh \\
&= \; -L(1, \eta_{-ua}) \, \mathrm{vol}(\mathbb{B}_{\infty,y}^\times \backslash \mathbb{B}_\infty^\times) \\
&\quad \cdot \sum_{h \in \mathbb{B}_{f,y}^\times \backslash \mathbb{B}_f^\times /U} \mathrm{vol}(\mathbb{B}_{f,y}^\times \backslash \mathbb{B}_{f,y}^\times hU) \; r(g, (h, h)) \phi_0(y, u).
\end{aligned}$$

Hence, it is reduced to verify

$$\begin{aligned}
&2 \frac{|B_y^\times \backslash \mathbb{B}_{f,y}^\times / (\mathbb{B}_{f,y}^\times \cap hUh^{-1})|}{[B_y^\times \cap (hUh^{-1}) : \mu_U]} \\
&= \; \mathrm{vol}(X_U) \, L(1, \eta_{-ua}) \, \mathrm{vol}(\mathbb{B}_{\infty,y}^\times \backslash \mathbb{B}_\infty^\times) \, \mathrm{vol}(\mathbb{B}_{f,y}^\times \backslash \mathbb{B}_{f,y}^\times hU).
\end{aligned}$$

Note $\mathrm{vol}(\mathbb{B}_{\infty,y}^\times \backslash \mathbb{B}_\infty^\times) = (2\pi^2)^d$ with $d = [F : \mathbb{Q}]$, and

$$\mathbb{B}_{f,y}^\times \backslash \mathbb{B}_{f,y}^\times hU = \left(\mathbb{B}_{f,y}^\times \backslash \mathbb{B}_{f,y}^\times hUh^{-1} \right) h = (\mathbb{B}_{f,y}^\times \cap hUh^{-1}) \backslash (hUh^{-1}) h.$$

The desired equality becomes

$$\mathrm{vol}(X_U) = \frac{2(2\pi^2)^{-d}}{L(1, \eta_{-ua})} \cdot \frac{|B_y^\times \backslash \mathbb{B}_{f,y}^\times / (\mathbb{B}_{f,y}^\times \cap hUh^{-1})|}{[B_y^\times \cap (hUh^{-1}) : \mu_U]} \cdot \frac{\mathrm{vol}(\mathbb{B}_{f,y}^\times \cap hUh^{-1})}{\mathrm{vol}(U)}.$$

For simplicity, denote

$$L = B_{y_0}, \;\; \mathbb{L}_f = \mathbb{A}_{L,f}, \;\; U_h = \mathbb{B}_{f,y}^\times \cap hUh^{-1}, \;\; U_Z = \mathbb{A}_f^\times \cap U_h = \mathbb{A}_f^\times \cap U.$$

Then η_{-ua} is exactly the quadratic character η_L associated to the CM extension L/F. What we need to show is just

$$\text{vol}(X_U) = \frac{2(2\pi^2)^{-d}}{L(1, \eta_L)} \cdot \frac{|L^\times \backslash \mathbb{L}_f^\times / U_h|}{[L^\times \cap U_h : \mu_U]} \cdot \frac{\text{vol}(U_h)}{\text{vol}(U)}.$$

We are going to need the following identities:

(1) $2L(1, \eta_L) = \text{vol}(\mathbb{L}^\times / L^\times \mathbb{A}^\times)$,

(2) $\text{vol}(\mathbb{L}^\times / L^\times \mathbb{A}^\times) = |\mathbb{A}_f^\times / F^\times U_Z|^{-1} \, \text{vol}(\mathbb{L}^\times / L^\times F_\infty^\times U_Z)$,

(3) $\text{vol}(\mathbb{L}^\times / L^\times F_\infty^\times U_Z) = 2^d \dfrac{|L^\times \backslash \mathbb{L}_f^\times / U_h|}{[L^\times \cap U_h : \mu_U]} \dfrac{\text{vol}(U_h)}{\text{vol}(U_Z)}$.

The first identity is just the result for the Tamagawa number of $SO(L, q) \cong L^\times / F^\times$. For the second identity, we have

$$\frac{\text{vol}(\mathbb{L}^\times / L^\times F_\infty^\times U_Z)}{\text{vol}(\mathbb{L}^\times / L^\times \mathbb{A}^\times)} = [L^\times \mathbb{A}^\times : L^\times F_\infty^\times U_Z] = |\mathbb{A}^\times / F^\times F_\infty^\times U_Z|.$$

For the third identity,

$$\frac{\text{vol}(\mathbb{L}^\times / L^\times F_\infty^\times U_Z)}{|L^\times \backslash \mathbb{L}_f^\times / U_h|} = \text{vol}\left(\frac{L^\times L_\infty^\times U_h}{L^\times F_\infty^\times U_Z} \right) = \text{vol}\left(\frac{L_\infty^\times U_h}{F_\infty^\times U_Z} \right) \frac{1}{[L^\times \cap U_h : \mu_U]}.$$

Use $\text{vol}(L_\infty^\times / F_\infty^\times) = 2^d$.

Applying (1), (2) and (3), the identity we need to prove is equivalent to

$$\text{vol}(X_U) = 4(4\pi^2)^{-d} \, |\mathbb{A}_f^\times / F^\times U_Z| \frac{\text{vol}(U_Z)}{\text{vol}(U)}. \tag{4.5.1}$$

This clean expression does not depend on L.

Recall from Proposition 4.2 that the Hodge bundle L_U has the same degree κ_U° on all geometrically connected components of X_U. It follows that

$$\text{vol}(X_U) = \kappa_U^\circ \cdot |\mathbb{A}_f^\times / F_+^\times q(U)|,$$

where $|\mathbb{A}_f^\times / F_+^\times q(U)|$ is just the number of geometrically connected components. Furthermore, by equation (4.3.2) in the proof of Proposition 4.2,

$$\kappa_U^\circ = \frac{2 w_U}{(4\pi^2)^d [\mu_U' : \mu_U^2] \text{vol}(U^1)}.$$

Thus (4.5.1) becomes

$$\frac{w_U}{[\mu_U' : \mu_U^2] \text{vol}(U^1)} \, |\mathbb{A}_f^\times / F_+^\times q(U)| = 2 |\mathbb{A}_f^\times / F^\times U_Z| \frac{\text{vol}(U_Z)}{\text{vol}(U)}.$$

Note that

$$\frac{|\mathbb{A}_f^\times / F_+^\times q(U)|}{|\mathbb{A}_f^\times / F^\times U_Z|} = [F^\times U_Z : F_+^\times q(U)] = [F^\times : F_+^\times] \frac{\mathrm{vol}(U_Z)}{\mathrm{vol}(q(U))} [\mu_U' : \mu_U].$$

The result follows from the following simple identities:

- $[F^\times : F_+^\times] = 2^d$,

- $[\mu_U : \mu_U^2] = 2^{d-1} w_U$,

- $\mathrm{vol}(U) = \mathrm{vol}(U^1) \cdot \mathrm{vol}(q(U))$.

$\square$

4.6   PULL-BACK FORMULA: NON-COMPACT CASE

In this section, we give a formula for $\Delta^* Z(g, \phi)_U$ and its degree in the case where X_U contains cusps. In other words, the open part X_U° is non-compact. Then we have $F = \mathbb{Q}$, $\Sigma = \{\infty\}$ and $\mathbb{B}_f = M_2(\mathbb{A}_f)$.

The extra computation for the pull-back comes from the cusps, and the extra computation for the degree comes from the non-holomorphic terms of the related Eisenstein series. The pull-back formula (4.4.3) can be modified in (4.4.6) with an extra term:

$$\deg \Delta^* Z(g, \phi)_U = -\mathrm{vol}(X_U) \, J(0, g, \phi)_U + \mathrm{vol}(X_U) \, B(g, \phi).$$

4.6.1   Pull-back as cycles

There is only one archimedean place, and the nearby quaternion algebra $B = M_2(\mathbb{Q})$. Recall that cusps on the upper half plane $\mathcal{H}$ form the set $\mathbb{P}^1(\mathbb{Q}) = \mathbb{Q} \cup \{\infty\}$ on which $B_+^\times = \mathrm{GL}_2(\mathbb{Q})_+$ acts transitively. Then the set of cusps on $X_U(\mathbb{C})$ is just the finite set

$$B_+^\times \backslash \mathbb{P}^1(\mathbb{Q}) \times \mathbb{B}_f^\times / U \cong \{\infty\} \times P(\mathbb{Q})_+ \backslash \mathbb{B}_f^\times / U.$$

Here $P(\mathbb{Q})_+$ is the set of upper triangular matrices with positive determinants, which is exactly the stabilizer of ∞ in $B_+^\times$. For any $h \in P(\mathbb{Q})_+ \backslash \mathbb{B}_f^\times / U$, we abbreviate the corresponding cusp $[\infty, h]_U$ as $\langle h \rangle_U$.

Resume the notations for CM points in the last section. In particular, $C(y)_U$ denotes a finite set of CM points for any y in $\mathbb{B}_{f,0}^{\mathrm{ad}}$. Note that Lemma 4.6 is still true in the current setting without any change. Lemma 4.7 becomes the following statement.

LEMMA 4.10. *The following are true for $x \in \mathbb{B}_f^\times$:*

(1) If $x \in F^\times U$, then $\Delta^* Z(x)_U = -\omega_{X_U}$.

(2) If $x \notin F^{\times} U$, then

$$\Delta^* Z(x)_U = \sum_{y \in \mathrm{Ad}(U) \backslash \mathbb{B}_{f,0}^{\mathrm{ad}} / F^{\times}} [F[y]^{\times} U \cap U x U : U] \, C(y)_U$$

$$+ \sum_{h \in P(\mathbb{Q})_+ \backslash \mathbb{B}_f^{\times} / U} \sum_{\substack{\gamma \in P(\mathbb{Q})_+ / (P(\mathbb{Q})_+ \cap h U h^{-1}) \\ h^{-1} \gamma h \in U x U}} \tau(\gamma) \, \langle h \rangle_U.$$

Here $\tau : P(\mathbb{R})_+ \to \mathbb{R}$ is the continuous function defined by

$$\tau \left(\begin{pmatrix} a & b \\ & d \end{pmatrix} \right) = \min(1, d/a).$$

PROOF. The result in (1) still follows from the definition of the canonical bundle as in the compact case. For (2), the contribution of non-cusps in $\Delta \cdot Z(x)_U$ is still given by Lemma 4.7. It remains to treat the contribution of cusps in $\Delta \cdot Z(x)_U$.

Let $\langle h \rangle_U$ be a cusp, represented by $h \in P(\mathbb{Q})_+ \backslash \mathbb{B}_f^{\times} / U$, in the intersection $\Delta \cdot Z(x)_U$. We need to compute the multiplicity at $\langle h \rangle_U$. Write

$$Z(x)_U \langle h \rangle_U = \sum_{y \in U x U / U} \langle h y \rangle_U.$$

We only need to consider cosets $y \in U x U / U$ with $\langle h y \rangle_U = \langle h \rangle_U$, which happens if and only if $h y U = \gamma h U$ for some $\gamma \in P(\mathbb{Q})_+$. It happens if and only if

$$y \in h^{-1} P(\mathbb{Q})_+ h U \cap U x U / U.$$

By the relation $y = h^{-1} \gamma h$, it is equivalent to

$$\gamma \in P(\mathbb{Q})_+ / (P(\mathbb{Q})_+ \cap h U h^{-1}) \quad \text{and} \quad h^{-1} \gamma h \in U x U.$$

Now we compute the multiplicity.

At a neighborhood of $\langle h \rangle_U$ in $X_U(\mathbb{C})$, we have

$$Z(x)_U [z, h]_U \simeq \sum_{y \in h^{-1} P(\mathbb{Q})_+ h U \cap U x U / U} [z, h y]_U.$$

We only sum over the set of y fixing $\langle h \rangle_U$, since other y does not contribute to the multiplicity at $\langle h \rangle_U$. Using $y = h^{-1} \gamma h$, we have $[z, h y]_U = [\gamma^{-1} z, h]_U$. It follows that

$$Z(x)_U [z, h]_U \simeq \sum_{\gamma} [\gamma^{-1} z, h]_U.$$

Consider the behavior of $z \to \infty$. In terms of the complex uniformization, a uniformizer at $\langle h \rangle_U$ is given by $q = e(z/r)$ where $r = [N(\mathbb{Z}) : N(\mathbb{Q}) \cap h U h^{-1}]$.

A local coordinate of $(\langle h \rangle_U, \langle h \rangle_U)$ in $X_U(\mathbb{C}) \times X_U(\mathbb{C})$ is just a pair $(q_1, q_2) = (e(z_1/r), e(z_2/r))$.

Write $\gamma = \begin{pmatrix} a & b \\ & d \end{pmatrix}$, and thus $\gamma^{-1} = \begin{pmatrix} a^{-1} & -ba^{-1}d^{-1} \\ & d^{-1} \end{pmatrix}$. Then

$$e(\gamma^{-1}z/r) = e\left(\frac{a^{-1}z - ba^{-1}d^{-1}}{d^{-1}r}\right) = q^{d/a}e\left(-\frac{b}{ar}\right).$$

It follows that a local equation of $Z(x)_U$ at $(\langle h \rangle_U, \langle h \rangle_U)$ is just

$$f(q_1, q_2) = \prod_\gamma (q_2 - e(-\frac{b}{ar})\, q_1^{d/a}).$$

It is a polynomial. The intersection multiplicity with the diagonal $q_1 = q_2$ is exactly

$$\sum_\gamma \min(1, d/a).$$

The result is proved. $\qquad \square$

By the lemma, Proposition 4.8 is modified as follows.

PROPOSITION 4.11. *Assume that* $F = \mathbb{Q}$ *and* $\Sigma = \{\infty\}$. *Assume that* $\phi = \phi^0 \otimes \phi_0$ *under the orthogonal decomposition* $\mathbb{B} = \mathbb{A} \oplus \mathbb{B}_0$. *Then for any* $g \in \widetilde{\mathrm{SL}}_2(\mathbb{A})$,

$$\Delta^* Z(g, \phi)_U \;=\; \sum_{u \in \mathbb{Q}^\times} \theta(g, u, \phi^0) C(g, u, \phi_0)_U + \sum_{u \in \mathbb{Q}^\times} W_0(g, u, \phi)\, L_U + D(g, \phi)_U.$$

Here the generating function

$$C(g, u, \phi_0)_U \;=\; -r(g)\phi_0(0, u)\, L_U$$
$$+ \sum_{y \in \mathrm{Ad}(U) \backslash \mathbb{B}_{f,0}^{\mathrm{ad}}} \frac{1}{[\mathbb{Q}[y]^\times \cap U : \mu_U]} r(g)\phi_0(y, u)\, C(y)_U$$

is as in Proposition 4.8, and

$$D(g, \phi)_U$$
$$= w_U \sum_{a \in \mathbb{Q}^\times} \sum_{h \in P(\mathbb{Q})_+ \backslash \mathbb{B}_f^\times / U} \sum_{\gamma \in P(\mathbb{Q})_+ / (P(\mathbb{Q})_+ \cap hUh^{-1})} r(g)\phi(h^{-1}\gamma h)_a\, \tau(\gamma)\, \langle h \rangle_U.$$

PROOF. Note that $\mu_U^2 = 1$ by $F = \mathbb{Q}$. We have a simplification

$$\sum_{u \in \mu_U^2 \backslash F^\times} = \sum_{u \in \mathbb{Q}^\times}.$$

Similar to the compact case, write

$$\Delta^* Z(\phi)_U = P + Q + R + S$$

with

$$P = -\sum_{\alpha \in \mathbb{Q}_+^\times \backslash \mathbb{A}_f^\times / q(U)} \sum_{u \in \mathbb{Q}^\times} \phi(0, \alpha^{-1} u)\, \Delta^* L_{K,\alpha},$$

$$Q = w_U \sum_{a \in \mathbb{Q}^\times} \sum_{x \in U \backslash (\mathbb{Q}^\times U)/U} \phi(x)_a\, \Delta^* Z(x)_U,$$

$$R = w_U \sum_{a \in \mathbb{Q}^\times} \sum_{x \in U \backslash (\mathbb{B}_f^\times - \mathbb{Q}^\times U)/U} \phi(x)_a\, \Delta^* Z(x)_U,$$

$$S = \sum_{\alpha \in \mathbb{Q}_+^\times \backslash \mathbb{A}_f^\times / q(U)} \sum_{u \in \mathbb{Q}^\times} W_0(u\alpha^{-1}, \phi)\, \Delta^* L_{K,\alpha}.$$

Here the extra part S comes from the constant part of $Z(g, \phi)_U$.

Similar to the compact case, it is easy to have

$$P = -\sum_{u \in \mathbb{Q}^\times} \phi(0, u)\, L_U,$$

$$Q = -\sum_{u \in \mathbb{Q}^\times} \sum_{x \in \mathbb{Q}^\times} \phi(x, u)\, \omega_{X_U},$$

$$S = \sum_{u \in \mathbb{Q}^\times} W_0(u, \phi)\, L_U.$$

The expression of R involves cusps. Recall that

$$\Delta^* Z(x)_U = \sum_{y \in \mathrm{Ad}(U) \backslash \mathbb{B}_{f,0}^{\mathrm{ad}} / \mathbb{Q}^\times} [\mathbb{Q}[y]^\times U \cap UxU : U]\, C(y)_U$$

$$+ \sum_{h \in P(\mathbb{Q})_+ \backslash \mathbb{B}_f^\times / U} \sum_{\substack{\gamma \in P(\mathbb{Q})_+ / (P(\mathbb{Q})_+ \cap hUh^{-1}) \\ h^{-1}\gamma h \in UxU}} \tau(\gamma)\, \langle h \rangle_U.$$

Accordingly, write

$$R = R_{\mathrm{cm}} + R_{\mathrm{cusp}}$$

where R_{cm} (resp. R_{cusp}) denotes the contribution of CM points (resp. cusps) in R. The expression for R_{cm} is still as before. Now we consider

$$R_{\mathrm{cusp}} = w_U \sum_{a \in \mathbb{Q}^\times} \sum_{x \in U \backslash (\mathbb{B}_f^\times - \mathbb{Q}^\times U)/U} \phi(x)_a$$

$$\sum_{h \in P(\mathbb{Q})_+ \backslash \mathbb{B}_f^\times / U} \sum_{\substack{\gamma \in P(\mathbb{Q})_+ / (P(\mathbb{Q})_+ \cap hUh^{-1}) \\ h^{-1}\gamma h \in UxU}} \tau(\gamma)\, \langle h \rangle_U.$$

By the substitution $x = h^{-1}\gamma h$,

$$R_{\mathrm{cusp}} = w_U \sum_{a \in \mathbb{Q}^\times} \sum_{h \in P(\mathbb{Q})_+ \backslash \mathbb{B}_f^\times / U} \sum_{\substack{\gamma \in P(\mathbb{Q})_+ / (P(\mathbb{Q})_+ \cap hUh^{-1}) \\ h^{-1}\gamma h \notin \mathbb{Q}^\times U}} \tau(\gamma)\, \phi(h^{-1}\gamma h)_a\, \langle h \rangle_U.$$

It is easy to have

$$R_{\mathrm{cusp}} = D(1, \phi)_U + R_{\mathrm{cusp},0}$$

with $D(1, \phi)_U$ given in the statement of the proposition and

$$R_{\mathrm{cusp},0} = -w_U \sum_{a \in \mathbb{Q}^\times} \sum_{h \in P(\mathbb{Q})_+ \backslash \mathbb{B}_f^\times / U} \sum_{\substack{\gamma \in P(\mathbb{Q})_+ / (P(\mathbb{Q})_+ \cap hUh^{-1}) \\ h^{-1}\gamma h \in \mathbb{Q}^\times U}} \tau(\gamma)\, \phi(h^{-1}\gamma h)_a\, \langle h \rangle_U.$$

The summation on γ is just $\gamma \in \mathbb{Q}^\times / (\mathbb{Q}^\times \cap hUh^{-1}) = \mathbb{Q}^\times / \mu_U$. Here $\mu_U \subset \{\pm 1\}$ with w_U. We further have $\tau(\gamma) = 1$ since γ is a scalar matrix. Thus

$$R_{\mathrm{cusp},0} = -\sum_{u \in \mathbb{Q}^\times} \sum_{z \in \mathbb{Q}^\times} \phi(z, u)_a \sum_{h \in P(\mathbb{Q})_+ \backslash \mathbb{B}_f^\times / U} \langle h \rangle_U.$$

The formula of the Hodge bundle in the non-compact case is

$$L_U = \omega_{X_U} + \sum_{y \in \mathrm{Ad}(U) \backslash \mathbb{B}_{f,0}^{\mathrm{ad}} / \mathbb{Q}^\times} \left(1 - \frac{1}{e_y} \right) C(y)_U + \sum_{h \in P(\mathbb{Q})_+ \backslash \mathbb{B}_f^\times / U} \langle h \rangle_U.$$

The extra part is the cusps, each of which has multiplicity one in the formula. It is easy to see that $R_{\mathrm{cusp},0}$ exactly serves as this extra part for the constant term in the original pull-back formula. More precisely, we have

$$P + Q + R_{\mathrm{cm}} + R_{\mathrm{cusp},0} = \sum_{u \in \mathbb{Q}^\times} \theta(1, u, \phi^0)\, C(1, u, \phi_0)_U.$$

$\square$

Remark. The generating function $C(g, u, \phi_0)_U$ is automorphic on the open part X_U°, but not automorphic on the compactification X_U any more. In particular, its degree is no longer automorphic.

4.6.2 Some coset identities

The degree of $C(g, u, \phi_0)_U$ is still given by Proposition 4.9. It remains to compute the degree of the cusp part $D(g, \phi)_U$. We first prove some expressions related to the set $P(\mathbb{Q})_+ \backslash \mathbb{B}_f^\times / U$ of cusps on X_U. The result even gives a way to rewrite the cusp part of Lemma 4.10.

LEMMA 4.12. *The following are true over $F = \mathbb{Q}$:*

(1) For any open compact subgroup C of $\mathbb{A}_f$,

$$\mathbb{A}_f = \mathbb{Q} + C, \quad \mathbb{A}_f / C = \mathbb{Q} / (\mathbb{Q} \cap C).$$

(2) *Let U be an open compact subgroup of $\mathrm{GL}_2(\mathbb{A}_f)$ contained in the maximal compact subgroup $U^0 = \mathrm{GL}_2(\widehat{\mathbb{Z}})$. Then*

$$P(\mathbb{Q})_+ \backslash \mathrm{GL}_2(\mathbb{A}_f)/U = N(\widehat{\mathbb{Z}})\backslash U^0/(U \cdot \{\pm 1\}).$$

(3) *Let $\Psi : \mathrm{GL}_2(\mathbb{A}_f) \to \mathbb{C}$ be a compactly supported function bi-invariant under U. Then*

$$\sum_{h \in P(\mathbb{Q})_+ \backslash \mathrm{GL}_2(\mathbb{A}_f)/U} \; \sum_{\gamma \in P(\mathbb{Q})_+/(P(\mathbb{Q})_+ \cap hUh^{-1})} \tau(\gamma) \; \Psi(h^{-1}\gamma h) \; \langle h \rangle_U$$

$$= \; \frac{1}{2} \sum_{h \in U^0/U} \langle h \rangle_U \sum_{x=(x_1,x_2) \in A(\mathbb{Q})_+}$$

$$\min(|x_1|_\infty, |x_2|_\infty) \int_{\mathbb{A}_f} \Psi\left(h^{-1}\begin{pmatrix} x_1 & b \\ & x_2 \end{pmatrix}h\right) db.$$

Here U^0 is as in (2), and τ is as in Lemma 4.10.

PROOF. The first identity in (1) is the standard strong approximation. The second identity in (1) follows from the first one.

Now we consider (2). We first claim that

$$P(\mathbb{Q})_+ \backslash \mathrm{GL}_2(\mathbb{A}_f)/U = (P(\mathbb{Q})_+ N(\mathbb{A}_f)) \backslash \mathrm{GL}_2(\mathbb{A}_f)/U.$$

It is equivalent to

$$P(\mathbb{Q})_+ N(\mathbb{A}_f)hU = P(\mathbb{Q})_+ hU, \quad \forall h \in \mathrm{GL}_2(\mathbb{A}_f).$$

Here both sides are viewed as subsets of $\mathrm{GL}_2(\mathbb{A}_f)$. It suffices to prove

$$N(\mathbb{A}_f)hUh^{-1} = N(\mathbb{Q})hUh^{-1}, \quad h \in \mathrm{GL}_2(\mathbb{A}_f).$$

In fact,

$$N(\mathbb{Q}) \cdot hUh^{-1} = N(\mathbb{Q}) \cdot (N(\mathbb{A}_f) \cap hUh^{-1}) \cdot hUh^{-1} = N(\mathbb{A}_f) \cdot hUh^{-1}.$$

Here the second equality follows from (1) by identifying $N \cong \mathbb{G}_a$. It proves the claimed result.

Go back to $P(\mathbb{Q})_+ \backslash \mathrm{GL}_2(\mathbb{A}_f)/U$. We will apply the Iwasawa decomposition $\mathrm{GL}_2(\mathbb{A}_f) = P(\mathbb{A}_f)U^0$. It is easy to see

$$P(\mathbb{Q})_+ N(\mathbb{A}_f)U^0 = N(\mathbb{A}_f)A(\mathbb{Q})_+ U^0 = N(\mathbb{A}_f)A(\mathbb{A}_f)U^0 = P(\mathbb{A}_f)U^0.$$

It follows that

$$
\begin{aligned}
& P(\mathbb{Q})_+ \backslash \mathrm{GL}_2(\mathbb{A}_f)/U \\
=\; & (P(\mathbb{Q})_+ N(\mathbb{A}_f)) \backslash P(\mathbb{A}_f)U^0/U \\
=\; & (P(\mathbb{Q})_+ N(\mathbb{A}_f)) \backslash P(\mathbb{Q})_+ N(\mathbb{A}_f)U^0/U \\
=\; & (P(\mathbb{Q})_+ N(\mathbb{A}_f) \cap U^0) \backslash U^0/U.
\end{aligned}
$$

It is easy to verify $P(\mathbb{Q})_+ N(\mathbb{A}_f) \cap U^0 = \{\pm 1\} \cdot N(\widehat{\mathbb{Z}})$. The result is proved.

Now we prove (3). We start with the left-hand side. First consider the summation on γ. For any matrix in $P(\mathbb{Q})_+ \cap hUh^{-1}$, the coefficients on its diagonal are units of $\mathbb{Z}$. They must be (simultaneously) 1 or -1. It follows that

$$P(\mathbb{Q})_+ \cap hUh^{-1} = \mu_U \cdot (N(\mathbb{Q}) \cap hUh^{-1}).$$

It follows that

$$\begin{aligned}
& P(\mathbb{Q})_+/(P(\mathbb{Q})_+ \cap hUh^{-1}) \\
& = \ (A(\mathbb{Q})_+/\mu_U) \times \big(N(\mathbb{Q})/(N(\mathbb{Q}) \cap hUh^{-1})\big).
\end{aligned}$$

Apply also the identity in (2). The left-hand side of (3) is equal to

$$\sum_{h \in N(\widehat{\mathbb{Z}})\backslash U^0/(U \cdot \{\pm 1\})} \ \sum_{x \in A(\mathbb{Q})_+/\mu_U} \ \sum_{n \in N(\mathbb{Q})/(N(\mathbb{Q}) \cap hUh^{-1})} \tau(x) \ \Psi(h^{-1}xnh) \ \langle h \rangle_U$$

$$= \ \frac{1}{2} \sum_{h \in N(\widehat{\mathbb{Z}})\backslash U^0/U} \ \sum_{x \in A(\mathbb{Q})_+} \ \sum_{n \in N(\mathbb{Q})/(N(\mathbb{Q}) \cap hUh^{-1})} \tau(x) \ \Psi(h^{-1}xnh) \ \langle h \rangle_U$$

$$= \ \frac{1}{2} \sum_{h \in N(\widehat{\mathbb{Z}})\backslash U^0/U} \ \sum_{x \in A(\mathbb{Q})_+} \ \sum_{n \in N(\mathbb{A}_f)/(N(\mathbb{A}_f) \cap hUh^{-1})} \tau(x) \ \Psi(h^{-1}xnh) \ \langle h \rangle_U.$$

Here the second identity uses the second result in (1).

On the other hand, the right-hand side of (3) is equal to

$$\frac{1}{2} \sum_{h \in U^0/U} \langle h \rangle_U \sum_{x=(x_1,x_2) \in A(\mathbb{Q})_+} \min(|x_1|_\infty, |x_2|_\infty)$$

$$\cdot |x_1|_{\mathbb{A}_f} \int_{\mathbb{A}_f} \Psi\left(h^{-1}\begin{pmatrix} x_1 & bx_1 \\ & x_2 \end{pmatrix} h\right) db$$

$$= \ \frac{1}{2} \sum_{h \in U^0/U} \langle h \rangle_U \sum_{x=(x_1,x_2) \in A(\mathbb{Q})_+} \min(1, x_2/x_1) \int_{N(\mathbb{A}_f)} \Psi(h^{-1}xnh) dn$$

$$= \ \frac{1}{2} \sum_{h \in U^0/U} \langle h \rangle_U \sum_{x \in A(\mathbb{Q})_+} \tau(x) \sum_{n \in N(\mathbb{A}_f)/N(\widehat{\mathbb{Z}})} \Psi(h^{-1}xnh).$$

It remains to verify that, for any $x \in A(\mathbb{Q})_+$,

$$\sum_{h \in N(\widehat{\mathbb{Z}})\backslash U^0/U} \ \sum_{n \in N(\mathbb{A}_f)/(N(\mathbb{A}_f) \cap hUh^{-1})} \Psi(h^{-1}xnh) \ \langle h \rangle_U$$

$$= \ \sum_{h \in U^0/U} \ \sum_{n \in N(\mathbb{A}_f)/N(\widehat{\mathbb{Z}})} \Psi(h^{-1}xnh) \ \langle h \rangle_U.$$

It is mentally easier to start with the right-hand side. In fact, the right-hand side is equal to

$$\sum_{h \in N(\widehat{\mathbb{Z}})\backslash U^0/U} \ \sum_{h' \in N(\widehat{\mathbb{Z}})hU/U} \ \sum_{n \in N(\mathbb{A}_f)/N(\widehat{\mathbb{Z}})} \Psi(h'^{-1}xnh') \ \langle h' \rangle_U.$$

Note that
$$N(\widehat{\mathbb{Z}})hU/U = (N(\widehat{\mathbb{Z}})/(N(\widehat{\mathbb{Z}}) \cap hUh^{-1})) \cdot h.$$

The triple sum becomes

$$\sum_{h \in N(\widehat{\mathbb{Z}})\backslash U^0/U} \sum_{n' \in N(\widehat{\mathbb{Z}})/(N(\widehat{\mathbb{Z}}) \cap hUh^{-1})} \sum_{n \in N(\mathbb{A}_f)/N(\widehat{\mathbb{Z}})} \Psi(h^{-1}n'^{-1}xnn'h) \langle n'h \rangle_U$$

$$= \sum_{h \in N(\widehat{\mathbb{Z}})\backslash U^0/U} [N(\widehat{\mathbb{Z}}) : (N(\widehat{\mathbb{Z}}) \cap hUh^{-1})] \sum_{n \in N(\mathbb{A}_f)/N(\widehat{\mathbb{Z}})} \Psi(h^{-1}xnh) \langle h \rangle_U$$

$$= \sum_{h \in N(\widehat{\mathbb{Z}})\backslash U^0/U} \sum_{n \in N(\mathbb{A}_f)/(N(\widehat{\mathbb{Z}}) \cap hUh^{-1})} \Psi(h^{-1}xnh) \langle h \rangle_U.$$

It finishes the proof since
$$N(\widehat{\mathbb{Z}}) \cap hUh^{-1} = N(\mathbb{A}_f) \cap hUh^{-1}$$

by the fact that $h \in U^0$. □

Remark. The result in Lemma 4.12 (3) "simplifies" the expression of Lemma 4.10 (2). In fact, set Ψ to be the characteristic function of UxU in Lemma 4.12 (3). We obtain

$$\Delta^* Z(x)_U = \sum_{y \in \mathrm{Ad}(U)\backslash \mathbb{B}_{f,0}^{\mathrm{ad}}/F^{\times}} [F[y]^{\times}U \cap UxU : U] \, C(y)_U$$

$$+ \frac{1}{2} \sum_{h \in U^0/U} \langle h \rangle_U \sum_{y=(y_1,y_2) \in A(\mathbb{Q})_+} \min(|y_1|_{\infty}, |y_2|_{\infty})$$

$$\cdot \int_{\mathbb{A}_f} 1_{UxU} \left(h^{-1} \begin{pmatrix} y_1 & b \\ & y_2 \end{pmatrix} h \right) db.$$

For example, let T_n be the standard Hecke correspondence on X_U, where n is a positive integer such that U is maximal at any prime factor p of n. Assume that n is not a perfect square so that $\Delta \cdot T_n$ is a proper intersection. Then the multiplicity of $\langle 1 \rangle_U$ in $\Delta^* T_n$ is exactly

$$\sum_{d|n,\ d>0} \min(d, n/d).$$

4.6.3 Degree of the pull-back

Denote by $V_{\mathrm{hyp}} = (\mathbb{Q}^2, q)$ with $q(x_1, x_2) = x_1 x_2$ the hyperbolic quadratic space over $\mathbb{Q}$. It is naturally embedded as a subspace of the matrix algebra (M_2, q) with $q = \det$ as diagonal matrices. There is a natural equivariant map from $\mathcal{S}(M_2(\mathbb{A}_f) \times \mathbb{A}_f^{\times})$ to $\mathcal{S}(V_{\mathrm{hyp}}(\mathbb{A}_f) \times \mathbb{A}_f^{\times})$, which already appears in Proposition 2.10.

The degree of the pull-back is best described using this map. Here we review it in more detail.

Equivariant maps

Let k be a non-archimedean local field. Define a map

$$\ell : \ \mathcal{S}(M_2(k) \times k^\times) \ \longrightarrow \ \mathcal{S}(V_{\text{hyp}}(k) \times k^\times)$$

by sending any $\phi \in \mathcal{S}(M_2(k) \times k^\times)$ to

$$(\ell\phi)((x_1, x_2), u) = |u|^{\frac{1}{2}} \int_{\mathrm{GL}_2(O_k)} \int_k \phi\left(h^{-1} \begin{pmatrix} x_1 & b \\ & x_2 \end{pmatrix} h, u\right) db \, dh.$$

In the case $x_1 \neq x_2$, a "more intrinsic" expression is

$$(\ell\phi)((x_1, x_2), u) = |u d_k^3|^{\frac{1}{2}} |x_1 - x_2| \int_{A(k)\backslash\mathrm{GL}_2(k)} \phi\left(h^{-1} \begin{pmatrix} x_1 & \\ & x_2 \end{pmatrix} h, u\right) dh.$$

It occurs in the proof of Proposition 2.10. Here $|d_k|^{\frac{1}{2}}$ appears as $\mathrm{vol}(O_k)$. In fact, by the Iwasawa decomposition, the second expression is equal to

$$|u|^{\frac{1}{2}} |x_1 - x_2| \int_{\mathrm{GL}_2(O_k)} \int_k \phi\left(h^{-1} \begin{pmatrix} x_1 & (x_1 - x_2)b \\ & x_2 \end{pmatrix} h, u\right) db \, dh$$

$$= \ |u|^{\frac{1}{2}} \int_{\mathrm{GL}_2(O_k)} \int_k \phi\left(h^{-1} \begin{pmatrix} x_1 & b \\ & x_2 \end{pmatrix} h, u\right) db \, dh.$$

Using the same formulae, we also have a map

$$\ell_0 : \mathcal{S}(M_2(k)_0 \times k^\times) \ \longrightarrow \ \mathcal{S}(V_{\text{hyp}}(k)_0 \times k^\times).$$

Here $M_2(k)_0$ is the subspace of trace-free elements, and $V_{\text{hyp}}(k)_0$ is the subspace of elements (x_1, x_2) of $V_{\text{hyp}}(k)$ with trace $x_1 + x_2 = 0$. We have $V_{\text{hyp}}(k)_0 \cong (k, q^-)$ with $q^-(z) = -z^2$. If $\phi = \phi^0 \otimes \phi_0$ is a decomposition respecting $M_2(k) = k \oplus M_2(k)_0$, then

$$\ell\phi = \phi^0 \otimes \ell_0\phi_0.$$

PROPOSITION 4.13. *(1) The map ℓ (resp. ℓ_0) is equivariant under the action of $\mathrm{GL}_2(k)$ (resp. $\widetilde{\mathrm{SL}}_2(k)$) as the symplectic similitude group via the Weil representation.*

(2) The images of the maps are

$$\mathrm{Im}(\ell_0) \ = \ \{\phi' \in \mathcal{S}(V_{\text{hyp}}(k)_0 \times k^\times) : \ \phi'((z, -z), u) = \phi'((-z, z), u)\},$$

$$\mathrm{Im}(\ell) \ = \ \{\phi' \in \mathcal{S}(V_{\text{hyp}}(k) \times k^\times) : \ \phi'((x_1, x_2), u) = \phi'((x_2, x_1), u)\}.$$

PROOF. We first consider (1). The following result is standard. Take ℓ for example. In the definition of ℓ, the action of h commutes with the action of $\mathrm{GL}_2(k)$ (as the symplectic group). Thus we only need to check

$$\ell'(r(g)\phi) = r(g)(\ell'\phi), \quad g \in \mathrm{GL}_2(k)$$

for

$$(\ell'\phi)((x_1, x_2), u) = |u|^{\frac{1}{2}} \int_k \phi \left(\begin{pmatrix} x_1 & b \\ & x_2 \end{pmatrix}, u \right) db.$$

By definition of the Weil representation, it is easy to check the results for $g = m(a), d(a)$ and $n(b)$. It remains to consider the action of the Weyl element w, which is essentially the Fourier transform. Write $(M_2, q) = V_{\text{hyp}} \oplus V'_{\text{hyp}}$ where $V'_{\text{hyp}} = (k^2, -q)$ is realized as the space of matrices of the form $\begin{pmatrix} & y_1 \\ y_2 & \end{pmatrix}$. It suffices to check the result for the case $\phi = \phi^{(1)} \otimes \phi^{(2)}$ with respect to the decomposition $(M_2, q) = V_{\text{hyp}} \oplus V'_{\text{hyp}}$. The action of w respects this decomposition. Thus it remains to check

$$\int_k \phi^{(2)} \left(\begin{pmatrix} & b \\ 0 & \end{pmatrix}, u \right) db = \int_k r(w)\phi^{(2)} \left(\begin{pmatrix} & b \\ 0 & \end{pmatrix}, u \right) db.$$

The right-hand side is equal to

$$\int_k \int_{V'_{\text{hyp}}(k)} \phi^{(2)} \left(\begin{pmatrix} & y_1 \\ y_2 & \end{pmatrix}, u \right) \psi(-u(by_2)) dy_1 dy_2 db$$

$$= \int_k \phi^{(2)} \left(\begin{pmatrix} & y_1 \\ 0 & \end{pmatrix}, u \right) dy_1.$$

Here we used the inversion formula to collapse a double integration.

Now we consider (2). It suffices to prove that for ℓ_0. Note that

$$\mathcal{S}(M_2(k)_0 \times k^\times) = \mathcal{S}(M - 2(k)_0) \otimes \mathcal{S}(k^\times),$$
$$\mathcal{S}(V_{\text{hyp}}(k)_0 \times k^\times) = \mathcal{S}(V_{\text{hyp}}(k)_0) \otimes \mathcal{S}(k^\times).$$

It is reduced to prove that

$$\ell_0(\mathcal{S}(M_2(k)_0)) = \{\phi' \in \mathcal{S}(V_{\text{hyp}}(k)_0) : \phi'(z, -z) = \phi'(-z, z)\}.$$

Here by abuse of notation, $\ell_0 : \mathcal{S}(M_2(k)_0) \to \mathcal{S}(V_{\text{hyp}}(k)_0)$ denotes the map given by

$$(\ell\phi_0)(z, -z) = \int_{\text{GL}_2(O_k)} \int_k \phi \left(h^{-1} \begin{pmatrix} z & b \\ & -z \end{pmatrix} h \right) db dh.$$

Use the alternative definition

$$(\ell\phi_0)(z, -z) = |d_k^3|^{\frac{1}{2}} |2z| \int_{A(k)\backslash\text{GL}_2(k)} \phi \left(h^{-1} \begin{pmatrix} z & \\ & -z \end{pmatrix} h \right) dh.$$

By a change of variable $h \to \begin{pmatrix} & 1 \\ 1 & \end{pmatrix} h$, we obtain

$$(\ell\phi_0)(z, -z) = (\ell\phi_0)(-z, z).$$

It follows that

$$\ell_0(\mathcal{S}(M_2(k)_0)) \subset \{\phi' \in \mathcal{S}(V_{\mathrm{hyp}}(k)_0) : \phi'(z, -z) = \phi'(-z, z)\}.$$

The other direction is implied by the fact that the right-hand side is an irreducible representation of $\widetilde{\mathrm{SL}}_2(k)$ under the Weil representation. In fact, it is the Howe lifting of the trivial representation of $O(V_{\mathrm{hyp}}(k)_0) = \{\pm 1\}$ to $\widetilde{\mathrm{SL}}_2(k)$. One can also check the irreducibility explicitly. $\qquad \square$

Go back to the global case. Define

$$\ell = \otimes_p \ell_p : \mathcal{S}(M_2(\mathbb{A}_f) \times \mathbb{A}_f^{\times}) \longrightarrow \mathcal{S}(V_{\mathrm{hyp}}(\mathbb{A}_f) \times \mathbb{A}_f^{\times}).$$

It is a well-defined equivariant map since

$$\ell_p(1_{M_2(\mathbb{Z}_p) \times \mathbb{Z}_p^{\times}}) = \frac{1}{\zeta_p(2)} 1_{V_{\mathrm{hyp}}(\mathbb{Z}_p) \times \mathbb{Z}_p^{\times}}$$

and the product converges absolutely.

Degree of the cusps

PROPOSITION 4.14. *Let* $U^0 = \mathrm{GL}_2(\widehat{\mathbb{Z}})$ *be the standard maximal compact subgroup of* $\mathbb{B}_f^{\times} = \mathrm{GL}_2(\mathbb{A}_f)$ *containing* U. *Then for any* $g \in \mathrm{GL}_2(\mathbb{A})$,

$$\deg D(g, \phi)_U$$
$$= \pi^2 \operatorname{vol}(X_U) \sum_{u \in \mathbb{Q}^{\times}} \sum_{x = (x_1, x_2) \in A(\mathbb{Q})_+} r(g) \ell \phi(x, u) |u|_{\infty}^{\frac{1}{2}} \min(|x_1|_{\infty}, |x_2|_{\infty}).$$

Here

$$r(g) \ell \phi(x, u) := r(g_f) \ell \phi_f(x, u) \cdot r(g_{\infty}) \phi_{\infty}(1, uq(x)).$$

PROOF. Recall from Proposition 4.11 that

$$D(g, \phi)_U$$
$$= w_U \sum_{a \in \mathbb{Q}^{\times}} \sum_{h \in P(\mathbb{Q})_+ \backslash \mathbb{B}_f^{\times}/U} \sum_{\gamma \in P(\mathbb{Q})_+/(P(\mathbb{Q})_+ \cap hUh^{-1})} r(g) \phi(h^{-1} \gamma h)_a \, \tau(\gamma) \, \langle h \rangle_U.$$

It suffices to compute the degree of the a-th coefficient

$$D_a(g, \phi)_U$$
$$= w_U \sum_{h \in P(\mathbb{Q})_+ \backslash \mathbb{B}_f^{\times}/U} \sum_{\gamma \in P(\mathbb{Q})_+/(P(\mathbb{Q})_+ \cap hUh^{-1})} r(g) \phi(h^{-1} \gamma h)_a \, \tau(\gamma) \, \langle h \rangle_U.$$

By Lemma 4.12 (3),

$$D_a(g, \phi)_U = \frac{w_U}{2} \sum_{h \in U^0/U} \langle h \rangle_U \sum_{x = (x_1, x_2) \in A(\mathbb{Q})_+} \min(|x_1|_{\infty}, |x_2|_{\infty})$$
$$\cdot \int_{\mathbb{A}_f} r(g) \phi \left(h^{-1} \begin{pmatrix} x_1 & b \\ & x_2 \end{pmatrix} h \right)_a \, db.$$

Taking degree, $\langle h \rangle_U$ becomes 1, and the summation over h becomes an integration on U^0. Therefore,

$$
\deg D_a(g, \phi)_U
$$
$$
= \frac{w_U}{2 \operatorname{vol}(U)} \sum_{x=(x_1,x_2) \in A(\mathbb{Q})_+} |u(x)|_\infty^{\frac{1}{2}} \min(|x_1|_\infty, |x_2|_\infty) \, r(g) \ell \phi(x)_a.
$$

Here $u(x) = a/(x_1 x_2)$.

Now we want to convert the coefficient in terms of $\operatorname{vol}(X_U)$. We claim that (for $F = \mathbb{Q}$)

$$
\operatorname{vol}(X_U) = \frac{w_U}{2\pi^2 \operatorname{vol}(U)}. \tag{4.6.1}
$$

Once it is true, then

$$
\deg D_a(g, \phi)_U
$$
$$
= \pi^2 \operatorname{vol}(X_U) \sum_{x=(x_1,x_2) \in A(\mathbb{Q})_+} |u(x)|_\infty^{\frac{1}{2}} \min(|x_1|_\infty, |x_2|_\infty) \, r(g) \ell \phi(x)_a.
$$

It is an expression as in the proposition.

It remains to verify (4.6.1). One can first verify it for maximal U, and derive the general case by comparing different U. Alternatively, it can be derived from (4.5.1) in the proof of Proposition 4.9 which asserts that

$$
\operatorname{vol}(X_U) = \pi^{-2} \, |\mathbb{Q}^\times \backslash \mathbb{A}_f^\times / U_Z| \frac{\operatorname{vol}(U_Z)}{\operatorname{vol}(U)}.
$$

It gives (4.6.1) since

$$
|\mathbb{Q}^\times \backslash \mathbb{A}_f^\times / U_Z| = \frac{\operatorname{vol}(\mathbb{Q}^\times \backslash \mathbb{A}_f^\times)}{\operatorname{vol}((U_Z \cap \mathbb{Q}^\times) \backslash U_Z)} = \frac{1/2}{\operatorname{vol}(U_Z)/w_U} = \frac{w_U}{2 \operatorname{vol}(U_Z)}.
$$

It finishes the proof. $\square$

Degree of the pull-back

Recall that in the compact case, Proposition 4.5 gives

$$
\deg \Delta^* Z(g, \phi)_U = -\operatorname{vol}(X_U) \, J(0, g, \phi)_U, \quad g \in \operatorname{GL}_2(\mathbb{A}).
$$

Here the mixed Eisenstein–theta series

$$
J(s, g, \phi)_U = \sum_{\gamma \in P(F) \backslash \operatorname{GL}_2(F)} \delta(\gamma g)^s \sum_{u \in \mu_U^2 \backslash F^\times} \sum_{x \in F} r(\gamma g) \phi(x, u).
$$

In the case $\phi = \phi^0 \otimes \phi_0$ under the orthogonal decomposition $\mathbb{B} = A \oplus \mathbb{B}_0$, one has

$$
J(s, g, u, \phi)_U = \sum_{u \in \mu_U^2 \backslash F^\times} \theta(g, u, \phi^0) E(s, g, u, \phi_0).
$$

In the non-compact case here, the equality is not true, and the difference of the two sides will be given by some summation on the hyperbolic space V_{hyp}.

THEOREM 4.15. *Assume that $F = \mathbb{Q}$ and $\Sigma = \{\infty\}$. For any $g \in \mathrm{GL}_2(\mathbb{A})$,*

$$\deg \Delta^* Z(g, \phi)_U \quad = \quad -\mathrm{vol}(X_U) \; J(0, g, \phi)_U + \mathrm{vol}(X_U) \; B(g, \phi),$$

where

$$
B(g, \phi): \quad = \quad \sum_{u \in \mathbb{Q}^\times} \sum_{x \in V_{\mathrm{hyp}}} r(g_f) \ell \phi_f(x, u) \; \boldsymbol{\beta}_{\phi_\infty}(g_\infty, x, u)
$$
$$
- \sum_{u \in \mathbb{Q}^\times} W'_{0,\infty}(0, g_\infty, u, \phi_\infty) \; |u|^{\frac{1}{2}}_{\mathbb{A}_f} \int_{\mathbb{A}_f} r(g_f) \ell \phi_f((0, x_2), u) dx_2.
$$

Here in the case $\phi_\infty = \phi_\infty^0 \otimes \phi_{0,\infty}$, the archimedean part

$$
\boldsymbol{\beta}_{\phi_\infty}(g, x, u) \quad = \quad -\frac{1}{2\sqrt{2}\gamma_u} W'_{-u(x_1 - x_2)^2/4}(0, g, u, \phi_{0,\infty}) \; r(g)\phi_\infty^0(1, u(x_1 + x_2)^2/4)
$$
$$
+ \pi^2 |u|^{\frac{1}{2}} \; \min(|x_1|, |x_2|) \; 1_{\mathbb{R}_+}(x_1 x_2) \; r(g)\phi_\infty(1, u x_1 x_2),
$$
$$
\forall \; x = (x_1, x_2) \in \mathbb{R}^2, \; u \in \mathbb{R}^\times, \; g \in \mathrm{GL}_2(\mathbb{R}).
$$

Here γ_u denotes the Weil index $\gamma(\mathbb{B}_{0,\infty}, uq) = e^{\frac{3\pi i}{4} \mathrm{sgn}(u)}$.

PROOF. It is easy to reduce the problem to the case $g \in \mathrm{SL}_2(\mathbb{A})$. Or equivalently, assume that $g \in \widetilde{\mathrm{SL}}_2(\mathbb{A})$ since the quadratic spaces in both sides are even-dimensional. Of course, in the proof we will meet odd-dimensional space since $\widetilde{\mathrm{SL}}_2(\mathbb{A})$ really makes a difference. By linearity, it suffices to prove the result for the split case $\phi = \phi^0 \otimes \phi_0$.

Recall that Proposition 4.11 gives

$$\Delta^* Z(g, \phi)_U$$
$$= \quad \sum_{u \in \mathbb{Q}^\times} \theta(g, u, \phi^0) C(g, u, \phi_0)_U + \sum_{u \in \mathbb{Q}^\times} W_0(0, g, u, \phi) \; L_U + D(g, \phi)_U.$$

The three terms on the right-hand side are the contributions respectively from the CM points, the Hodge bundle and the cusps.

The generating function

$$C(g, u, \phi_0)_U = -r(g)\phi_0(0, u) \; L_U$$
$$+ \sum_{y \in \mathrm{Ad}(U) \backslash \mathbb{B}_{f,0}^{\mathrm{ad}}} \frac{1}{[F[y]^\times \cap U : \mu_U]} r(g)\phi_0(y, u) \; C(y)_U$$

has the same expression as in Proposition 4.8. In particular, the proof of Proposition 4.9 applies to $\deg C(g, u, \phi_0)_U$ and gives

$$\deg C(g, u, \phi_0)_U = -\mathrm{vol}(X_U) \; r(g)\phi_0(0, u)$$
$$+ \mathrm{vol}(X_U) \sum_{a \in \mathbb{Q}^\times} L(1, \eta_{-ua}) \int_{\mathbb{B}_{x_a}^\times \backslash \mathbb{B}^\times} r(g, (h, h))\phi_0(x_a, u) dh.$$

Then Proposition 2.10 exactly gives

$$\deg C(g, u, \phi_0)_U = -\text{vol}(X_U)\, E^{\text{hol}}(0, g, u, \phi_0).$$

Here $E^{\text{hol}}(0, g, u, \phi_0)$ is the holomorphic part of $E(0, g, u, \phi_0)$.
  It follows that

$$\deg \Delta^* Z(g, \phi)_U + \text{vol}(X_U)\, J(0, g, \phi)_U$$
$$= \deg D(g, \phi) - \text{vol}(X_U) \sum_{u \in \mathbb{Q}^\times} W_0(0, g, u, \phi)$$
$$+ \text{vol}(X_U) \sum_{u \in \mathbb{Q}^\times} \theta(g, u, \phi^0) E^{\text{nhol}}(0, g, u, \phi_0).$$

We are going to rewrite these three terms on the right-hand side.
  By Proposition 4.14,

$$\deg D(g, \phi)_U$$
$$= \pi^2\, \text{vol}(X_U) \sum_{u \in \mathbb{Q}^\times} \sum_{x=(x_1,x_2) \in A(\mathbb{Q})_+} |u|_\infty^{\frac{1}{2}} \min(|x_1|_\infty, |x_2|_\infty)\, r(g)\ell\phi(x, u).$$

On the other hand, by Proposition 2.10,

$$E^{\text{nhol}}(0, g, u, \phi_0) = -\sum_{z \in \mathbb{Q}} \frac{1}{2\sqrt{2}\gamma_u} W'_{-uz^2,\infty}(0, g_\infty, u, \phi_{0,\infty})\, r(g_f)\ell_0\phi_{0,f}(z, u).$$

It follows that

$$\sum_{u \in \mathbb{Q}^\times} \theta(g, u, \phi^0) E^{\text{nhol}}(0, g, u, \phi)$$
$$= -\sum_{u \in \mathbb{Q}^\times} \sum_{x \in \mathbb{Q}} \sum_{z \in \mathbb{Q}} \frac{1}{2\sqrt{2}\gamma_u} W'_{-uz^2,\infty}(0, g_\infty, u, \phi_{0,\infty})$$
$$r(g_\infty)\phi_\infty^0(x, u)\, r(g_f)\ell\phi_f(x + z, x - z, u)$$
$$= -\sum_{u \in \mathbb{Q}^\times} \sum_{x_1,x_2 \in \mathbb{Q}} \frac{1}{2\sqrt{2}\gamma_u} W'_{-u(x_1-x_2)^2/4,\infty}(0, g_\infty, u, \phi_{0,\infty})$$
$$r(g_\infty)\phi_\infty^0(\frac{x_1 + x_2}{2}, u)\, r(g_f)\ell\phi_f(x_1, x_2, u).$$

It remains to treat $W_0(0, g, u, \phi)$. By Proposition 2.9,

$$W_0(0, g, u, \phi) = W'_{0,v}(0, g, u, \phi_v) \prod_{v \nmid \infty} W^\circ_{0,v}(0, g, u, \phi_v)$$

where for any $v \neq \infty$,

$$W^\circ_{0,v}(0, g, u, \phi_v) = |u|_v \int_{\text{SL}_2(\mathbb{Z}_v)} \int_{\mathbb{Q}_v} \int_{\mathbb{Q}_v} r(g)\phi_v \left(\begin{pmatrix} 0 & x \\ 0 & y \end{pmatrix} h_1, u \right) dx\, dy\, dh_1.$$

We claim that

$$W_{0,v}^{\circ}(0, g, u, \phi_v) = |u|_v^{\frac{1}{2}} \int_{\mathbb{Q}_v} r(g) \ell \phi_v((0, y), u) dy.$$

For that it suffices to verify

$$\int_{\mathrm{SL}_2(\mathbb{Z}_v)} \int_{\mathbb{Q}_v} \int_{\mathbb{Q}_v} r(g) \phi_v \left(\begin{pmatrix} 0 & x \\ 0 & y \end{pmatrix} h_1, u \right) dx dy dh_1$$

$$= \int_{\mathrm{GL}_2(\mathbb{Z}_v)} \int_{\mathbb{Q}_v} \int_{\mathbb{Q}_v} r(g) \phi_v \left(h^{-1} \begin{pmatrix} 0 & x \\ 0 & y \end{pmatrix} h, u \right) dx dy dh.$$

Start with the right-hand side. Write $h^{-1} = \begin{pmatrix} a & b \\ c & d \end{pmatrix}$, then we have

$$h^{-1} \begin{pmatrix} 0 & x \\ 0 & y \end{pmatrix} = \begin{pmatrix} 0 & ax + by \\ 0 & cx + dy \end{pmatrix}.$$

A change of variable $(x', y') = (ax + by, cx + dy)$ transfers the right-hand side to

$$\int_{\mathrm{GL}_2(\mathbb{Z}_v)} \int_{\mathbb{Q}_v} \int_{\mathbb{Q}_v} r(g) \phi_v \left(\begin{pmatrix} 0 & x \\ 0 & y \end{pmatrix} h, u \right) dx dy dh.$$

It only remains to change the domain of the integration from $\mathrm{GL}_2(\mathbb{Z}_v)$ to $\mathrm{SL}_2(\mathbb{Z}_v)$. Write $h \in \mathrm{GL}_2(\mathbb{Z}_v)$ as $d(a)h'$ with $a \in \mathbb{Z}_v^\times$ and $h' \in \mathrm{SL}_2(\mathbb{Z}_v)$. The last triple integral becomes

$$\int_{\mathrm{SL}_2(\mathbb{Z}_v)} \int_{\mathbb{Z}_v^\times} \int_{\mathbb{Q}_v} \int_{\mathbb{Q}_v} r(g) \phi_v \left(\begin{pmatrix} 0 & ax \\ 0 & ay \end{pmatrix} h', u \right) dx dy da dh$$

$$= \int_{\mathrm{SL}_2(\mathbb{Z}_v)} \int_{\mathbb{Q}_v} \int_{\mathbb{Q}_v} r(g) \phi_v \left(\begin{pmatrix} 0 & x \\ 0 & y \end{pmatrix} h', u \right) dx dy dh.$$

It finishes the proof. □

What is needed for Proposition 4.4

Theorem 4.15 asserts

$$\deg \Delta^* Z(g, \phi)_U = -\mathrm{vol}(X_U) J(0, g, \phi)_U + \mathrm{vol}(X_U) B(g, \phi),$$

with the "extra series" $B(g, \phi)$ a new part in the non-compact case. Recall that Proposition 4.4 is proved in the compact case by Proposition 4.5. To finish it in the non-compact case, it remains to prove

$$\int_{\mathrm{GL}_2(\mathbb{Q}) \backslash \mathrm{GL}_2(\mathbb{A})/Z(\mathbb{A})}^* \varphi(g) \, B(g, \phi) \, dg = 0, \quad \forall \, \varphi \in \sigma, \; \phi \in \overline{\mathcal{S}}(\mathbb{B} \times \mathbb{A}^\times). \quad (4.6.2)$$

We can assume that ϕ_∞ is the standard Gaussian since $\overline{\mathcal{S}}(\mathbb{B}_\infty \times \mathbb{R}^\times)$ is generated by the standard one under the action of the Lie algebra $\mathrm{gl}_2(\mathbb{R})$.

Before considering the integration, a basic question is to explain the modularity of

$$
\begin{aligned}
B(g,\phi) = & \sum_{u \in \mathbb{Q}^\times} \sum_{x \in V_{\mathrm{hyp}}} r(g_f)\ell\phi_f(x,u)\, \beta_{\phi_\infty}(g_\infty,x,u) \\
& - \sum_{u \in \mathbb{Q}^\times} W'_{0,\infty}(0,g_\infty,u,\phi_\infty)\, |u|_{\mathbb{A}_f}^{\frac{1}{2}} \int_{\mathbb{A}_f} r(g_f)\ell\phi_f((0,x_2),u)dx_2.
\end{aligned}
$$

It is automorphic since it is the difference of two automorphic forms by Theorem 4.15. It is reasonable to expect a more direct interpretation, and such an interpretation should lead to a proof of (4.6.2).

The double sum in the series $B(g,\phi)$ looks like a theta series (on V_{hyp}), except that the archimedean part $\beta_{\phi_\infty}(g_\infty,x,u)$ is not defined in terms of the Weil representation. However, as we will see in the following section, the archimedean part surprisingly transfers according to the Weil representation by the work of Hirzebruch–Zagier [HZ] on Hilbert modular surfaces. It makes the double sum in $B(g,\phi)$ appear as a theta series. The theta series is "singular" in that the archimedean part, even transferring according to the Weil representation, is much worse than a Schwartz function. It is also different from pseudo-theta series coming from derivatives and local heights we will see in this book. Our alternative approach is to use partial Fourier transform to rewrite the series in a simpler form. Then we eventually write the whole series $B(g,\phi)$, including the sum with the constant part, as a Poincaré series up to analytic continuation. Then the modularity is proved and equation (4.6.2) follows immediately.

4.7 INTERPRETATION: NON-COMPACT CASE

As described at the end of the last section, the goal of this section is to re-interpret the series $B(g,\phi)$ and finish the proof of Theorem 3.22.

4.7.1 The archimedean part

Here we rewrite the archimedean part $\beta_{\phi_\infty}(g_\infty,x,u)$ of $B(g,\phi)$ by translating a result of Hirzebruch–Zagier [HZ].

The work of Hirzebruch–Zagier

Hirzebruch–Zagier [HZ] studied the intersection numbers of Hecke operators on Hilbert modular surfaces. Though written in a different language, their treatment is very similar to our case. In terms of the language of this book, the main result of [HZ] is roughly as follows.

Fix a real quadratic field K over $\mathbb{Q}$ (of prime discriminant), and denote by

$$M^\circ = \mathrm{SL}_2(O_K)\backslash\mathcal{H}^2$$

the Hilbert modular surface associated to K. There is a generating function

$$Z_{\mathrm{HZ}}(z) = \sum_{N \geq 0} T_N e^{2\pi i N \tau}, \quad z \in \mathcal{H}.$$

Each coefficient T_N is an explicit special curve in M analogous to the classical Hecke correspondence on $X(1) \times X(1)$. In particular, T_1 is just the image of the modular curve $X(1)$ diagonally embedded in M.

Denote by M the minimal desingularization of the compactification $M^\circ \cup \{\text{cusps}\}$. There is a canonical extension T_N^c of T_N to a divisor in M, and thus there is a canonical extension $Z_{\mathrm{HZ}}^c(z)$ of $Z_{\mathrm{HZ}}(z)$ to M. The paper computed the intersection number of T_1^c with $Z_{\mathrm{HZ}}^c(z)$. One main result of the paper is the expression

$$T_1^c \cdot Z_{\mathrm{HZ}}^c(z) \approx \theta(z)\mathcal{F}(z) + \mathcal{W}(z).$$

Here θ is the classical holomorphic theta series of weight $1/2$, $\mathcal{F}(z)$ is Zagier's almost holomorphic Eisenstein series of weight $3/2$, and $\mathcal{W}(z)$ is a (non-holomorphic) theta function defined by the real quadratic space K. Here we write "$\approx$" since the equality is up to certain linear operations.

Then [HZ] concluded that $T_1^c \cdot Z_{\mathrm{HZ}}^c(z)$ is a modular form of weight two in $z \in \mathcal{H}$. It also obtained the modularity of $T_N^c \cdot Z_{\mathrm{HZ}}^c(z)$. At that time, the modularity of $Z_{\mathrm{HZ}}^c(z)$ or $Z_{\mathrm{HZ}}(z)$ (even as a cohomology class) was not known. The result of [HZ] is a landmark for the succeeding works [KM1, KM2, KM3, GKZ, Bor, Zha, YZZ] on modularity of generating series of special cycles on Shimura varieties.

The counterparts of

$$K, \quad M, \quad T_1^c, \quad Z_{\mathrm{HZ}}^c(z), \quad \theta(z), \quad \mathcal{F}(z), \quad \mathcal{W}(z)$$

in this book are respectively

$$\mathbb{Q}^2, \quad X_U \times X_U, \quad \Delta, \quad Z(g, \phi)_U, \quad \theta(g, \phi^0), \quad E(0, g, \phi_0), \quad B(g, \phi).$$

The archimedean part of $B(g, \phi)$ is essentially the same as that of $\mathcal{W}(z)$. So we can apply the corresponding result of [HZ] to it.

The archimedean part

Assume that the archimedean parts ϕ_∞, $\phi_{0,\infty}$ and ϕ_∞^0 are standard. Here we treat the archimedean part

$$\beta_{\phi_\infty}(g, x, u)$$
$$= -\frac{1}{2\sqrt{2}\gamma_u} W'_{-u(x_1-x_2)^2/4}(0, g, u, \phi_{0,\infty}) \, r(g)\phi_\infty^0(1, u(x_1+x_2)^2/4)$$
$$+ \pi^2 |u|^{\frac{1}{2}} \min(|x_1|, |x_2|) \, 1_{\mathbb{R}_+}(x_1 x_2) \, r(g)\phi_\infty(1, u x_1 x_2).$$

Recall that in Corollary 2.12 we have obtained

$$W'_{a,\infty}(0,1,u,\phi_0) \;=\; \gamma_u \, 8\sqrt{2}\pi^2 e^{-2\pi a}\beta(-4\pi a), \quad u > 0,\; a \le 0.$$

Here β, as introduced in [HZ], is given by

$$\beta(z) = \frac{1}{16\pi}\int_1^\infty e^{-zt}t^{-\frac{3}{2}}\,dt, \quad \mathrm{Re}(z) \ge 0.$$

It follows that for $g = 1$, $u > 0$,

$$\boldsymbol{\beta}_{\phi_\infty}(1,x,u)$$
$$= \; \pi^2 e^{-2\pi u x_1 x_2}\left(u^{\frac{1}{2}}\min(|x_1|,|x_2|)\,1_{\mathbb{R}_+}(x_1 x_2) - 4\beta(\pi u(x_1 - x_2)^2)\right).$$

If $u < 0$, then $\boldsymbol{\beta}_{\phi_\infty}(1,x,u) = 0$ automatically.
For simplicity, denote

$$\boldsymbol{\beta}(x,u) := \boldsymbol{\beta}_{\phi_\infty}(1,x,u), \quad (x,u) \in V_{\mathrm{hyp}}(\mathbb{R}) \times \mathbb{R}^\times.$$

It is a continuous function on $V_{\mathrm{hyp}}(\mathbb{R}) \times \mathbb{R}^\times$, integrable in x for any fixed u. It is not a Schwartz function in the usual sense because it is not differentiable around $x_1 x_2 = 0$. Nonetheless, by the formulae of the Weil representation, the function $r(g)\boldsymbol{\beta}$ is a well-defined continuous function on $V_{\mathrm{hyp}}(\mathbb{R}) \times \mathbb{R}^\times$ for any $g \in \mathrm{GL}_2(\mathbb{R})$.

The following is a surprising result of [HZ] which makes their series $\mathcal{W}(z)$ a theta series. We state it in the current language.

PROPOSITION 4.16. *Assume that ϕ_∞ is the standard Gaussian. For any $g \in \mathrm{GL}_2(\mathbb{R})$ and $(x,u) \in V_{\mathrm{hyp}}(\mathbb{R}) \times \mathbb{R}^\times$,*

$$\boldsymbol{\beta}_{\phi_\infty}(g,x,u) = r(g)\boldsymbol{\beta}(x,u).$$

PROOF. It is essentially Proposition 1 in §2.3 of [HZ]. The equality holds for $g = 1$ by definition and we need to extend it to the general case. It is easy to verify that

$$\begin{aligned}
\boldsymbol{\beta}_{\phi_\infty}(n(b)g,x,u) &= \boldsymbol{\beta}_{\phi_\infty}(g,x,u)\,\psi(buq(x)), & b \in \mathbb{R},\\
\boldsymbol{\beta}_{\phi_\infty}(m(a)g,x,u) &= \boldsymbol{\beta}_{\phi_\infty}(g,ax,u)\,|a|, & a \in \mathbb{R}^\times,\\
\boldsymbol{\beta}_{\phi_\infty}(d(a)g,x,u) &= \boldsymbol{\beta}_{\phi_\infty}(g,x,a^{-1}u)\,|a|^{-\frac{1}{2}}, & a \in \mathbb{R}^\times.
\end{aligned}$$

The formulae match the following formula

$$\begin{aligned}
r(n(b)g)\boldsymbol{\beta}(x,u) &= r(g)\boldsymbol{\beta}(x,u)\,\psi(buq(x)), & b \in \mathbb{R},\\
r(m(a)g)\boldsymbol{\beta}(x,u) &= r(g)\boldsymbol{\beta}(ax,u)\,|a|, & a \in \mathbb{R}^\times,\\
r(d(a)g)\boldsymbol{\beta}(x,u) &= r(g)\boldsymbol{\beta}(x,a^{-1}u)\,|a|^{-\frac{1}{2}}, & a \in \mathbb{R}^\times.
\end{aligned}$$

It follows that if the proposition is true for some g, then it is true for all elements of $P(\mathbb{R})g$.

By the Bruhat decomposition

$$\mathrm{GL}_2(\mathbb{R}) = P(\mathbb{R}) \cup P(\mathbb{R})wN(\mathbb{R}),$$

we only need to verify

$$\beta_{\phi_\infty}(wn(b), x, u) = r(wn(b))\beta(x, u), \quad \forall\, b \in \mathbb{R}.$$

We can assume that $u > 0$ since both sides are zero for $u < 0$. We need to prove that the Fourier transform of $r(n(b))\beta(x, u)$ is exactly equal to $\beta_{\phi_\infty}(wn(b), x, u)$.

It is easy to have

$$
\begin{aligned}
& r(n(b))\beta(x, u) \\
={} & \beta(x, u)\,\psi(buq(x)) \\
={} & \pi^2 e^{2\pi i u(b+i)x_1 x_2}\left(u^{\frac{1}{2}}\min(|x_1|, |x_2|)\,1_{\mathbb{R}_+}(x_1 x_2) - 4\beta(\pi u(x_1 - x_2)^2)\right).
\end{aligned}
$$

Now we compute $\beta_{\phi_\infty}(wn(b), x, u)$. The function $\beta_{\phi_\infty}(g, x, u)$ is of weight two under the right action of $\mathrm{SO}(2, \mathbb{R})$ on g. Use the Iwasawa decomposition

$$
\begin{pmatrix} & 1 \\ -1 & \end{pmatrix}\begin{pmatrix} 1 & b \\ & 1 \end{pmatrix} = \begin{pmatrix} 1 & -b/r^2 \\ & 1 \end{pmatrix}\begin{pmatrix} 1/r & \\ & r \end{pmatrix}\begin{pmatrix} -b/r & 1/r \\ -1/r & -b/r \end{pmatrix}
$$

with $r - \sqrt{b^2 + 1}$. We obtain

$$
\begin{aligned}
& \beta_{\phi_\infty}(wn(b), x, u) \\
={} & \frac{(-b+i)^2}{r^2}\beta_{\phi_\infty}(n(-b/r^2)m(1/r), x, u) \\
={} & \pi^2 \frac{(-b+i)^2}{r^4}e^{2\pi i u r^{-2}(-b+i)x_1 x_2} \\
& \cdot \left(u^{\frac{1}{2}}\min(|x_1|, |x_2|)\,1_{\mathbb{R}_+}(x_1 x_2) - 4r\,\beta(\pi u r^{-2}(x_1 - x_2)^2)\right).
\end{aligned}
$$

By a simple change of variable, it suffices to assume that $u = 1$. Denoting $z = b + i$, in terms of the notation of Proposition 1 in [HZ, §2.3], we simply have

$$
\begin{aligned}
r(n(b))\beta(x, u) &= -2\pi^2\,W_z(x_1, x_2), \\
\beta_{\phi_\infty}(wn(b), x, u) &= -2\pi^2\,z^{-2}\,W_{-1/z}(x_1, x_2).
\end{aligned}
$$

Then the result we need is exactly the result of Proposition 1 in [HZ, §2.3]. $\square$

Poisson summation formula

Go back to the series $B(g, \phi)$. Assume that ϕ_∞ is standard. By Proposition 4.16, we have

$$
\begin{aligned}
B(g, \phi) \quad = \quad & \sum_{u \in \mathbb{Q}^\times} \sum_{x \in V_{\text{hyp}}} r(g_f) \ell \phi_f(x, u) \; r(g_\infty) \beta(x, u) \\
& - \sum_{u \in \mathbb{Q}^\times} W'_{0,\infty}(0, g_\infty, u, \phi_\infty) \, |u|_{\mathbb{A}_f}^{\frac{1}{2}} \int_{\mathbb{A}_f} r(g_f) \ell \phi_f((0, x_2), u) dx_2.
\end{aligned}
$$

A rough thought is that the double sum

$$
\sum_{u \in \mathbb{Q}^\times} \sum_{x \in V_{\text{hyp}}} r(g_f) \ell \phi_f(x, u) \; r(g_\infty) \beta(x, u)
$$

is a theta series on V_{hyp}. Then the proof of (4.6.2) would follow the general fact that the (global) theta lifting of any cuspidal automorphic representation of $\mathrm{GL}_2(\mathbb{A})$ to $\mathrm{GSO}(V_{\text{hyp}}(\mathbb{A})) = \mathbb{A}^\times \times \mathbb{A}^\times$ is zero. However, this argument is NOT true since it simply ignores the singularity of $\beta(x, u)$ and the presence of the second term in $B(g, \phi)$.

In fact, it is very easy to see that the second term

$$
\sum_{u \in \mathbb{Q}^\times} W'_{0,\infty}(0, g_\infty, u, \phi_\infty) \, |u|_{\mathbb{A}_f}^{\frac{1}{2}} \int_{\mathbb{A}_f} r(g_f) \ell \phi_f((0, x_2), u) dx_2,
$$

an expression for the intertwining part $W_0(0, g, \phi)$ of the Eisenstein series $E(0, g, \phi)$, is not automorphic in g (unless it is identically zero). We know that $B(g, \phi)$ is automorphic. So the double sum is not automorphic in general. It cannot be treated as a theta series in the usual way. The problem is caused by the slow decay of β and the singularity of β around $x_1 x_2 = 0$. They violate the Poisson summation formula used in the proof of the modularity of theta series.

Next, we recall a version of the Poisson summation formula. Let (V, q) be a quadratic space over $\mathbb{Q}$. For an integrable function $\Phi : V(\mathbb{A}) \to \mathbb{C}$, recall its Fourier transform

$$
\widehat{\Phi}(x) = \int_{V(\mathbb{A})} \Phi(y) \psi(\langle x, y \rangle) dy.
$$

Here the integration uses the self-dual Haar measure, and the pairing

$$
\langle x, y \rangle = q(x + y) - q(x) - q(y)
$$

is as usual. The Poisson summation formula, which only holds for suitable Φ, is

$$
\sum_{x \in V} \Phi(x) = \sum_{x \in V} \widehat{\Phi}(x).
$$

In particular, it holds for the usual Schwartz–Bruhat functions on $V(\mathbb{A})$, i.e., linear combinations of pure tensors $\otimes_v \Phi_v$ where $\Phi_p \in \mathcal{S}(V(\mathbb{Q}_p))$ is locally constant and compactly supported for every finite place p, and $\Phi_\infty \in \mathcal{S}(V(\mathbb{R}))$ is infinitely differentiable and all its partial derivatives (of any orders) have the property of rapid decay.

In the Poisson summation formula, the condition of being Schwartz–Bruhat is very strong. One can relax them in many ways. The following result improves it at the archimedean part.

LEMMA 4.17. *The Poisson summation formula holds for Φ if $\Phi = \otimes_v \Phi_v$ where $\Phi_p \in \mathcal{S}(V(\mathbb{Q}_p))$ for every finite place p, and Φ_∞ is a continuous and integrable function on $V(\mathbb{R})$ satisfying the decay condition*

$$|\Phi_\infty(x)| + |\widehat{\Phi}_\infty(x)| < C(1+|x|)^{-\dim V - \epsilon}$$

for some $C > 0$ and $\epsilon > 0$. Here $|\cdot|$ is the norm induced from the usual norm on $\mathbb{R}^{\dim V}$ by any choice of an $\mathbb{R}$-linear identification $V(\mathbb{R}) \simeq \mathbb{R}^{\dim V}$.

PROOF. It can be easily reduced to the classical setting. Let $V_{\mathbb{Z}}$ be any $\mathbb{Z}$-lattice of V on which $\psi_\infty \circ q$ is trivial. By linearity, we can assume that $\Phi_p = 1_{a_p + p^{e_p} V_{\mathbb{Z}_p}}$ for every finite place p. Of course, $a_p = 0$ and $e_p = 0$ for almost all p. By a translation of the form $x \to x + a$ for some $a \in V$, we can assume that all $a_p = 0$. By a contraction of the form $x \to kx$ for some $k \in \mathbb{Q}^\times$, we can assume that all $k = 0$. Then $\Phi_p = 1_{V_{\mathbb{Z}_p}}$ for every finite place p. Thus

$$\sum_{x \in V} \Phi(x) = \sum_{x \in V_{\mathbb{Z}}} \Phi_\infty(x).$$

It is reduced to the Poisson summation for the lattice $V_{\mathbb{Z}}$ in the real vector space $V(\mathbb{R})$. See [Gra, Theorem 3.1.17] for example. It is easy to see that the proof goes through a general (non-degenerate) norm q on $V(\mathbb{R})$. $\qquad\square$

Go back to the function $\beta(x, u)$. Write

$$\beta(x, u) = \beta_1(x, u) - \beta_2(x, u)$$

where

$$\beta_1(x, u) = \pi^2 e^{-2\pi u x_1 x_2} u^{\frac{1}{2}} \min(|x_1|, |x_2|) \, 1_{\mathbb{R}_+}(u) \, 1_{\mathbb{R}_+}(x_1 x_2),$$
$$\beta_2(x, u) = 4\pi^2 e^{-2\pi u x_1 x_2} \beta(\pi u (x_1 - x_2)^2) \, 1_{\mathbb{R}_+}(u).$$

Fix $u > 0$. It is easy to see that

$$\beta_2(x, u) < 4\pi^2 e^{-\pi u (x_1^2 + x_2^2)}.$$

It decays very fast. The function $\beta_1(x, u)$ does not satisfy the growth condition in the lemma. In fact, taking $x = (t, t^{-1})$ with $t \to \infty$, then

$$\beta_1(x, u) = \pi^2 e^{-2\pi u} u^{\frac{1}{2}} t^{-1} \sim |x|^{-1}.$$

It explains in some extent the failure of the Poisson summation formula for $\beta(\cdot, u)$.

In [HZ], the series $\mathcal{W}(z)$ is essentially a multiple of

$$\sum_{x \in O_K} r(g_\infty)\boldsymbol{\beta}(x, 1).$$

Here g_∞ is the unique element of the parabolic subgroup $P^1(\mathbb{R})$ with $g_\infty(i) = z$. It does not have the counterpart of the constant part appearing in $B(g, \phi)$. Then [HZ] uses the Poisson summation formula on O_K to conclude the modularity of $\mathcal{W}(z)$. One can verify that the Poisson summation formula holds in that case, even though the function $\boldsymbol{\beta}$ does not satisify the decay condition in Lemma 4.17. These different properties reflect the differences between the anisotropic K and the isotropic $\mathbb{Q}^2$.

4.7.2 Partial Fourier transforms

Our idea is to use partial Fourier transforms to write $B(g, \phi)$ in a simpler form to make the modularity transparent. Partial Fourier transforms work here by two reasons:

- The function $\boldsymbol{\beta}((x_1, x_2), u)$, viewed as a function in x_2 by fixing (x_1, u), satisfies the decay condition in Lemma 4.17. Thus the Poisson summation formula for x_2 is valid, which gives a partial Poisson summation formula for x.

- Under the partial Fourier transform, the action of GL_2 via the Weil representation is transferred to the regular action. This intertwining property greatly simplifies the series.

We will illustrate the idea by the example of the usual theta series (defined by Schwartz–Bruhat functions).

Schwartz–Bruhat case

Let k be a local field, and V_{hyp} be the hyperbolic quadratic space over k. Let $S(V_{\mathrm{hyp}}(k) \times k^\times)$ be the space of Schwartz–Bruhat functions. If k is non-archimedean, it is just $\mathcal{S}(V_{\mathrm{hyp}}(k) \times k^\times)$. If k is archimedean, it is much bigger than the space $\mathcal{S}(V_{\mathrm{hyp}}(k) \times k^\times)$ we use more often. See §2.1 for more details. We only use the space $S(V_{\mathrm{hyp}}(k) \times k^\times)$ in this subsection.

For any $\Phi \in S(V_{\mathrm{hyp}}(k) \times k^\times)$, its partial Fourier transform (on the second variable) is defined by

$$\widetilde{\Phi}((x_1, x_2), u) = |u|^{\frac{1}{2}} \int_k \Phi((x_1, y), u)\psi(ux_2 y)dy.$$

Here the measure dy is the Haar measure on k in §1.6 and the factor $|u|^{\frac{1}{2}}$ makes it self-dual.

Recall that $\mathrm{GL}_2(k)$ acts on $S(V_{\mathrm{hyp}}(k) \times k^\times)$ by the Weil representation r. The regular action ρ of $\mathrm{GL}_2(k)$ on $S(V_{\mathrm{hyp}}(k) \times k^\times)$ is given by

$$\rho(g)\Phi(x, u) = \Phi(xg, \det(g)^{-1} u).$$

Here the action xg is the usual right multiplication of the matrix g on the row vector $x = (x_1, x_2)$. The following result is classical.

LEMMA 4.18. *For any $g \in \mathrm{GL}_2(k)$ and $\Phi \in S(V_{\mathrm{hyp}}(k) \times k^\times)$,*

$$\widetilde{r(g)\Phi} = \rho(g)\widetilde{\Phi}.$$

In other words, the partial Fourier transform defines a $\mathrm{GL}_2(k)$-intertwining map

$$(S(V_{\mathrm{hyp}}(k) \times k^\times), \ r) \longrightarrow (S(V_{\mathrm{hyp}}(k) \times k^\times), \ \rho).$$

PROOF. Use the definition of Weil representation to check the identity for $m(a), n(b), d(a)$ and w. See also [JL, Proposition 1.6] for the case of $S(V_{\mathrm{hyp}}(k))$. $\qquad\square$

The partial Fourier transform can be used to write a theta series on V_{hyp} as the sum of a residue form with a Poincaré series. Let F be a number field and $\Phi \in S(V_{\mathrm{hyp}}(\mathbb{A}) \times \mathbb{A}^\times)$ be a Schwartz–Bruhat function. We are going to rewrite the theta series

$$\theta(g, \Phi) = \sum_{u \in F^\times} \sum_{(x_1, x_2) \in V_{\mathrm{hyp}}(F)} r(g)\Phi((x_1, x_2), u), \quad g \in \mathrm{GL}_2(\mathbb{A}).$$

Use the Poisson summation on x_2 by fixing (x_1, u). We get

$$\sum_{x_2 \in F} r(g)\Phi((x_1, x_2), u) = \sum_{x_2 \in F} \widetilde{r(g)\Phi}((x_1, x_2), u).$$

Then the lemma permits us to change $\widetilde{r(g)\Phi}$ to $\rho(g)\widetilde{\Phi}$. It follows that

$$\theta(g, \Phi) = \sum_{u \in F^\times} \sum_{(x_1, x_2) \in F^2} \rho(g)\widetilde{\Phi}((x_1, x_2), u).$$

Consider the right action of $\mathrm{GL}_2(F)$ on $F^2 \times F^\times$ defined by

$$g : ((x_1, x_2), u) \longmapsto ((x_1, x_2)g, \det(g)^{-1}u).$$

There are two orbits, represented respectively by $((0,0), 1)$ and $((0,1), 1)$. The stabilizers of these two representatives are respectively $\mathrm{SL}_2(F)$ and $N(F)$. It follows that

$$\theta(g, \Phi) = \sum_{u \in F^\times} \widetilde{\Phi}((0,0), \det(g)^{-1}u) + \sum_{\gamma \in N(F)\backslash\mathrm{GL}_2(F)} \rho(\gamma g)\widetilde{\Phi}((0,1), 1).$$

Both terms on the right-hand side are automorphic for $g \in \mathrm{GL}_2(\mathbb{A})$. The first term is a residue form in the sense that it is an automorphic form for $\det(g) \in \mathrm{GL}_1(\mathbb{A})$. The second term is a Poincaré series in that

$$\rho(ng)\widetilde{\Phi}((0,1),1) = \rho(ng)\widetilde{\Phi}((0,1),1), \quad n \in N(\mathbb{A}).$$

It is well-known that both of them are orthogonal to cusp forms.

Archimedean part

Go back to the function $\beta((x_1, x_2), u)$. For any $g \in \mathrm{GL}_2(\mathbb{R})$, the function $r(g)\beta$ is continuous and integrable in x_2. So we can still define the partial Fourier transform

$$\widetilde{r(g)\beta}((x_1, x_2), u) = |u|^{\frac{1}{2}} \int_{\mathbb{R}} r(g)\beta((x_1, y), u)\psi(ux_2 y)dy.$$

The action ρ of $\mathrm{GL}_2(\mathbb{R})$ on $\widetilde{r(g)\beta}$ is defined as in the Schwartz case.

We will see that the only discontinuous points of the function $\widetilde{r(g)\beta}(x, u)$ are given by $x = 0$. Thus Lemma 4.18 is true for β as long as $x \neq 0$.

LEMMA 4.19. *The following are true for any $g \in \mathrm{GL}_2(\mathbb{R})$ and $u \in \mathbb{R}^\times$.*

(1) *For any $(x_1, x_2) \neq (0, 0)$,*

$$\widetilde{\beta}((x_1, x_2), u) = \frac{1 - e^{-\pi u(x_1^2 + x_2^2)}}{4u(x_1 - ix_2)^2} 1_{\mathbb{R}_+}(u).$$

(2) *The function $\widetilde{r(g)\beta}(x, u)$ is continuous at all points (x, u) with $x \neq 0$.*

(3) *The function $r(g)\beta((x_1, x_2), u)$, viewed as a function of $x_2 \in \mathbb{R}$ for any fixed (x_1, u), satisfies the decay condition of Φ_∞ in Lemma 4.17.*

(4) *If $x \neq 0$, then*

$$\widetilde{r(g)\beta}(x, u) = \rho(g)\widetilde{\beta}(x, u).$$

(5) *If $x = 0$, then*

$$\widetilde{r(g)\beta}(0, u) = \frac{\pi(ci - d)}{4(ci + d)} 1_{\mathbb{R}_+}(u \det(g)),$$

$$W_0'(0, g, u, \phi_\infty) = \frac{\pi(ci - d)}{2(ci + d)} 1_{\mathbb{R}_+}(u \det(g)).$$

Here we write $g = \begin{pmatrix} a & b \\ c & d \end{pmatrix}$.

PROOF. Result (1) is obtained by explicit computations. We only sketch it here. Write

$$\beta(x, u) = \beta_1(x, u) - \beta_2(x, u)$$

where

$$
\begin{aligned}
\beta_1(x, u) &= \pi^2 e^{-2\pi u x_1 x_2} \, u^{\frac{1}{2}} \min(|x_1|, |x_2|) \, 1_{\mathbb{R}_+}(u) \, 1_{\mathbb{R}_+}(x_1 x_2), \\
\beta_2(x, u) &= 4\pi^2 e^{-2\pi u x_1 x_2} \, \beta(\pi u (x_1 - x_2)^2) \, 1_{\mathbb{R}_+}(u).
\end{aligned}
$$

Explicit computation gives

$$
\begin{aligned}
\widetilde{\beta_1}((x_1, x_2), u) &= \frac{1 - e^{-2\pi u x_1 (x_1 - i x_2)}}{4u(x_1 - i x_2)^2} 1_{\mathbb{R}_+}(u), \\
\widetilde{\beta_2}((x_1, x_2), u) &= \frac{e^{-\pi u (x_1^2 + x_2^2)} - e^{-2\pi u x_1 (x_1 - i x_2)}}{4u(x_1 - i x_2)^2} 1_{\mathbb{R}_+}(u).
\end{aligned}
$$

Here the second identity is obtained by replacing β by its definition

$$\beta(z) = \frac{1}{16\pi} \int_1^\infty e^{-zt} t^{-\frac{3}{2}} dt$$

and changing the order of the integration. It gives

$$\widetilde{\beta}((x_1, x_2), u) = \frac{1 - e^{-\pi u (x_1^2 + x_2^2)}}{4u(x_1 - i x_2)^2} 1_{\mathbb{R}_+}(u).$$

It proves (1).

The function $\widetilde{\beta}$ is continuous away from $(0, 0)$. Expand the exponential function. We have

$$\widetilde{\beta}((x_1, x_2), u) = \frac{\pi}{4} \frac{x_1 + i x_2}{x_1 - i x_2} 1_{\mathbb{R}_+}(u) + O(x_1^2 + x_2^2), \quad (x_1, x_2) \to (0, 0). \quad (4.7.1)$$

It follows that all the directional limits exist. Then it verifies (2), (3) and (4) for $g = 1$.

We first prove (2) and (3) for general g. By Proposition 4.16, β has weight two under the action r of $O(2, \mathbb{R})$. Thus by the Iwasawa decomposition, we only need to consider the case $g \in P(\mathbb{R})$. It is routine to verify that

$$\widetilde{r(g)\beta} = \rho(g)\widetilde{\beta}, \quad \forall g \in P(\mathbb{R}).$$

It follows that we only need to prove the results for $\rho(g)\widetilde{\beta}$ with $g \in P(\mathbb{R})$. They can be easily reduced to $g = 1$.

Now we prove (4) for general g. Note that both $\widetilde{r(g)\beta}(x, u)$ and $\rho(g)\widetilde{\beta}(x, u)$ are continuous at $x \neq 0$. It suffices to prove that, for each g and u,

$$\widetilde{r(g)\beta}(x, u) = \rho(g)\widetilde{\beta}(x, u)$$

for almost all $x \in \mathbb{R}^2$. The proof is similar to Lemma 4.18. To see the appearance of "almost all," we sketch it here. The result for $g = 1$ is trivial. It is easy to see that if the result is true for g, then it is also true for all elements in the coset $P(\mathbb{R})g$. So it suffices to prove the result for the Weyl element $g = w$. Assume that $u > 0$ since both sides are zero if $u < 0$. By definition,

$$
\begin{aligned}
\widetilde{r(w)\beta}&((x_1, x_2), u) \\
&= |u|^{\frac{1}{2}} \int_{\mathbb{R}} \widetilde{r(w)\beta}((x_1, y), u) \ \psi(ux_2 y) dy \\
&= |u|^{\frac{3}{2}} \int_{\mathbb{R}} \int_{\mathbb{R}} \int_{\mathbb{R}} \beta((z, t), u) \ \psi(ux_1 t) \ \psi(uyz) \ \psi(ux_2 y) dz dt dy \\
&= |u| \int_{\mathbb{R}} \int_{\mathbb{R}} \widetilde{\beta}((z, x_1), u) \ \psi(uyz) \ \psi(ux_2 y) dz dy.
\end{aligned}
$$

Here the powers of $|u|$ come out by different ways of normalizing the Haar measures. Using the inversion formula to contract the double integral, we obtain

$$
\widetilde{r(w)\beta}((x_1, x_2), u) = \widetilde{\beta}((-x_2, x_1), u) = \widetilde{\rho(w)\beta}((x_1, x_2), u).
$$

It holds at all (continuous points) $(x_1, x_2) \neq (0, 0)$.

It remains to prove (5). We first compute $\widetilde{r(g)\beta}(0, u)$. By definition, $\widetilde{r(g)\beta}$ is the partial Fourier transform for x_2, so it is continuous in x_2. It follows that

$$
\widetilde{r(g)\beta}(0, u) = \lim_{x_2 \to 0} \widetilde{r(g)\beta}((0, x_2), u).
$$

By (4), for $x_2 \neq 0$, we have

$$
\widetilde{r(g)\beta}((0, x_2), u) = \widetilde{\rho(g)\beta}((0, x_2), u) = \widetilde{\beta}((cx_2, dx_2), \det(g)^{-1}u).
$$

Taking the limit $x_2 \to 0$, by equation (4.7.1), we have

$$
\widetilde{r(g)\beta}(0, u) = \frac{\pi}{4} \cdot \frac{c + id}{c - id} 1_{\mathbb{R}_+}(u \det(g)) = \frac{\pi}{4} \cdot \frac{ci - d}{ci + d} 1_{\mathbb{R}_+}(u \det(g)).
$$

On the other hand, Proposition 2.11 gives

$$
W_0'(0, 1, u, \phi_\infty) = -\frac{\pi}{2} 1_{\mathbb{R}_+}(u).
$$

It is easy to verify that

$$
W_0'(0, \gamma g, u, \phi_\infty) = W_0'(0, g, \det(\gamma)^{-1}u, \phi_\infty), \quad \forall \gamma \in P(\mathbb{R}).
$$

Therefore, the weight-two property yields that

$$
W_0'(0, g, u, \phi_\infty) = -\frac{\pi}{2} e^{2i\theta} 1_{\mathbb{R}_+}(u \det(g)).
$$

Here θ comes from the Iwasawa decomposition

$$\begin{pmatrix} a & b \\ c & d \end{pmatrix} = \begin{pmatrix} a' & b' \\ & d' \end{pmatrix} \begin{pmatrix} \cos(\theta) & \sin(\theta) \\ -\sin(\theta) & \cos(\theta) \end{pmatrix}.$$

But it is easy to verify

$$e^{2i\theta} = -\frac{ci - d}{ci + d}.$$

It finishes the proof.   □

4.7.3   Interpretation of the extra series

In this subsection we will write $B(g, \phi)$ as a Poincaré series, and finish the proof of Theorem 3.22 in the non-compact case.

Partial Poisson summation

Recall from Theorem 4.15 that

$$\begin{aligned}
B(g, \phi) &= \sum_{u \in \mathbb{Q}^\times} \sum_{x \in V_{\mathrm{hyp}}} r(g_f) \ell \phi_f(x, u) \, r(g_\infty) \boldsymbol{\beta}(x, u) \\
&\quad - \sum_{u \in \mathbb{Q}^\times} W'_{0,\infty}(0, g_\infty, u, \phi_\infty) \, |u|_{\mathbb{A}_f}^{\frac{1}{2}} \int_{\mathbb{A}_f} r(g_f) \ell \phi_f((0, x_2), u) dx_2.
\end{aligned}$$

In the second summation, the non-archimedean part is exactly $\widetilde{r(g)\ell\phi_f}(0, u)$ by definition, and the archimedean part is equal to $2\, \widetilde{r(g_\infty)}\boldsymbol{\beta}(0, u)$ by Lemma 4.19. It follows that

$$\begin{aligned}
B(g, \phi) &= \sum_{u \in \mathbb{Q}^\times} \sum_{x \in V_{\mathrm{hyp}}} r(g_f) \ell \phi_f(x, u) \, r(g_\infty) \boldsymbol{\beta}(x, u) \\
&\quad - 2 \sum_{u \in \mathbb{Q}^\times} \widetilde{r(g_f) \ell \phi_f}(0, u) \, \widetilde{r(g_\infty)}\boldsymbol{\beta}(0, u).
\end{aligned}$$

We consider a more general situation. For any $\eta_f \in \mathcal{S}(V_{\mathrm{hyp}}(\mathbb{A}_f) \times \mathbb{A}_f^\times)$, denote $\eta = \eta_f \otimes \boldsymbol{\beta}$ and

$$B(g, \eta) := \sum_{u \in \mathbb{Q}^\times} \sum_{x \in V_{\mathrm{hyp}}} r(g)\eta(x, u) - \sum_{u \in \mathbb{Q}^\times} \widetilde{r(g)\eta}(0, u) - \sum_{u \in \mathbb{Q}^\times} \widetilde{r(g)\eta}'(0, u).$$

Here $\widetilde{r(g)\eta}'$ denotes the partial Fourier transform with respect to x_1. We will prove that $B(g, \eta)$ is automorphic and "orthogonal" to all cusp forms with trivial central character at infinity. The modularity explains the reason for changing half of $\widetilde{r(g)\eta}$ to $\widetilde{r(g)\eta}'$.

For clarification, recall that for $\eta_p \in \mathcal{S}(V_{\mathrm{hyp}}(\mathbb{Q}_p) \times \mathbb{Q}_p^\times)$ we have the following three Fourier transforms:

$$\widetilde{\eta}_p((x_1, x_2), u) = |u|_p^{\frac{1}{2}} \int_{\mathbb{Q}_p} \eta_p((x_1, y), u) \, \psi_p(x_2 y) \, dy,$$

$$\widetilde{\eta}_p'((x_1, x_2), u) = |u|_p^{\frac{1}{2}} \int_{\mathbb{Q}_p} \eta_p((y, x_2), u) \, \psi_p(x_1 y) \, dy,$$

$$\widehat{\eta}_p((x_1, x_2), u) = |u|_p \int_{\mathbb{Q}_p} \int_{\mathbb{Q}_p} \eta_p((y_1, y_2), u) \, \psi_p(x_2 y_1 + x_1 y_2) \, dy_1 dy_2.$$

The powers of $|u|_p$ are to change the measures to the self-dual measures. The composition of any two transforms above is equal to the third one composed with a transformation of the form $(x_1, x_2) \to (\pm x_1, \pm x_2)$ or $(x_1, x_2) \to (\pm x_2, \pm x_1)$. The definitions are also valid for $\boldsymbol{\beta}$ in the archimedean case.

The special case $B(g, \ell \phi_f \otimes \boldsymbol{\beta})$ is exactly the original series $B(g, \phi)$. In fact, $\widetilde{r(g_f)\ell\phi_f}'(0, u) = \widetilde{r(g_f)\ell\phi_f}(0, u)$ follows from the symmetry

$$\ell(r(g_f)\phi_f)((x_1, x_2), u) = \ell(r(g_f)\phi_f)((x_2, x_1), u).$$

See Proposition 4.13 for the symmetry. The identity $\widetilde{r(g_\infty)\boldsymbol{\beta}}'(0, u) = \widetilde{r(g_\infty)\boldsymbol{\beta}}(0, u)$ also follows from the symmetry of $r(g_\infty)\boldsymbol{\beta}$. We will see that the equality $\widetilde{r(g_f)\ell\phi_f}'(0, u) = \widetilde{r(g_f)\ell\phi_f}(0, u)$ is the only special property we need in the consideration.

Go back to the study of the general $B(g, \eta)$. To ease the notation, consider

$$B(g, u, \eta): = \sum_{x \in V_{\mathrm{hyp}}} r(g)\eta(x, u) - \widetilde{r(g)\eta}(0, u) - \widetilde{r(g)\eta}'(0, u).$$

Here $u \in \mathbb{Q}^\times$. Write the summation x into $x_1 \in \mathbb{Q}$ and $x_2 \in \mathbb{Q}$. Apply the Poisson summation formula in Lemma 4.17 to the sum $x_2 \in \mathbb{Q}$. We obtain

$$B(g, u, \eta)$$
$$= \sum_{x_1 \in \mathbb{Q}} \sum_{x_2 \in \mathbb{Q}} \widetilde{r(g)\eta}((x_1, x_2), u) - \widetilde{r(g)\eta}((0, 0), u) - \widetilde{r(g)\eta}'((0, 0), u)$$
$$= \sum_{x_1 \in \mathbb{Q}} {\sum_{x_2 \in \mathbb{Q}}}' \widetilde{r(g)\eta}((x_1, x_2), u) - \widetilde{r(g)\eta}'((0, 0), u).$$

Here the summation

$$\sum_{x_2 \in \mathbb{Q}}' = \sum_{x_2 \in \mathbb{Q} - (\{x_1\} \cap \{0\})}.$$

The double summation excludes the point $(x_1, x_2) = (0, 0)$.

By Lemma 4.18 and Lemma 4.19, change the action r to the action ρ in the double sum. As for the $(0, 0)$-term, it is easy to have

$$\widetilde{r(g)\eta_f}'((0, 0), u) = \widetilde{r(g_f)\eta_f}((0, 0), u) = \widetilde{\rho(g_f)\widetilde{\eta}_f}((0, 0), u).$$

Therefore,

$$B(g, u, \eta)$$
$$= \sum_{x_1 \in \mathbb{Q}} {\sum_{x_2 \in \mathbb{Q}}}' \rho(g)\widetilde{\eta}((x_1, x_2), u) - \widehat{\rho(g_f)\widetilde{\eta}_f}((0,0), u) \ \widetilde{r(g_\infty)}\boldsymbol{\beta}((0,0), u).$$

Interpretation by Poincaré series

The last double sum for $B(g, u, \eta)$ is not absolutely convergent, while each single sum is absolutely convergent in the sense that

$${\sum_{x_2 \in \mathbb{Q}}}' |\rho(g)\widetilde{\eta}((x_1, x_2), u)| < \infty$$

and

$$\sum_{x_1 \in \mathbb{Q}} \left| {\sum_{x_2 \in \mathbb{Q}}}' \rho(g)\widetilde{\eta}((x_1, x_2), u) \right| < \infty.$$

As we will see, the situation resembles the classical almost holomorphic weight-two Eisenstein series

$$G_2(z) = \sum_{m \in \mathbb{Z}} {\sum_{n \in \mathbb{Z}}}' \frac{1}{(mz+n)^2} - \frac{\pi}{\text{Im}(z)}.$$

We will actually reduce the problem to the classical case.

Ignore the convergence for the moment. Then the summation $(x_1, x_2) \in \mathbb{Q}^2 - \{(0,0)\}$ is a single orbit of $(0,1)$ under the right action of $\text{SL}_2(\mathbb{Q})$. The stabilizer of $(0,1)$ in $\text{SL}_2(\mathbb{Q})$ is exactly $N(\mathbb{Q})$. It follows that the double sum in $B(g, u, \eta)$ can be formally transformed to

$$\sum_{\gamma \in N(\mathbb{Q}) \backslash \text{SL}_2(\mathbb{Q})} \rho(\gamma g)\widetilde{\eta}((0,1), u).$$

The standard way to rigorize the above idea is to introduce an s-variable and consider the analytic continuation. Define

$$B(s, g, u, \eta) := \sum_{\gamma \in N(\mathbb{Q}) \backslash \text{SL}_2(\mathbb{Q})} \delta_\infty(\gamma g)^s \, \rho(\gamma g)\widetilde{\eta}((0,1), u), \quad g \in \text{GL}_2(\mathbb{A}).$$

Here δ_∞ only picks up the archimedean part. The series is a Poincaré series in that

$$\delta_\infty(ng)^s \, \rho(ng)\widetilde{\eta}((0,1), u) = \delta_\infty(g)^s \, \rho(g)\widetilde{\eta}((0,1), u), \quad n \in N(\mathbb{A}).$$

It is not exactly an Eisenstein series in the sense that

$$\delta_\infty(g)^s \, \rho(g)\widetilde{\eta}((0,1), u)$$

is not a principal series. We will see that the series $B(s, g, u, \eta)$ is absolutely convergent for $\text{Re}(s) > 0$, and thus defines an automorphic form for $g \in \text{SL}_2(\mathbb{A})$.

In the absolute convergence range $\text{Re}(s) > 0$, it is easy to recover the expression

$$B(s, g, u, \eta) = \sum_{x_1 \in \mathbb{Q}} \sideset{}{'}\sum_{x_2 \in \mathbb{Q}} \delta_\infty(\gamma_{(x_1, x_2)} g)^s \rho(g) \widetilde{\eta}((x_1, x_2), u).$$

Here $\gamma_{(x_1, x_2)}$ denotes any matrix in $\text{SL}_2(\mathbb{Q})$ of the form $\begin{pmatrix} * & * \\ x_1 & x_2 \end{pmatrix}$, and $\delta_\infty(\gamma_{(x_1, x_2)} g)$ does not depend on the choice of the matrix. In fact, explicit computation gives

$$\delta_\infty(\gamma_{(x_1, x_2)} g) = \frac{|\text{Im}(z)|^{1/2}}{|x_1 z + x_2|}.$$

Here we denote $g_\infty = \begin{pmatrix} a & b \\ c & d \end{pmatrix}$ and $z = \dfrac{ai + b}{ci + d}$.

Now we introduce the counterpart $B(s, g, \eta)$ of the original series $B(g, \eta)$. We might define it to be

$$\sum_{u \in \mathbb{Q}^\times} B(s, g, u, \eta) = \sum_{\gamma \in N(\mathbb{Q}) \backslash \text{GL}_2(\mathbb{Q})} |\det(\gamma)|_\infty^{-\frac{s}{2}} \delta_\infty(\gamma g)^s \, \rho(\gamma g) \widetilde{\eta}((0, 1), 1).$$

But it would not be automorphic on $g \in \text{GL}_2(\mathbb{A})$ due to the extra power of $|\det(\gamma)|_\infty$ on the right-hand side. To solve the problem, we multiple the sum by a similar power of $|\det(g)|_\infty$. To keep the central character of the sum trivial at infinity, we divide it by the same power of $|\det(g)|_\mathbb{A}$. Therefore, define

$$B(s, g, \eta) := |\det(g)|_{\mathbb{A}_f}^{\frac{s}{2}} \sum_{u \in \mathbb{Q}^\times} B(s, g, u, \eta).$$

Here

$$|\det(g)|_{\mathbb{A}_f} = \prod_{p < \infty} |\det(g)|_p.$$

Note that the summation on u is actually a finite summation (depending on η_f and g_f). Then

$$B(s, g, \eta) = \sum_{\gamma \in N(\mathbb{Q}) \backslash \text{GL}_2(\mathbb{Q})} |\det(\gamma g)|_{\mathbb{A}_f}^{\frac{s}{2}} \delta_\infty(\gamma g)^s \, \rho(\gamma g) \widetilde{\eta}((0, 1), 1).$$

It is automorphic on $g \in \text{GL}_2(\mathbb{A})$ and has trivial central character at infinity.

THEOREM 4.20. *The series $B(s, g, u, \eta)$ and $B(s, g, \eta)$ are absolutely convergent for $\text{Re}(s) > 0$, and have meromorphic continuations to all $s \in \mathbb{C}$. In particular, both of them are holomorphic at $s = 0$ with*

$$B(0, g, u, \eta) = B(g, u, \eta), \quad B(0, g, \eta) = B(g, \eta).$$

PROOF. It suffices to prove the results for $B(s, g, u, \eta)$. We reduce it to the standard Eisenstein series of weight two for congruence subgroups in the classical language.

Start with the archimedean part. By Lemma 4.19,

$$\widetilde{\beta}((x_1, x_2), u) = \frac{1 - e^{-\pi u(x_1^2 + x_2^2)}}{4u(x_1 - ix_2)^2} 1_{\mathbb{R}_+}(u).$$

It has two parts

$$\alpha_1(x_1, x_2) = \frac{1}{4u(x_1 - ix_2)^2} 1_{\mathbb{R}_+}(u),$$

$$\alpha_2(x_1, x_2) = -\frac{e^{-\pi u(x_1^2 + x_2^2)}}{4u(x_1 - ix_2)^2} 1_{\mathbb{R}_+}(u).$$

We accordingly have

$$B(s, g, u, \eta) = B^1(s, g) + B^2(s, g)$$

where

$$B^1(s, g) = \sideset{}{'}\sum_{x_1 \in \mathbb{Q}} \sum_{x_2 \in \mathbb{Q}} \delta_\infty(\gamma_{(x_1, x_2)} g)^s \, \rho(g)(\widetilde{\eta}_f \otimes \alpha_2)((x_1, x_2), u),$$

$$B^2(s, g) = \sideset{}{'}\sum_{x_1 \in \mathbb{Q}} \sum_{x_2 \in \mathbb{Q}} \delta_\infty(\gamma_{(x_1, x_2)} g)^s \, \rho(g)(\widetilde{\eta}_f \otimes \alpha_2)((x_1, x_2), u).$$

For the theorem, it suffices to assume $u_\infty > 0$. Then we also assume $\det(g_\infty) > 0$ since every term is zero if $\det(g_\infty) < 0$. Denote $g_\infty = \begin{pmatrix} a & b \\ c & d \end{pmatrix}$ and $z = \dfrac{ai + b}{ci + d}$. Then

$$\delta_\infty(\gamma_{(x_1, x_2)} g) = \frac{\operatorname{Im}(z)^{1/2}}{|x_1 z + x_2|}.$$

By this, it is easy to see that $B^2(s, g)$ is absolutely convergent (and thus holomorphic) for any $s \in \mathbb{C}$. In particular,

$$B^2(0, g) = \sideset{}{'}\sum_{x_1 \in \mathbb{Q}} \sum_{x_2 \in \mathbb{Q}} \rho(g)(\widetilde{\eta}_f \otimes \alpha_2)((x_1, x_2), u)$$

is the contribution of α_2 in $B(g, u, \eta)$.

Now consider $B^1(s, g)$, which has the convergence trouble. We first have

$$\rho(g_\infty)\alpha_1(x_1, x_2) = -\frac{(ad - bc)}{4u(ci + d)^2(x_1 z + x_2)^2},$$

$$\widetilde{r(g_\infty)\beta}((0, 0), u) = \frac{\pi(ci - d)}{4(ci + d)}.$$

The second identity is still from Lemma 4.19. It follows that

$$
B^1(s, g)
$$
$$
= -\frac{(ad - bc)}{4u(ci + d)^2} \sum_{x_1 \in \mathbb{Q}} \sideset{}{'}\sum_{x_2 \in \mathbb{Q}} \frac{\mathrm{Im}(z)^{s/2}}{(x_1 z + x_2)^2 |x_1 z + x_2|^s} \rho(g_f) \widetilde{\eta}_f((x_1, x_2), u).
$$

Note that g and u are fixed. Write $\Phi_f(x_1, x_2) = \rho(g_f) \widetilde{\eta}_f((x_1, x_2), u)$ for simplicity. It is a Schwartz function on $\mathcal{S}(V_{\mathrm{hyp}}(\mathbb{A}_f) \times \mathbb{A}_f^\times)$. Then

$$
B^1(s, g) = -\frac{(ad - bc)}{4u(ci + d)^2} B^{11}(s, g)
$$

where

$$
B^{11}(s, g) = \sum_{x_1 \in \mathbb{Q}} \sideset{}{'}\sum_{x_2 \in \mathbb{Q}} \frac{\mathrm{Im}(z)^{s/2}}{(x_1 z + x_2)^2 |x_1 z + x_2|^s} \Phi_f(x_1, x_2).
$$

It suffices to prove similar results for $B^{11}(s, g)$, i.e., $B^{11}(s, g)$ is absolutely convergent for $\mathrm{Re}(s) > 0$ and has a meromorphic continuation to $s \in \mathbb{C}$ with

$$
B^{11}(0, g)
$$
$$
= \sum_{x_1 \in \mathbb{Q}} \sideset{}{'}\sum_{x_2 \in \mathbb{Q}} \frac{1}{(x_1 z + x_2)^2} \Phi_f(x_1, x_2) - \frac{\pi}{\mathrm{Im}(z)} \int_{\mathbb{A}_f} \int_{\mathbb{A}_f} \Phi_f(x_1, x_2) dx_1 dx_2.
$$

Here the double integral comes from the identity

$$
\widehat{\rho(g_f) \widetilde{\eta}}_f((0, 0), u) = |u|_{\mathbb{A}_f} \int_{\mathbb{A}_f} \int_{\mathbb{A}_f} \rho(g_f) \widetilde{\eta}_f((x_1, x_2), u) dx_1 dx_2.
$$

By additivity, we can assume that $\Phi_f = \otimes_p \Phi_p$ with $\Phi_p = 1_{r_p + p^{e_p} \mathbb{Z}_p^2}$ for some $r_p \in \mathbb{Q}_p^2$ at every finite place p. Of course, $r_p = 0$ and $e_p = 0$ for almost all p. By a change of variable of the form $(x_1, x_2) \to c(x_1, x_2)$ with $c \in \mathbb{Q}^\times$, we can assume that all $e_p \geq 0$ and $r_p \in \mathbb{Z}_p^2$. It follows that there are $r \in \mathbb{Z}^2$ and $N \in \mathbb{Z}_+$ such that $\Phi_p = 1_{r + N \mathbb{Z}_p^2}$ for every p.

In this case, we simply have

$$
B^{11}(s, g) = \sum_{x_1 \in r_1 + N\mathbb{Z}} \sideset{}{'}\sum_{x_2 \in r_2 + N\mathbb{Z}} \frac{\mathrm{Im}(z)^{s/2}}{(x_1 z + x_2)^2 |x_1 z + x_2|^s}.
$$

Here we write $r = (r_1, r_2)$. We need to prove the meromorphic continuation and the identity

$$
B^{11}(0, g) = \sum_{x_1 \in r_1 + N\mathbb{Z}} \sideset{}{'}\sum_{x_2 \in r_2 + N\mathbb{Z}} \frac{1}{(x_1 z + x_2)^2} - \frac{\pi}{N^2 \, \mathrm{Im}(z)}.
$$

It is exactly the result for the classical Eisenstein series of weight two for the principal congruence subgroup $\Gamma(N)$. See [Sch, §III-2, §V-3] for example. □

Proof of Proposition 4.4

Now it is easy to finish the proof of Proposition 4.4. Recall that it is reduced to equation (4.6.2), which asserts that

$$\int_{GL_2(\mathbb{Q})\backslash GL_2(\mathbb{A})/Z(\mathbb{A})}^{*} \varphi(g) \; B(g, \phi) \; dg = 0, \quad \forall \, \varphi \in \sigma, \; \phi \in \overline{\mathcal{S}}(\mathbb{B} \times \mathbb{A}^{\times}).$$

We can assume that ϕ_{∞} is standard by the action of the Lie algebra $sl_2(\mathbb{R})$.

By Theorem 4.20, it suffices to prove that, for $\mathrm{Re}(s) > 0$,

$$\int_{GL_2(\mathbb{Q})\backslash GL_2(\mathbb{A})/Z(\mathbb{A})}^{*} \varphi(g) \; B(s, g, \eta) \; dg = 0. \tag{4.7.2}$$

Here for any $g \in GL_2(\mathbb{A})$,

$$B(s, g, \eta) \;\; = \sum_{\gamma \in N(\mathbb{Q})\backslash GL_2(\mathbb{Q})} |\det(\gamma g)|_{\mathbb{A}_f}^{\frac{s}{2}} \; \delta_{\infty}(\gamma g)^s \; \rho(\gamma g)\widetilde{\eta}((0,1), 1).$$

The proof is similar to the proof of the orthogonality between Eisenstein series and cusp forms.

By definition of regularized integration in §1.6, the left-hand side of (4.7.2) is equal to

$$\int_{GL_2(\mathbb{Q})\backslash GL_2(\mathbb{A})/Z(\mathbb{A})} \fint_{Z(\mathbb{Q})\backslash Z(\mathbb{A}_f)} \varphi(zg) \; B(s, zg, \eta) \; dz \, dg$$

$$= \;\; 2 \int_{GL_2(\mathbb{Q})\backslash GL_2(\mathbb{A})/Z(\mathbb{R})} \varphi(g) \; B(s, g, \eta) \; dg.$$

By the expression of $B(s, g, \eta)$, the last integral is transformed to

$$\int_{N(\mathbb{Q})\backslash GL_2(\mathbb{A})/Z(\mathbb{R})} \varphi(g) \; |\det(g)|_{\mathbb{A}_f}^{\frac{s}{2}} \; \delta_{\infty}(g)^s \; \rho(g)\widetilde{\eta}((0,1), 1) \; dg.$$

Like a principle series,

$$|\det(g)|_{\mathbb{A}_f}^{\frac{s}{2}} \; \delta_{\infty}(g)^s \; \rho(g)\widetilde{\eta}((0,1), 1)$$

is invariant under the left action of $N(\mathbb{A})$. The integration is zero by the cuspidality condition

$$\int_{N(\mathbb{Q})\backslash N(\mathbb{A})} \varphi(ng)dn = 0.$$

It finishes the proof.

Chapter Five

Assumptions on the Schwartz Function

In this chapter and the rest of this book, we assume all the geometric assumptions in §3.6.5.

In this chapter, we impose some assumptions on the Schwartz function $\phi \in \overline{\mathcal{S}}(\mathbb{V} \times \mathbb{A}^\times)$, which we will keep from the rest of this book. These assumptions greatly simplify the computations, but imply the kernel identity for all ϕ.

In §5.1, we restate the kernel identity in terms of un-normalized kernel functions $Z(g, \phi, \chi)_U$ and $I'(0, g, \phi, \chi)_U$. It depends on U, but we always fix a U from now on. The rest of this book is to work on this version.

In §5.2, we state the assumptions. It is the key section of this chapter. We also claim in Theorem 5.7 that we can "add" these assumptions to the kernel identity without losing the generality.

In §5.3 and §5.4 we study two different classes of degenerate Schwartz functions in the assumptions. The goals are to prove Theorem 5.7.

5.1 RESTATING THE KERNEL IDENTITY

Let $\phi = \phi_f \otimes \phi_\infty \in \overline{\mathcal{S}}(\mathbb{V} \times \mathbb{A}^\times)$ be a Schwartz function. Assume that the archimedean part ϕ_∞ is standard, and that the finite part ϕ_f is invariant under the action of $K = U \times U$ for some open compact subgroup U of $\mathbb{B}_f^\times$.

5.1.1 Rewriting the analytic kernel

Since we change the space from $\mathcal{S}(\mathbb{B} \times \mathbb{A}^\times)$ to $\overline{\mathcal{S}}(\mathbb{B} \times \mathbb{A}^\times)$, the definition of the kernel function needs modification. The definition of $I(s, g, \phi)_U$ below are in the same flavor of holomorphic theta series and Eisenstein series in §4.1.

Define a mixed theta-Eisenstein series

$$I(s, g, \phi)_U := \sum_{u \in \mu_U^2 \backslash F^\times} \sum_{\gamma \in P^1(F) \backslash \mathrm{SL}_2(F)} \delta(\gamma g)^s \sum_{x_1 \in E} r(\gamma g) \phi(x_1, u). \qquad (5.1.1)$$

Here $\mu_U = F^\times \cap U$ as before. The series converges by the same reason as that for (4.1.4). It is well-defined and gives an automorphic form of $\mathrm{GL}_2(\mathbb{A})$. As in the case of $\overline{\mathcal{S}}(\mathbb{B} \times \mathbb{A}^\times)$, it is a finite linear combination of products of theta series and Eisenstein series. We do not need this fact for the moment.

If $w_U = 1$, we can rewrite it as

$$I(s, g, \phi)_U = \sum_{\gamma \in P(F) \backslash \mathrm{GL}_2(F)} \delta(\gamma g)^s \sum_{(x_1, u) \in \mu_U \backslash (E \times F^\times)} r(\gamma g) \phi(x_1, u).$$

Here the action of μ_U on $E \times F^\times$ is still $\alpha \circ (x_1, u) \mapsto (\alpha x_1, \alpha^{-2} u)$. In general, we need a factor $w_U = |U \cap \{1, -1\}|$ before $r(\gamma g) \phi(x_1, u)$ for any $(x_1, u) \in \mu_U \backslash (E^\times \times F^\times)$.

We further define the twisted average

$$I(s, g, \chi, \phi)_U := \int_{T(F) \backslash T(\mathbb{A})/Z(\mathbb{A})}^{*} I(s, g, r(t, 1)\phi)_U \, \chi(t) dt. \qquad (5.1.2)$$

Here $T = E^\times$ denotes the torus over F, $Z = F^\times$ denotes the sub-torus, and the regularized integration is introduced in (1.6.1). Note that both $\chi(t)$ and $I(s, g, r(t, 1)\phi)_U$ are invariant under both $T(F)T(F_\infty)$ as functions of $t \in T(\mathbb{A})$, so that the regularized integration is well-defined.

The definitions of $I(s, g, \phi)_U$ and $I(s, g, \chi, \phi)_U$ depend on the choice of U. In fact, if we have a different choice $U' \subset U$, then

$$I(s, g, \phi)_{U'} = [\mu_U^2 : \mu_{U'}^2] I(s, g, \phi)_U, \quad I(s, g, \chi, \phi)_{U'} = [\mu_U^2 : \mu_{U'}^2] I(s, g, \chi, \phi)_U.$$

In most of the computations, we will fix a level U.

5.1.2 Recall the height series

We first recall the definition of the generating function and the height series. Recall that the generating series

$$Z(g, \phi)_U = Z_0(g, \phi)_U + Z_*(g, \phi)_U, \quad g \in \mathrm{GL}_2(\mathbb{A}).$$

The constant term and the non-constant part are respectively

$$Z_0(g, \phi)_U = -\sum_{\alpha \in F_+^\times \backslash \mathbb{A}_f^\times / q(U)} \sum_{u \in \mu_U^2 \backslash F^\times} E_0(0, g, \alpha^{-1} u, \phi) \, L_{K,\alpha},$$

$$Z_*(g, \phi)_U = w_U \sum_{a \in F^\times} \sum_{x \in K \backslash \mathbb{B}_f^\times} r(g) \phi(x)_a \, Z(x)_U.$$

Recall that the height series are as follows:

$$Z(g, (h_1, h_2), \phi)_U = \langle Z(g, \phi)_U \, [h_1]_U^\circ, \, [h_2]_U^\circ \rangle_{\mathrm{NT}}, \quad h_1, h_2 \in \mathbb{B}^\times;$$

$$Z(g, \chi, \phi)_U = \int_{T(F) \backslash T(\mathbb{A})/Z(\mathbb{A})}^{*} Z(g, (t, 1), \phi)_U \, \chi(t) dt.$$

Their normalizations are given by

$$\widetilde{Z}(g, (h_1, h_2), \phi) = \frac{2^{[F:\mathbb{Q}]-1} h_F}{[O_F^\times : \mu_U^2] \sqrt{|D_F|}} Z(g, (h_1, h_2), \phi)_U,$$

$$\widetilde{Z}(g, \chi, \phi) = \frac{2^{[F:\mathbb{Q}]-1} h_F}{[O_F^\times : \mu_U^2] \sqrt{|D_F|}} Z(g, \chi, \phi)_U.$$

The normalized series are independent of U.

5.1.3 Rewriting the kernel identity

In terms of the kernel functions depending on U, Theorem 3.21 becomes the following result.

THEOREM 5.1. *Let $\phi \in \overline{\mathcal{S}}(\mathbb{V} \times \mathbb{A}^\times)$ be a Schwartz function bi-invariant under the action of U for some open compact subgroup U of $\mathbb{B}_f^\times$. Then for any $\varphi \in \sigma$,*

$$(I'(0, \cdot, \chi, \phi)_U, \ \varphi)_{\mathrm{Pet}} = 2\,(Z(\cdot, \chi, \phi)_U, \ \varphi)_{\mathrm{Pet}}.$$

Remark. In general, the ideal identity $I'(0, g, \chi, \phi)_U = 2Z(g, \chi, \phi)_U$ is not true. By a philosophy of Kudla, an identity of the form

$$I'(0, g, r(t_1, t_2)\phi)_U \approx 2Z(g, (t_1, t_2), \phi)_U, \quad t_1, t_2 \in T(\mathbb{A})$$

is called an arithmetic Siegel–Weil formula. However, such an equality in terms of our definitions is too good to be true even up to holomorphic projection. To have such an identity, one has to modify the definition of the right-hand side by some "natural" way. For more examples of arithmetic Siegel–Weil formulae, we refer to [KRY1, KRY2, Ku3].

Here we check that Theorem 5.1 is equivalent to Theorem 3.21. In fact, they are just stated under different normalizations. Let Φ and ϕ be as in both theorems. We will show the equalities in two theorems are equivalent assuming $\phi = \overline{\Phi}$. It suffices to show that

$$I(s, g, \chi, \Phi) = \frac{2^{[F:\mathbb{Q}]-1} h_F}{[O_F^\times : \mu_U^2]\sqrt{|D_F|}} I(s, g, \chi, \phi)_U. \tag{5.1.3}$$

Recall that

$$I(s, g, \chi, \Phi) = \int_{T(F)\backslash T(\mathbb{A})} I(s, g, r(t, 1)\Phi) \, \chi(t)dt,$$

$$I(s, g, \chi, \phi)_U = \int_{T(F)\backslash T(\mathbb{A})/Z(\mathbb{A})}^{*} I(s, g, r(t, 1)\phi)_U \, \chi(t)dt.$$

Here

$$I(s, g, \Phi) = \sum_{\gamma \in P^1(F)\backslash \mathrm{SL}_2(F)} \sum_{u \in F^\times} \delta(\gamma g)^s \sum_{x_1 \in E} r(\gamma g)\Phi(x_1, u),$$

$$I(s, g, \phi)_U = \sum_{\gamma \in P^1(F)\backslash \mathrm{SL}_2(F)} \sum_{u \in \mu_U^2 \backslash F^\times} \delta(\gamma g)^s \sum_{x_1 \in E} r(\gamma g)\phi(x_1, u).$$

We see that the proof of (5.1.3) is exactly the same as (4.4.4). We will not repeat it here.

5.2   THE ASSUMPTIONS AND BASIC PROPERTIES

5.2.1   Assumptions on the Schwartz function

Let S_F be the set of all places of F. We first have a disjoint union

$$S_F = S_\infty \cup S_{\text{nonsplit}} \cup S_{\text{split}},$$

where

- S_∞ is the set of archimedean places of F.

- S_{nonsplit} is the set of non-archimedean places of F nonsplit in E.

- S_{split} is the set of non-archimedean places of F split in E.

We further decompose it as a disjoint union

$$S_F = S_\infty \cup S_1 \cup S_2 \cup (S_{\text{nonsplit}} - S_1) \cup (S_{\text{split}} - S_2),$$

where we pick S_1 and S_2 as follows:

- S_1 is a finite subset of S_{nonsplit}, containing all places in S_{nonsplit} that are either ramified over $\mathbb{Q}$, or ramified in E, or ramified in $\mathbb{B}$, or ramified in σ, or ramified in χ. Assume that S_1 has at least two elements.

- S_2 consists of two places in S_{split} at which σ and χ are unramified.

ASSUMPTION 5.2. *The Schwartz function* $\phi = \otimes \phi_v \in \overline{\mathcal{S}}(\mathbb{B} \times \mathbb{A}^\times)$ *is a pure tensor, and* ϕ_v *is standard for any* $v \in S_\infty$. *See §4.1 for details of standard Schwartz functions.*

ASSUMPTION 5.3. *For any* $v \in S_1$, ϕ_v *lies in the space*

$$\overline{\mathcal{S}}^1(\mathbb{B}_v \times F_v^\times) := \{\phi_v \in \overline{\mathcal{S}}(\mathbb{B}_v \times F_v^\times) :$$
$$\phi_v(x, u) = 0 \text{ if } v(uq(x)) \geq -v(d_v) \text{ or } v(uq(x_2)) \geq -v(d_v)\}.$$

Here d_v *is the local different of* F *at* v, *and* x_2 *denotes the orthogonal projection of* x *in* $\mathbb{V}_{2,v} = E_v \mathbf{j}_v$.

ASSUMPTION 5.4. *For any* $v \in S_2$, ϕ_v *lies in the space*

$$\overline{\mathcal{S}}^2(\mathbb{B}_v \times F_v^\times) := \{\phi_v \in \overline{\mathcal{S}}(\mathbb{B}_v \times F_v^\times) :$$
$$r(g)\phi_v(0, u) = 0, \quad \forall \, g \in \text{GL}_2(F_v), \ u \in F_v^\times\}.$$

ASSUMPTION 5.5. *For any* $v \in S_{\text{nonsplit}} - S_1$, *assume that* ϕ_v *is the standard characteristic function of* $O_{\mathbb{B}_v} \times O_{F_v}^\times$.

ASSUMPTION 5.6. *Let* $U = \prod_v U_v$ *be an open compact subgroup of* $\mathbb{B}_f^\times$ *satisfying*

- ϕ is invariant under the action of $K = U \times U$.

- χ is invariant under the action of $U_T := U \cap T(\mathbb{A}_f)$.

- U_v is of the form $(1 + \varpi_v^r O_{\mathbb{B}_v})^{\times}$ for some $r \geq 0$ for every finite place v.

- U_v is maximal for all $v \in S_{\mathrm{nonsplit}} - S_1$ and $v \in S_2$.

- U does not contain -1.

- U is sufficiently small such that each connected component of the complex points of X_U is an unramified quotient of $\mathcal{H}$ by the complex uniformization.

We explain the last condition. Let $\tau : F \hookrightarrow \mathbb{C}$ be an embedding. Then

$$X_{U,\tau}(\mathbb{C}) = \coprod_{h \in B(\tau)_+^{\times} \backslash \mathbb{B}_f^{\times}/U} \Gamma_h \backslash \mathcal{H} \cup \{\mathrm{cusps}\}.$$

Here $\Gamma_h = B(\tau)_+^{\times} \cap hUh^{-1}$ is a discrete subgroup of $B(\tau)_+^{\times}$. Now we require that the quotient map $\mathcal{H} \to \Gamma_h \backslash \mathcal{H}$ is unramified everywhere for every τ and every h.

Notice that we do not impose any condition at places in $S_{\mathrm{split}} - S_2$. The crucial assumptions that simplify the computations are Assumption 5.3 and Assumption 5.4. In Theorem 5.7 below, we will see that those assumptions do not lose the generality of the derivative formula. The global argument using the theorem to extend the derivative formula to the general case is in the last chapter.

Assumption 5.3 will always be combined with the assumption $g \in 1_{S_1} \mathrm{GL}_2(\mathbb{A}^{S_1})$. Its major effects are as follows:

- Kill the constant term of $E'(0, g, u, \phi_1)$.

- Kill the self-intersections of CM points in the height series $Z(g, (t_1, t_2), \phi)$.

- Kill the i_v-part of the local height pairings of CM points in $Z(g, (t_1, t_2), \phi)$ for $v \in S_{\mathrm{split}}$.

- Kill the logarithmic singularities coming from both the derivatives and the local heights at v. Consequently, both k_{ϕ_v} and m_{ϕ_v} are non-singular for $v \in S_1$ so that the v-part kernel functions can be approximated by theta series.

The major effects of Assumption 5.4 are as follows:

- Kill the absolute constant term of the analytic kernel $I'(0, g, \chi, \phi)_U$ so that it satisfies the growth of the holomorphic projection.

- Kill the constant term of the generating series $Z(g, \phi)$.

- Kill the arithmetic intersections coming for the Hodge classes in the height series $Z(g, (t_1, t_2), \phi)$.

- Kill the j_v-part of the local height pairing of CM points in $Z(g, (t_1, t_2), \phi)$ for $v \in S_{\mathrm{nonsplit}}$.

5.2.2 Sufficiency of the assumptions

The main result in this chapter asserts that it suffices to prove the formula under the above assumptions.

THEOREM 5.7. *Theorem 5.1 is true for all (ϕ, U) if and only if it is true for all (ϕ, U) satisfying Assumptions 5.2–5.6.*

Let (ϕ, U) be as in Theorem 5.1. Take $\Phi \in \mathcal{S}(\mathbb{B} \times \mathbb{A}^\times)$ with $\overline{\Phi} = \phi$. Then Theorem 5.1 for (ϕ, U) is equivalent to Theorem 3.21 for Φ. It follows that we can always shrink U to meet Assumption 5.6. By Lemma 3.23 (2), Theorem 5.7 is a consequence of the following result.

PROPOSITION 5.8. *There exists a Schwartz function $\Phi = \otimes_v \Phi_v \in \mathcal{S}(\mathbb{B} \times \mathbb{A}^\times)$ satisfying the following two conditions:*

- *The Schwartz function $\phi = \overline{\Phi} \in \overline{\mathcal{S}}(\mathbb{B} \times \mathbb{A}^\times)$ satisfies Assumptions 5.2–5.5.*

- *The pairing $\alpha(\theta(\phi_f \otimes \varphi_f)) \neq 0$ for some $\varphi \in \sigma$.*

Here $\theta = \otimes_v \theta_v$ is the product of the local theta lifting

$$\theta_v : \sigma_v \otimes \mathcal{S}(\mathbb{B}_v \otimes F_v^\times) \longrightarrow \pi_v \otimes \widetilde{\pi}_v$$

normalized in (2.2.3). The bilinear form $\alpha = \otimes_v \alpha_v$ is a product of local bilinear forms $\alpha_v : \pi_v \otimes \widetilde{\pi}_v \to \mathbb{C}$ given by

$$\alpha_v(f_{1,v}, f_{2,v}) = \frac{L(1, \eta_v)L(1, \pi_v, \mathrm{ad})}{\zeta_v(2)L(\frac{1}{2}, \pi_v, \chi_v)} \int_{F_v^\times \backslash E_v^\times} (\pi_v(t) f_{1,v}, f_{2,v}) \, \chi_v(t) dt.$$

The proposition is actually local. Namely, we only need to find (Φ_v, φ_v) for each place v such that $\phi_v = \overline{\Phi}_v$ satisfies the relevant assumption and $\alpha(\theta_v(\Phi_v \otimes \varphi_v)) \neq 0$. Here we take the convention that $\overline{\Phi}_v = \Phi_v$ if v is non-archimedean.

The archimedean case is trivial. In fact, for any archimedean place v, one can take Φ_v to be any Schwartz function with $\overline{\Phi}_v$ standard, and $\varphi_v \in \sigma_v$ such that $\theta_v(\overline{\Phi}_v \otimes \varphi_v) \neq 0$. Since π is one-dimensional, this implies $\alpha(\theta_v(\overline{\Phi}_v \otimes \varphi_v)) \neq 0$.

In the case $v \in S_{\mathrm{nonsplit}} - S_1$, by Assumption 5.5 ϕ_v has to be standard. In this unramified case, $\alpha(\theta_v(\phi_v \otimes \varphi_v)) = 1$ if we take φ_v to be the spherical vector. In the case $v \in S_{\mathrm{split}} - S_2$, there is no restriction on ϕ_v so the proposition is also true. It remains to treat the case $v \in S_1$ and the case $v \in S_2$. They are exactly given in the next two sections.

5.2.3   Some simple properties of the assumptions

In the following, we state some simple results on the descriptions of $\overline{S}^1(\mathbb{B}_v \times F_v^\times)$ and $\overline{S}^2(\mathbb{B}_v \times F_v^\times)$. They will be useful when applying the assumptions above. The first result is related to Assumption 5.3.

LEMMA 5.9. *Let $v \in S_{\text{nonsplit}}$ and $\phi_v \in \overline{S}(\mathbb{B}_v \times F_v^\times)$. Assume that there is a constant $c > 0$ such that $\phi_v(x, u) = 0$ for all (x, u) with $v(uq(x_2)) \geq c$. Then we can write*

$$\phi_v = \sum_{i=1}^r \phi_{1,v}^i \otimes \phi_{2,v}^i.$$

Here for each i, $\phi_{1,v}^i \in \overline{S}(E_v \times F_v^\times)$ and $\phi_{2,v}^i \in \overline{S}(E_v \mathfrak{j} \times F_v^\times)$ satisfy $\phi_{2,v}^i(x_2, u) = 0$ for all (x_2, u) with $v(uq(x_2)) \geq c$.

PROOF. We can find non-trivial open subgroups of $A_1 \subset E_v$, $A_2 \subset O_{E_v}^\times$, $A_3 \subset O_{F_v}^\times$ such that ϕ_v is constant on any coset of $A_1 \backslash E_v \times A_2 \backslash E_v \mathfrak{j} \times A_3 \backslash F_v^\times$. Here A_1 acts by addition, and A_2, A_3 act by multiplication. It follows that we have linear combination

$$\phi_v = \sum_{a_1 \in A_1 \backslash E_v} \sum_{a_2 \in A_2 \backslash E_v \mathfrak{j}} \sum_{b \in A_3 \backslash F_v^\times} \phi_v(a_1 + a_2, b) \, 1_{(a_1+A_1) \times (A_3 b)} \otimes 1_{(A_2 a_2) \times (A_3 b)}.$$

It is essentially a finite sum. The coefficient of $1_{(a_1+A_1) \times (A_3 b)} \otimes 1_{(A_2 a_2) \times (A_3 b)}$ is nonzero only if $v(bq(a_2)) < c$. It proves the result since $v(A_3 b \, q(A_2 a_2)) = v(bq(a_2))$. □

The following result is related to Assumption 5.4. Indeed it gives an alternative description of $\overline{S}^2(\mathbb{B}_v \times F_v^\times)$.

LEMMA 5.10. *Let v be a non-archimedean place of F and $\phi_v \in \overline{S}(\mathbb{B}_v \times F_v^\times)$. The following are equivalent:*

(1) *The value*

$$r(g)\phi_v(0, u) = 0, \quad \forall\, g \in \mathrm{GL}_2(F_v),\ u \in F_v^\times.$$

(2) *The average*

$$\int_{\mathbb{B}_v(a)} r(g, h)\phi_v(x, u) dx = 0,$$

$$\forall\, g \in \mathrm{GL}_2(F_v),\ h \in \mathbb{B}_v^\times \times \mathbb{B}_v^\times,\ a \in F_v^\times,\ u \in F_v^\times.$$

PROOF. Recall that $\mathbb{B}_v(a) = \{x \in \mathbb{B}_v : q(x) = a\}$ is a homogeneous space of $\mathbb{B}_v^1 = \mathbb{B}_v(1)$, and endowed with the measure induced from the Haar measure of

$\mathbb{B}_v^1$. We first observe that we can assume $h = 1$ in (2). In fact,

$$
\int_{\mathbb{B}_v(a)} r(g,h)\phi_v(x,u)dx = \int_{\mathbb{B}_v(a)} r(g)\phi_v(h^{-1}x,\nu(h)u)dx
$$

$$
= \int_{\mathbb{B}_v(a')} r(g)\phi_v(x,u')dx
$$

with $a' = a\nu(h)^{-1}$ and $u' = u\nu(h)$.

The local Siegel-Weil gives

$$
\int_{\mathbb{B}_v(a)} r(g)\phi_v(x,u)dx = \int_{F_v} r(wn(b)g)\phi_v(0,u)\psi_v(-ab)db.
$$

It follows that (1) implies (2). Conversely, (2) implies (1) by viewing the right-hand side as the Fourier transform of $r(wn(b)g)\phi_v(0,u)$ as a function of $b \in F_v$. $\qquad\square$

Remark. Consider the case ϕ_v is a local component of a global ϕ. The Siegel Eisenstein series $E(0, g, u, \phi)$ is introduced before. Then (1) would imply the vanishing of its constant term, and (2) would imply the vanishing of all other terms.

5.3   DEGENERATE SCHWARTZ FUNCTIONS I

The goal in this section is to prove Proposition 5.8 for $v \in S_1$. We will prove slightly more general results below.

Let v be a non-archimedean place of F nonsplit in E. Recall that we have introduced

$$
\overline{\mathcal{S}}^1(\mathbb{B}_v \times F_v^\times) = \{\phi_v \in \overline{\mathcal{S}}(\mathbb{B}_v \times F_v^\times) :
$$
$$
\phi_v(x,u) = 0 \text{ if } v(uq(x)) \geq -v(d_v) \text{ or } v(uq(x_2)) \geq -v(d_v)\}.
$$

The goal here is to show that this space "generates" the whole space in the consideration of the final result. The following is the main result of this section.

PROPOSITION 5.11. *Let σ_v be an infinite dimensional irreducible representation of $\mathrm{GL}_2(F_v)$. Then for any nonzero $\mathrm{GL}_2(F_v)$-equivariant homomorphism $\overline{\mathcal{S}}(\mathbb{B}_v \times F_v^\times) \to \sigma_v$, the image of $\overline{\mathcal{S}}^1(\mathbb{B}_v \times F_v^\times)$ in σ_v is nonzero.*

It implies Proposition 5.8 for $v \in S_1$. In fact, fix any $f_{1,v} \otimes f_{2,v} \in \pi_v \otimes \widetilde{\pi}_v$ with $\alpha(f_{1,v} \otimes f_{2,v}) \neq 0$. Consider the nonzero homomorphism

$$
\alpha : \overline{\mathcal{S}}(\mathbb{B}_v \times F_v^\times) \longrightarrow \sigma_v, \quad \phi_v \longmapsto \theta'(\phi_v \otimes f_{1,v} \otimes f_{2,v}).
$$

Here

$$
\theta' : \overline{\mathcal{S}}(\mathbb{B}_v \times F_v^\times) \otimes \pi_v \otimes \widetilde{\pi}_v \longrightarrow \sigma_v
$$

is the theta lifting in the opposite direction. It is unique up to scalars. By the above proposition, we can find φ_v lies in the image of α. In other words, we have $\theta(\phi_v \otimes \varphi_v) = c f_{1,v} \otimes f_{2,v}$ for some $c \neq 0$. Then (ϕ_v, φ_v) satisfies Proposition 5.8.

In the following, we prove Proposition 5.11. For convenience, we introduce

$$\overline{\mathcal{S}}^1_{\text{weak}}(\mathbb{B}_v \times F_v^\times) := \{\phi_v \in \overline{\mathcal{S}}(\mathbb{B}_v \times F_v^\times) : \phi_v|_{(\mathbb{B}_v^{\text{sing}} \cup E_v) \times F_v^\times} = 0\}.$$

Here $\mathbb{B}_v^{\text{sing}} = \{x \in \mathbb{B}_v : q(x) = 0\}$. By compactness, for any $\phi_v \in \overline{\mathcal{S}}^1_{\text{weak}}(\mathbb{B}_v \times F_v^\times)$, there exists a constant c such that $\phi_v(x, u) = 0$ if $v(uq(x)) > c$. The same result holds for $uq(x_2)$ in the nonsplit case. Then it is easy to see that $\overline{\mathcal{S}}^1(\mathbb{B}_v \times F_v^\times)$ generates $\overline{\mathcal{S}}^1_{\text{weak}}(\mathbb{B}_v \times F_v^\times)$ under the action of the group $m(F_v^\times) \subset \mathrm{GL}_2(F_v)$ of elements $\begin{pmatrix} a & \\ & a^{-1} \end{pmatrix}$. Thus the former can be viewed as an effective version of the latter. In particular, it suffices to prove the proposition for $\overline{\mathcal{S}}^1_{\text{weak}}(\mathbb{B}_v \times F_v^\times)$.

We will prove a more general result. For simplicity, let F be a nonarchimedean local field and let (V, q) be a non-degenerate quadratic space over F of even dimension. Then we have the Weil representation of $\mathrm{GL}_2(F)$ on $\overline{\mathcal{S}}(V \times F^\times)$, the space of Schwartz–Bruhat functions on $V \times F^\times$. Let

$$\alpha : \overline{\mathcal{S}}(V \times F^\times) \to \sigma$$

be a surjective morphism to an irreducible and admissible representation σ of $\mathrm{GL}_2(F)$. We will prove the following result which obviously implies Proposition 5.11.

PROPOSITION 5.12. *Let W be a proper subspace of V of even dimension. Assume that σ is not one dimensional, and that in the case $W \neq 0$, W is nondegenerate, and that its orthogonal complement W' is anisotropic. Then there is a function $\phi \in \overline{\mathcal{S}}(V \times F^\times)$ with a nonzero image in σ such that the support $\mathrm{supp}(\phi)$ of ϕ contains only elements (x, u) such that $q(x) \neq 0$ and that*

$$W(x) := W + Fx$$

is non-degenerate of dimension $\dim W + 1$.

Let us start with the following proposition, which allows us to modify any test function to a function with support at points $(x, u) \in V \times F^\times$ with components x of nonzero norm $q(x) \neq 0$.

PROPOSITION 5.13. *Let $\phi \in \overline{\mathcal{S}}(V \times F^\times)$ be an element with a nonzero image in σ. Then there is a function $\widetilde{\phi} \in \overline{\mathcal{S}}(V \times F^\times)$ with a nonzero image in σ such that*

$$\mathrm{supp}(\widetilde{\phi}) \subset \mathrm{supp}(\phi) \cap \left(V_{q \neq 0} \times F^\times\right).$$

The key to prove this proposition is the following lemma. It is well-known but we give a proof for readers' convenience.

LEMMA 5.14. *Let σ be an irreducible admissible representation of $\mathrm{GL}_2(F)$ whose dimension is greater than one. Then the only vector in σ invariant under the action of the unipotent group $N(F)$ is zero.*

PROOF. Let v be such an invariant vector. By smoothness, it is also fixed by some compact open subgroup U of $\mathrm{SL}_2(F)$. Then v is invariant under the subgroup generated by $N(F)$ and U. It is easy to see that $U - P^1(F)$ is non-empty, and let $\gamma \in U - P^1(F)$ be one element. A basic fact asserts that $\mathrm{SL}_2(F)$ is generated by $N(F)$ and γ as long as γ is not in $P^1(F)$. It follows that v is invariant under $\mathrm{SL}_2(F)$.

If $v \neq 0$, then by irreducibility σ is generated by v under the action of $\mathrm{GL}_2(F)$. It follows that all elements of σ are invariant under $\mathrm{SL}_2(F)$. Thus the representation σ factors through the determinant map, which implies it must be one-dimensional. Contradiction! $\qquad\square$

PROOF OF PROPOSITION 5.13. Applying the lemma above, we obtain an element $b \in F$ such that

$$\sigma(n(b))\alpha(\phi) - \alpha(\phi) \neq 0.$$

The left-hand side is equal to $\alpha(\widetilde{\phi})$ with

$$\widetilde{\phi} = r(n(b))\phi - \phi.$$

By definition, we have

$$\widetilde{\phi}(x, u) = (\psi(buq(x)) - 1)\phi(x).$$

Thus such a $\widetilde{\phi}$ has support

$$\mathrm{supp}(\widetilde{\phi}) \subset \mathrm{supp}(\phi) \cap \left(V_{q\neq 0} \times F^{\times}\right).$$

$\qquad\square$

PROOF OF PROPOSITION 5.12. If $W = 0$, then the result is implied by Proposition 5.13. Thus we assume that $W \neq 0$. We have an orthogonal decomposition $V = W \oplus W'$, and an identification

$$\overline{\mathcal{S}}(V \times F^{\times}) = \overline{\mathcal{S}}(W \times F^{\times}) \otimes \overline{\mathcal{S}}(W' \times F^{\times}).$$

The action of $\mathrm{GL}_2(F)$ is given by actions on $\overline{\mathcal{S}}(W \times F^{\times})$ and $\overline{\mathcal{S}}(W' \times F^{\times})$ respectively. Choose any $\phi \in \overline{\mathcal{S}}(V \times F^{\times})$ such that $\alpha(\phi) \neq 0$. We may assume that $\phi = f \otimes f'$ is a pure tensor.

Since W' is anisotropic, $\overline{\mathcal{S}}(W'_{q\neq 0} \times F^{\times})$ is a subspace in $\overline{\mathcal{S}}(W' \times F^{\times})$ with quotient $\overline{\mathcal{S}}(F^{\times})$. The quotient map is given by evaluation at $(0, u)$. Thus

$$\overline{\mathcal{S}}(W' \times F^{\times}) = \overline{\mathcal{S}}(W'_{q\neq 0} \times F^{\times}) + r(w)\overline{\mathcal{S}}(W'_{q\neq 0} \times F^{\times})$$

as w acts as the Fourier transform up to a scale multiple. In this way, we may write

$$f' = f'_1 + r(w)f'_2, \qquad f'_i \in \overline{\mathcal{S}}(W'_{q \neq 0} \times F^\times).$$

Then we have a decomposition

$$\phi = \phi_1 + r(w)\phi_2, \quad \phi_1 := f \otimes f'_1, \ \phi_2 := r(w^{-1})f \otimes f'_2.$$

One of $\alpha(\phi_i) \neq 0$, and the support of this ϕ_i consists of points (x, u) such that $W(x)$ is non-degenerate. Applying Proposition 5.13 to this ϕ_i, we get the function we want. □

5.4   DEGENERATE SCHWARTZ FUNCTIONS II

The goal in this section is to prove Proposition 5.8 for $v \in S_2$. It is a consequence of the following Proposition 5.15.

Let v be a non-archimedean place of F split in $\mathbb{B}$. The space we are considering is

$$\overline{\mathcal{S}}^2(\mathbb{B}_v \times F_v^\times) = \{\phi_v \in \overline{\mathcal{S}}(\mathbb{B}_v \times F_v^\times) :$$
$$r(g)\phi_v(0, u) = 0, \quad \forall\, g \in \mathrm{GL}_2(F_v), \ u \in F_v^\times\}.$$

For simplicity, we abbreviate $(F_v, E_v, \mathbb{B}_v, \sigma_v, \pi_v)$ as (F, E, B, σ, π). In the following result, we only need the assumptions that B is a matrix algebra, both σ and χ are unramified, and that σ is unitary. We do not need E to be split over F in this local case.

PROPOSITION 5.15. *Then there exists a degenerate Schwartz function $\phi \in \overline{\mathcal{S}}^2(B \times F^\times)$ such that the following are true:*

(1) The function ϕ is invariant under $\mathrm{GL}_2(O_F) \times O_B^\times \times O_B^\times$.

(2) The pairing $\alpha(\theta(\phi \otimes \varphi)) \neq 0$ for any nonzero φ in the one-dimensional space $\pi^{\mathrm{GL}_2(O_F)}$.

PROOF. Since all the data are unramified, the lemma can be verified by explicit computations. However, here we provide an explicit ϕ and check it without involved computations.

We first take ϕ_0 to be the standard characteristic function of $O_B \times O_F^\times$. Then (1) and (2) are satisfied. The truth of (2) follows from the unramified theta lifting

$$\theta(\phi_0 \otimes \varphi) = f_1 \otimes f_2$$

where f_1 and f_2 are respectively nonzero vectors in the one-dimensional spaces $\pi^{O_B^\times}$ and $\widetilde{\pi}^{O_B^\times}$. The problem is that ϕ_0 does not lie in $\overline{\mathcal{S}}^2(B \times F^\times)$. We solve this problem by the action of the spherical Hecke algebra.

Let $L \in C_c^\infty(B^\times)$ be a locally constant and compactly supported function on $B^\times$. Then L acts on π by the usual Hecke actions, and it acts on $\phi' \in \overline{\mathcal{S}}(B \times F^\times)$ by

$$(L\phi')(x, u) := \int_{B^\times} \phi'(h^{-1}x, q(h)u) \ L(h)dh.$$

The integration uses the Haar measure with $\mathrm{vol}(O_B^\times) = 1$. Note that the action of L commutes with the action of $g \in \mathrm{GL}_2(F)$. It is immediate that

$$r(g)(L\phi')(0, u) = (L \ r(g)\phi')(0, u) = \int_{B^\times} r(g)\phi'(0, q(h)u) \ L(h)dh.$$

Assume that ϕ' is invariant under the left (or right) action of $O_B^\times$, and that the support of L is contained in $O_B^\times B^1 O_B^\times = B^1 O_B^\times$. Then $\phi'(0, q(h)u) = \phi'(0, u)$ for any h in the support of L. In that case,

$$r(g)(L\phi')(0, u) = \deg(L) \ r(g)\phi'(0, u),$$

where

$$\deg(L) = \int_{B^\times} L(x)dx.$$

In summary, the difference $L\phi' - \deg(L)\phi'$ lies in $\overline{\mathcal{S}}^2(B \times F^\times)$ as long as L is supported on $B^1 O_B^\times$ and ϕ' is invariant under the left (or right) action of $O_B^\times$.

Go back to the standard ϕ_0. We will find some L such that

$$\phi := L\phi_0 - \deg(L)\phi_0$$

satisfies the conditions of the proposition. In fact, assume that L is in the spherical Hecke algebra

$$\mathcal{H}_{O_B^\times} = C_c^\infty(O_B^\times \backslash B^\times / O_B^\times).$$

It makes Lf computable and ϕ satisfy (1).

The algebra $\mathcal{H}_{O_B^\times}$ acts on the one-dimensional space $\pi^{O_B^\times}$ via a character

$$\lambda_\pi : \mathcal{H}_{O_B^\times} \to \mathbb{C}.$$

As usual, denote by $T_\wp$ the characteristic function of

$$O_B^\times \begin{pmatrix} \varpi & \\ & 1 \end{pmatrix} O_B^\times.$$

Then we have

$$\lambda_\pi(T_\wp) = N^{\frac{1}{2}}(\alpha + \beta).$$

Here N is the cardinality of the residue field of F, and α and β are the Satake parameters of σ.

Therefore,

$$\alpha(\theta(L\phi \otimes \varphi)) = \alpha(Lf_1 \otimes f_2) = \lambda_\sigma(L)\, \alpha(f_1 \otimes f_2).$$

It follows that

$$\alpha(\theta(\phi \otimes \varphi)) = (\lambda_\sigma(L) - \deg(L))\, \alpha(f_1 \otimes f_2).$$

It suffices to find $L \in \mathcal{H}_{O_B^\times}$ such that $\lambda_\sigma(L) \neq \deg(L)$. Set $L(x) := (T_\varphi)^2(\varpi x)$ for any $x \in B^\times$. It is supported on $B^1 O_B^\times$, and the above inequality becomes

$$\frac{1}{\alpha\beta}(N^{\frac{1}{2}}(\alpha + \beta))^2 \neq (N+1)^2.$$

Equivalently, we want

$$\frac{\alpha}{\beta} \neq N, \frac{1}{N}.$$

It is true because σ is unramified, unitary and infinite-dimensional. In fact, any such representation arises from the obvious case $|\alpha| = |\beta| = 1$ or from a complementary series. In the latter case, the absolute values $|\alpha|, |\beta|$ are exactly in the open interval $(N^{-\frac{1}{2}}, N^{\frac{1}{2}})$. $\qquad\square$

Chapter Six

Derivative of the Analytic Kernel

Let $\phi = \phi_f \otimes \phi_\infty \in \overline{\mathcal{S}}(\mathbb{V} \times \mathbb{A}^\times)$ be a Schwartz function. Assume that the archimedean part ϕ_∞ is standard, and that the finite part ϕ_f is invariant under the action of $K = U \times U$ for some open compact subgroup U of $\mathbb{B}_f^\times$.

Recall that in §5.1 we have introduced the series

$$I(s, g, \chi, \phi)_U = \int_{T(F)\backslash T(\mathbb{A})/Z(\mathbb{A})}^* I(s, g, r(t, 1)\phi)_U \, \chi(t)dt,$$

where

$$I(s, g, \phi)_U = \sum_{u \in \mu_U^2 \backslash F^\times} \sum_{\gamma \in P^1(F)\backslash \mathrm{SL}_2(F)} \delta(\gamma g)^s \sum_{x_1 \in E} r(\gamma g)\phi(x_1, u).$$

In this chapter, we compute the derivative $I'(0, g, \chi, \phi)_U$ and its holomorphic projection $\mathcal{P}r I'(0, g, \chi, \phi)_U$. We assume all the assumptions in §5.2, which significantly simplify the results. The main content of this section is various local formulae. We usually fix U and abbreviate $I(s, g, \phi)_U$ and $I(s, g, \chi, \phi)_U$ as $I(s, g, \phi)$ and $I(s, g, \chi, \phi)$, or even as $I(s, g)$ and $I(s, g, \chi)$.

In §6.1, we decompose the kernel function $I'(0, g, \phi)$ into a sum of infinitely many local terms $I'(0, g, \phi)(v)$ indexed by places v of F nonsplit in E. Each local term is a period integral of some kernel function $\mathcal{K}^{(v)}(g, (t_1, t_2))$.

In §6.2, we deal with the v-part $I'(0, g, \phi)(v)$ for non-archimedean v. An explicit formula is given in the unramified case, and an approximation is given in the ramified case assuming the Schwartz function is degenerate.

In §6.2, we show an explicit result of the v-part $I'(0, g, \phi)(v)$ for archimedean v.

In §6.4, we review a general formula of holomorphic projection, and estimate the growth of the kernel function in order to apply the formula.

In §6.5, we compute the holomorphic projection $\mathcal{P}r I'(0, g, \chi)$ of the kernel $I'(0, g, \chi)$. When compared with the geometric kernel function, we use $\mathcal{P}r I'(0, g, \chi)$ instead of $I'(0, g, \chi)$.

6.1 DECOMPOSITION OF THE DERIVATIVE

Let $\phi \in \overline{\mathcal{S}}(\mathbb{V} \times \mathbb{A}^\times)$ be bi-invariant under the action of an open compact subgroup U of $\mathbb{B}_f^\times$. Here $\mathbb{V} = (\mathbb{B}, q)$ is determined by a set Σ of places odd cardinality containing all the archimedean places. Recall that

$$I(s, g, \phi)_U = \sum_{u \in \mu_U^2 \backslash F^\times} \sum_{\gamma \in P^1(F) \backslash \mathrm{SL}_2(F)} \delta(\gamma g)^s r(\gamma g) \phi_2(0, u) \sum_{x_1 \in E} r(\gamma g) \phi_1(x_1, u).$$

In this section we decompose the derivative $I'(0, g, \phi)_U$ into a sum over places of F where the derivative is taken. As in the case of $\mathcal{S}(\mathbb{V} \times \mathbb{A}^\times)$, the series $I(s, g, \phi)_U$ is a finite linear combination of products of theta series and Eisenstein series. So the derivative is moved to the Eisenstein series.

Recall that we have the orthogonal decomposition $\mathbb{V} = \mathbb{V}_1 \oplus \mathbb{V}_2$, where $\mathbb{V}_1 = E_\mathbb{A}$ and $\mathbb{V}_2 = E_\mathbb{A} j$. It yields a decomposition

$$\mathcal{S}(\mathbb{V} \times \mathbb{A}^\times) = \mathcal{S}(\mathbb{V}_1 \times \mathbb{A}^\times) \otimes \mathcal{S}(\mathbb{V}_2 \times \mathbb{A}^\times).$$

More precisely, any $\phi_1 \in \mathcal{S}(\mathbb{V}_1 \times \mathbb{A}^\times)$ and $\phi_2 \in \mathcal{S}(\mathbb{V}_2 \times \mathbb{A}^\times)$ gives $\phi_1 \otimes \phi_2 \in \mathcal{S}(\mathbb{V} \times \mathbb{A}^\times)$ defined by

$$(\phi_1 \otimes \phi_2)(x_1 + x_2, u) := \phi_1(x_1, u)\phi_2(x_2, u).$$

Any element of $\mathcal{S}(\mathbb{V} \times \mathbb{A}^\times)$ is a finite linear combination of functions of the form $\phi_1 \otimes \phi_2$. More importantly, the decomposition preserves Weil representation in the sense that

$$r(g, (t_1, t_2))(\phi_1 \otimes \phi_2)(x, u) = r_1(g, (t_1, t_2))\phi_1(x_1, u) \; r_2(g, (t_1, t_2))\phi_2(x_2, u)$$

for any $(g, (t_1, t_2)) \in \mathrm{GL}_2(\mathbb{A}) \times E_\mathbb{A}^\times \times E_\mathbb{A}^\times$. Here we write r_1, r_2 for the Weil representation associated to the vector spaces $\mathbb{V}_1, \mathbb{V}_2$. The group $E_\mathbb{A}^\times \times E_\mathbb{A}^\times$ acts on $\mathbb{V}_\ell$ by $(t_1, t_2) \circ x_\ell = t_1 x_\ell t_2^{-1}$. It is compatible with the action on $\mathbb{V}$.

By linearity, we may reduce the computation to the decomposable case $\phi = \phi_1 \otimes \phi_2$ and we further assume that $\phi_2 = \otimes_v \phi_{2,v}$ is a pure tensor. It follows that

$$I(s, g, \phi)_U = \sum_{u \in \mu_U^2 \backslash F^\times} \theta(g, u, \phi_1) \, E(s, g, u, \phi_2),$$

where for any $g \in \mathrm{GL}_2(\mathbb{A})$, the theta series and the Eisenstein series are given by

$$\theta(g, u, \phi_1) = \sum_{x_1 \in E} r(g)\phi_1(x_1, u),$$

$$E(s, g, u, \phi_2) = \sum_{\gamma \in P^1(F) \backslash \mathrm{SL}_2(F)} \delta(\gamma g)^s r(\gamma g)\phi_2(0, u).$$

It suffices to study the behavior of Eisenstein series at $s = 0$. Let us start with the standard Fourier expansion

$$E(s, g, u, \phi_2) = \delta(g)^s r(g)\phi_2(0, u) + \sum_{a \in F} W_a(s, g, u, \phi_2).$$

Here the Whittaker function for $a \in F$, $u \in F^\times$ is given by

$$W_a(s, g, u, \phi_2) = \int_{\mathbb{A}} \delta(wn(b)g)^s \ r(wn(b)g)\phi_2(0, u)\psi(-ab)db.$$

For each place v of F, we also introduce the local Whittaker function for $a \in F_v$, $u \in F_v^\times$ by

$$W_{a,v}(s, g, u, \phi_{2,v}) = \int_{F_v} \delta(wn(b)g)^s \ r(wn(b)g)\phi_{2,v}(0, u)\psi_v(-ab)db.$$

In the following we will suppress the dependence of the series on ϕ, ϕ_1, ϕ_2 and U.

6.1.1 Vanishing of the central value

Before taking the derivative, we recall the fact that $E(0, g, u) = 0$ and examine the local reason for that. It will inspire us to arrange the derivative according to Kudla's philosophy. The results below are considered in §2.5.1 in a slightly different setting.

We use slightly different normalizations of the local Whittaker functions by the Weil index $\gamma_{u,v} = \gamma(\mathbb{V}_{2,v}, uq)$. For $a \in F_v^\times$, denote

$$W_{a,v}^\circ(s, g, u) = \gamma_{u,v}^{-1} W_{a,v}(s, g, u).$$

Normalize the intertwining part by

$$W_{0,v}^\circ(s, g, u) = \gamma_{u,v}^{-1} \frac{L(s+1, \eta_v)}{L(s, \eta_v)} |D_v|^{-\frac{1}{2}} |d_v|^{-\frac{1}{2}} W_{0,v}(s, g, u).$$

Here we use the convention that $D_v = d_v = 1$ if v is archimedean. The normalizing factor $\dfrac{L(s+1, \eta_v)}{L(s, \eta_v)}$ has a zero at $s = 0$ when E_v is split, and is equal to π^{-1} at $s = 0$ when v is archimedean.

Globally, denote

$$W_a^\circ(s, g, u) = \prod_v W_{a,v}^\circ(s, g_v, u).$$

Then

$$W_a^\circ(s, g, u) = -W_a(s, g, u), \quad a \in F^\times, \ u \in F^\times.$$

It follows from the incoherence condition $\prod_v \gamma_{u,v} = 1$. By the functional equation for $L(s, \eta)$, we also have

$$W_0^\circ(0, g, u) = -W_0(0, g, u), \quad u \in F^\times.$$

PROPOSITION 6.1. *(1) In the sense of analytic continuation for $s \in \mathbb{C}$,*

$$W_{0,v}^{\circ}(0, g, u) = r(g)\phi_{2,v}(0, u).$$

Therefore, the global

$$W_0^{\circ}(0, g, \phi) = r(g)\phi(0, u).$$

Furthermore, for almost all places v,

$$W_{0,v}^{\circ}(s, g, u) = \delta_v(g)^{-s} r(g)\phi_{2,v}(0, u).$$

(2) Assume $a \in F_v^{\times}$.

(a) If a is not represented by $(\mathbb{V}_{2,v}, uq)$, then $W_{a,v}^{\circ}(0, g, u) = 0$.

(b) Assume that there exists $x_a \in \mathbb{V}_{2,v}$ satisfying $uq(x_a) = a$. Then

$$W_{a,v}^{\circ}(0, g, u) = \frac{1}{L(1, \eta_v)} \int_{E_v^1} r(g, h)\phi_{2,v}(x_a, u)dh.$$

Here the integration uses the Haar measure on E_v^1 normalized in §1.6.2.

If $g \in \mathrm{SL}_2(\mathbb{A})$, it follows from Proposition 2.7 for the quadratic space $(\mathbb{V}_2, uq)$. The general case is reduced to $\mathrm{SL}_2(\mathbb{A})$ by an action of $d(c)$.

PROPOSITION 6.2. *For any $\phi_2 \in \overline{\mathcal{S}}(\mathbb{V}_2 \times \mathbb{A}^{\times})$, one has $E(0, g, u) = 0$ for any $g \in \mathrm{GL}_2(\mathbb{A})$ and $u \in \mathbb{A}^{\times}$. It follows that $I(0, g, \phi) = 0$ identically.*

This is Proposition 2.8. We recall the proof a little bit. The vanishing of the constant term $E_0(0, g, u)$ is immediate. Let $a \in F^{\times}$, and consider the Whittaker function

$$W_a(0, g, u) = \prod_v W_{a,v}(0, g, u).$$

The local result asserts that $W_{a,v}(0, g, u) \neq 0$ only if a is represented by $(\mathbb{V}_{2,v}, uq)$. Then the key is the following simple result.

LEMMA 6.3. *For any $a, u \in F^{\times}$, there is a place v of F such that a is not represented by $(\mathbb{V}_{2,v}, uq)$.*

This result is very important to us, so we repeat its proof here. Denote by B the (global) quaternion algebra over F generated by E and j with relations

$$j^2 = -au^{-1}, \quad jt = \bar{t}j, \ \forall \, t \in E.$$

If au^{-1} is represented by some element x_v of $(\mathbb{V}_{2,v}, q)$ for all places v, then the map $j \mapsto x_v$ gives an isomorphism $B_{\mathbb{A}} \cong \mathbb{B}$. It contradicts the incoherence assumption on $\mathbb{B}$.

6.1.2   Decomposition of the derivative

For an element $u \in F^\times$ and a place v of F, denote by $F_u(v)$ the set of $a \in F^\times$ represented by $(\mathbb{V}_2(\mathbb{A}^v), uq)$. Then $F_u(v)$ is non-empty only if E isnonsplit at v. Note that Lemma 6.3 implies that such an a is not represented by $(\mathbb{V}_{2,v}, uq)$. By Proposition 6.1, $W^\circ_{a,v}(0, g, u) = 0$ for any $a \in F_u(v)$.

Fix a place v of Fnonsplit in E. For any $a \in F_u(v)$, taking the derivative yields

$$W^{\circ\prime}_a(0, g, u) = W^\circ_{a,v}{}'(0, g, u) W^{\circ,v}_a(0, g, u).$$

It follows that

$$E'(0, g, u) = E'_0(0, g, u) - \sum_{v \text{ nonsplit}} \sum_{a \in F_u(v)} W^\circ_{a,v}{}'(0, g, u) W^{\circ,v}_a(0, g, u).$$

Here the constant term

$$E_0(s, g, u) = \delta(g)^s r(g) \phi_2(0, u) + W_0(s, g, u).$$

In fact, if $a \in F^\times$ does not belong to $F_u(v)$ for any v, then $W^\circ_{a,w}(s, g, u)$ vanishes at $s = 0$ for at least two places w, which implies $W^{\circ\prime}_a(0, g, u) = 0$.

NOTATION 6.4. *Let v be a place*nonsplit *in E. If $\phi = \phi_1 \otimes \phi_2$ and $\phi_2 = \otimes_v \phi_{2,v}$, denote the v-part by*

$$E'(0, g, u, \phi_2)(v) := \sum_{a \in F_u(v)} W^\circ_{a,v}{}'(0, g, u, \phi_2) W^{\circ,v}_a(0, g, u, \phi_2),$$

$$I'(0, g, \phi)(v) := \sum_{u \in \mu_U^2 \backslash F^\times} \theta(g, u, \phi_1) E'(0, g, u, \phi_2)(v).$$

The definition for $I'(0, g, \phi)(v)$ extends to general ϕ by linearity.

By definition, we have a decomposition

$$I'(0, g) = - \sum_{v \text{ nonsplit}} I'(0, g)(v) + \sum_{u \in \mu_U^2 \backslash F^\times} \theta(g, u) E'_0(0, g, u).$$

We will show later that the "extra part," the second sum, is essentially zero by the assumptions in §5.2. But we first take care of $I'(0, g)(v)$ for any fixed non-split v.

Fix a place v of Fnon-split in E. Denote by $B = B(v)$ the nearby quaternion algebra. Then we have a splitting $B = E + Ej$. Let $V = (B, q)$ be the corresponding quadratic space with the reduced norm q, and $V = V_1 \oplus V_2$ be the corresponding orthogonal decomposition. We identify the quadratic spaces $V_{2,w} = \mathbb{V}_{2,w}$ unless $w = v$. Then we have

$$F_u(v) = uq(V_2) - \{0\}.$$

In fact, any $a \in F_u(v)$ is represented by $(V_{2,v}, uq)$ since $uq(V_{2,v}) - \{0\}$ and $uq(\mathbb{V}_{2,v}) - \{0\}$ are two (only) different cosets of $F_v^\times/q(E_v^\times)$. On the other hand, a is represented by $(V_2(\mathbb{A}^v), uq)$ by definition. Then the Hasse principle implies that a is represented by (V_2, uq).

In the following, the integral $\fint$ uses the Haar measure of total volume one as introduced in §1.6.

PROPOSITION 6.5. *For any place v nonsplit in E,*

$$I'(0, g, \phi)(v) = 2\fint_{Z(\mathbb{A})T(F)\backslash T(\mathbb{A})} \mathcal{K}_\phi^{(v)}(g, (t, t)) dt,$$

where for $t_1, t_2 \in T(\mathbb{A})$ and $g \in \mathrm{GL}_2(\mathbb{A})$,

$$
\begin{aligned}
\mathcal{K}_\phi^{(v)}(g, (t_1, t_2)) &= \mathcal{K}_{r(t_1,t_2)\phi}^{(v)}(g) \\
&= \sum_{u \in \mu_U^2 \backslash F^\times} \sum_{y \in V - V_1} k_{r(t_1,t_2)\phi_v}(g, y, u) r(g, (t_1, t_2)) \phi^v(y, u).
\end{aligned}
$$

Here $k_{\phi_v}(g, y, u)$ is linear in ϕ_v. In the case $\phi_v = \phi_{1,v} \otimes \phi_{2,v}$ under the orthogonal decomposition, it is given by

$$k_{\phi_v}(g, y, u) = \frac{L(1, \eta_v)}{\mathrm{vol}(E_v^1)} r(g) \phi_{1,v}(y_1, u) W_{uq(y_2),v}^{\circ}{}'(0, g, u, \phi_{2,v})$$

for any $y = y_1 + y_2 \in V_v$ with $y_2 \neq 0$ under the orthogonal decomposition $V = V_1 + V_2$.

PROOF. By linearity, it suffices to treat the case $\phi_v = \phi_{1,v} \otimes \phi_{2,v}$. By Proposition 6.1,

$$
\begin{aligned}
&E'(0, g, u)(v) \\
={}& \sum_{y_2 \in E^1 \backslash (V_2 - \{0\})} W_{uq(y_2),v}^{\circ}{}'(0, g, u) W_{uq(y_2)}^{\circ,v}(0, g, u) \\
={}& \frac{1}{L^v(1, \eta)} \sum_{y_2 \in E^1 \backslash (V_2 - \{0\})} W_{uq(y_2),v}^{\circ}{}'(0, g, u) \int_{E^1(\mathbb{A}^v)} r(g) \phi_2^v(y_2 \tau, u) d\tau \\
={}& \frac{1}{\mathrm{vol}(E_v^1) L^v(1, \eta)} \sum_{y_2 \in E^1 \backslash (V_2 - \{0\})} \int_{E^1(\mathbb{A})} W_{uq(y_2\tau),v}^{\circ}{}'(0, g, u) r(g) \phi_2^v(y_2 \tau, u) d\tau \\
={}& \frac{1}{\mathrm{vol}(E_v^1) L^v(1, \eta)} \int_{E^1 \backslash E^1(\mathbb{A})} \sum_{y_2 \in V_2 - \{0\}} W_{uq(y_2\tau),v}^{\circ}{}'(0, g, u) r(g) \phi_2^v(y_2 \tau, u) d\tau.
\end{aligned}
$$

Therefore, we have the following expression for $I'(0,g)(v)$:

$$I'(0,g)(v) = \frac{1}{\mathrm{vol}(E_v^1)L^v(1,\eta)} \sum_{u\in\mu_U^2\backslash F^\times} \sum_{y_1\in V_1} r(g)\phi_1(y_1,u)$$

$$\cdot \int_{E^1\backslash E^1(\mathbb{A})} \sum_{y_2\in V_2-\{0\}} W_{uq(y_2\tau),v}^{\circ\;\prime}(0,g,u)r(g)\phi_2^v(y_2\tau,u)d\tau.$$

Combine the two sums for y_1 and y_2 to obtain

$$\int_{Z(\mathbb{A})T(F)\backslash T(\mathbb{A})} \sum_{u\in\mu_U^2\backslash F^\times} \sum_{\substack{y=y_1+y_2\in V \\ y_2\neq 0}}$$

$$r(g)\phi_{1,v}(y_1,u)W_{uq(y_2t^{-1}\bar{t}),v}^{\circ\;\prime}(0,g,u)r(g)\phi^v(t^{-1}yt,u)dt.$$

By definition of k_{ϕ_v} and $\mathcal{K}_\phi^{(v)}$, we have

$$I'(0,g)(v)$$

$$= \frac{1}{L(1,\eta)} \int_{Z(\mathbb{A})T(F)\backslash T(\mathbb{A})} \sum_{u\in\mu_U^2\backslash F^\times} \sum_{x\in V-V_1} k_{\phi_v}(g,t^{-1}yt,u)r(g)\phi^v(t^{-1}yt,u)dt$$

$$= \frac{1}{L(1,\eta)} \int_{Z(\mathbb{A})T(F)\backslash T(\mathbb{A})} \mathcal{K}_\phi^{(v)}(g,(t,t))dt.$$

Since $\mathrm{vol}(Z(\mathbb{A})T(F)\backslash T(\mathbb{A})) = 2L(1,\eta)$, we get the result. Here we have used the relation $k_{\phi_v}(g,t^{-1}yt,u) = k_{r(t,t)\phi_v}(g,y,u)$ which follows from the definition. See also the lemma below. $\qquad\square$

LEMMA 6.6. *The function $k_{\phi_v}(g,y,u)$ behaves like Weil representation under the action of $P(F_v)$ and $E_v^\times \times E_v^\times$. Namely,*

$$\begin{aligned}
k_{\phi_v}(m(a)g,y,u) &= |a|^2 k_{\phi_v}(g,ay,u), \quad a\in F_v^\times,\\
k_{\phi_v}(n(b)g,y,u) &= \psi(buq(y))k_{\phi_v}(g,y,u), \quad b\in F_v,\\
k_{\phi_v}(d(c)g,y,u) &= |c|^{-1}k_{\phi_v}(g,y,c^{-1}u), \quad c\in F_v^\times,\\
k_{r(t_1,t_2)\phi_v}(g,y,u) &= k_{\phi_v}(g,t_1^{-1}yt_2,q(t_1t_2^{-1})u), \quad (t_1,t_2)\in E_v^\times\times E_v^\times.
\end{aligned}$$

PROOF. These identities follow from the definition of Weil representation and some simple transformation of integrals. It suffices to assume that $\phi_v = \phi_{1,v}\otimes\phi_{2,v}$ by linearity. Then it follows from similar results for the Whittaker function $W_{a,v}(s,g,u)$. We omit the proof. $\qquad\square$

Now we take care of the contribution from the constant term $E_0'(0,g,u)$. It will simply vanish in some degenerate cases.

PROPOSITION 6.7. *Under Assumption 5.3,*

$$I'(0,g,\phi) = - \sum_{v\text{ nonsplit}} I'(0,g,\phi)(v), \quad \forall g\in P(F_{S_1})\mathrm{GL}_2(\mathbb{A}^{S_1}).$$

PROOF. By the assumption, $\phi_v(E_v, F_v^\times) = 0$ for any $v \in S_1$. By Lemma 5.9, we can simply assume that $\phi = \phi_1 \otimes \phi_2$ with $\phi_{2,v}(0, F_v^\times) = 0$ for any $v \in S_1$.

We will check that

$$E_0'(0, g, u) = \log \delta(g) r_2(g) \phi_2(0, u) + W_0'(0, g, u)$$

vanishes at $g \in P(F_{S_1})\mathrm{GL}_2(\mathbb{A}^{S_1})$. It is immediate that $r_2(g)\phi_2(0, u) = 0$ for $g \in P(F_{S_1})\mathrm{GL}_2(\mathbb{A}^{S_1})$. It remains to treat $W_0'(0, g, u)$.

Take the derivative on

$$W_0(s, g, u) = -\frac{L(s, \eta)}{L(s+1, \eta)} W_0^\circ(s, g, u) \prod_v |D_v|^{\frac{1}{2}} |d_v|^{\frac{1}{2}}$$

$$= -\frac{L(s, \eta)/L(0, \eta)}{L(s+1, \eta)/L(1, \eta)} \prod_v W_{0,v}^\circ(s, g, u).$$

We obtain

$$W_0'(0, g, u) = -\frac{d}{ds}\Big|_{s=0} \left(\log \frac{L(s, \eta)}{L(s+1, \eta)}\right) W_0^\circ(0, g, u)$$
$$- \sum_v W_{0,v}^\circ{}'(0, g, u) \prod_{v' \neq v} W_{0,v'}^\circ(0, g, u).$$

By Proposition 6.1, we get

$$W_0'(0, g, u) = -\frac{d}{ds}\Big|_{s=0} \left(\log \frac{L(s, \eta)}{L(s+1, \eta)}\right) r(g) \phi_2(0, u)$$
$$- \sum_v W_{0,v}^\circ{}'(0, g, u) r(g^v) \phi_2^v(0, u).$$

Then $r(g)\phi_2(0, u) = 0$ as above, and $r(g^v)\phi_2^v(0, u) = 0$ for any v since it has a factor $r(g_{v'})\phi_{2,v'}(0, u) = 0$ for any $v' \in S_1 - \{v\}$. □

6.2   NON-ARCHIMEDEAN COMPONENTS

Assume that v is a non-archimedean placenonsplit in E. Resume the notations in the last section. We now consider the local kernel function $k_{\phi_v}(g, y, u)$, which has the expression

$$k_{\phi_v}(g, y, u) = \frac{L(1, \eta_v)}{\mathrm{vol}(E_v^1)} r(g) \phi_{1,v}(y_1, u) W_{uq(y_2),v}^\circ{}'(0, g, u, \phi_{2,v}),$$

$$y = y_1 + y_2 \in V_v - V_{1v},$$

if $\phi_v = \phi_{1,v} \otimes \phi_{2,v}$.

6.2.1   Main results

Let v be a non-archimedean placenonsplit in E, and B_v be the quaternion division algebra over F_v non-isomorphic to $\mathbb{B}_v$.

PROPOSITION 6.8. *The following results are true:*

(1) Assume that $v \in S_{\mathrm{nonsplit}} - S_1$ is unramified as in Assumption 5.5. Then

$$k_{\phi_v}(1, y, u) = 1_{O_{B_v} \times O_{F_v}^\times}(y, u) \frac{v(q(y_2)) + 1}{2} \log N_v.$$

(2) Assume that $v \in S_1$ so that $\phi_v \in \overline{\mathcal{S}}^1(\mathbb{B}_v \times F_v^\times)$ satisfies Assumption 5.3. Then $k_{\phi_v}(1, y, u)$ extends to a Schwartz function of $(y, u) \in B_v \times F_v^\times$.

We consider its consequences. We first look at (1). In that unramified case, it is easy to see that

$$k_{\phi_v}(1, y, u) = k_{\phi_v}(g, y, u), \quad \forall\, g \in \mathrm{GL}_2(O_{F_v}).$$

Then by Iwasawa decomposition and Lemma 6.6, we know $k_{r(t_1,t_2)\phi_v}(g, y, u)$ explicitly for all $(g, (t_1, t_2))$. It will cancel the local height of CM points at v.

Now we consider a place v that does not satisfy the conditions in (1). Then the computation of k_{ϕ_v} may be very complicated or useless. It is better to consider the whole series

$$\mathcal{K}_\phi^{(v)}(g, (t_1, t_2)) = \sum_{u \in \mu_U^2 \backslash F^\times} \sum_{y \in V - V_1} k_{r(t_1,t_2)\phi_v}(g, y, u)\, r(g, (t_1, t_2))\phi^v(y, u).$$

It looks like a theta series. We call it a *pseudo-theta series*. It has a strong connection with the usual theta series.

In (2), we have shown that $k_{\phi_v}(y, u) := k_{\phi_v}(1, y, u)$ extends to a Schwartz function for $(y, u) \in V_v \times F_v^\times$ under Assumption 5.3. We did this because we want to compare the above pseudo-theta series with the usual theta series

$$\theta(g, (t_1, t_2), k_{\phi_v} \otimes \phi^v)$$
$$= \sum_{u \in \mu_U^2 \backslash F^\times} \sum_{y \in V} r(g, (t_1, t_2))k_{\phi_v}(y, u)\, r(g, (t_1, t_2))\phi^v(y, u).$$

It seems that these two series have a good chance to equal if $g_v = 1$. In fact, it is supported by the equality

$$r(t_1, t_2)k_{\phi_v}(y, u) = k_{r(t_1,t_2)\phi_v}(1, y, u)$$

shown in Lemma 6.6.

Another difficulty for them to be equal is that the summations of y are over slightly different spaces. This problem is also solved by Assumption 5.3. In fact, the assumption implies that $r(t_1, t_2)\phi^v(y, u) = 0$ for all $y \in E$, since it has a factor $r(t_1, t_2)\phi_{v'}(y, u) = 0$ for any $v' \in S_1 - \{v\}$. Therefore, the two series are equal if $g_{S_1} = 1$. We can also extend the equality to $P(F_{S_1})\mathrm{GL}_2(\mathbb{A}^{S_1})$ by Lemma 6.6.

COROLLARY 6.9. *Let $v \in S_1$ so that $\phi_v \in \overline{\mathcal{S}}^1(\mathbb{B}_v \times F_v^\times)$ satisfies Assumption 5.3. Then*

$$\mathcal{K}_\phi^{(v)}(g, (t_1, t_2)) = \theta(g, (t_1, t_2), k_{\phi_v} \otimes \phi^v)$$

for all

$$(g, (t_1, t_2)) \in P(F_{S_1})\mathrm{GL}_2(\mathbb{A}^{S_1}) \times T(\mathbb{A}) \times T(\mathbb{A}).$$

In that situation, we say $\mathcal{K}_\phi^{(v)}$ is *approximated* by $\theta(k_{\phi_v} \otimes \phi^v)$. They are usually not equal, unless we know the modularity of the pseudo-theta series.

6.2.2   The computation

To prove Proposition 6.8, we first show a formula for the Whittaker function $W_{a,v}^\circ(s, 1, u)$ in the most general case.

PROPOSITION 6.10. *Let v be any non-archimedean place of F.*

(1) For any $a \in F_v$,

$$W_{a,v}^\circ(s, 1, u) = |d_v|^{\frac{1}{2}}(1 - N_v^{-s}) \sum_{n=0}^\infty N_v^{-ns+n} \int_{D_n(a)} \phi_{2,v}(x_2, u)d_ux_2,$$

where d_ux_2 is the self-dual measure of $(\mathbb{V}_{2,v}, uq)$ and

$$D_n(a) = \{x_2 \in \mathbb{V}_{2,v} : uq(x_2) \in a + p_v^n d_v^{-1}\}.$$

(2) Assume that $\phi_{2,v}(x_2, u) = 0$ if $v(uq(x_2)) > -v(d_v)$. Then there is a constant $c > 0$ such that $W_{a,v}^\circ(s, 1, u) = 0$ identically for all $a \in F_v$ satisfying $v(a) > c$ or $v(a) < -c$.

PROOF. We first compute (1). Recall that

$$W_{a,v}(s, g, u) = \int_{F_v} \delta(wn(b)g)^s \, r(wn(b)g)\phi_{2,v}(0, u)\psi_v(-ab)db.$$

Expand the action of w in terms of the Fourier transform. We obtain

$$W_{a,v}(s, g, u) = \gamma_{u,v} \int_{F_v} \delta(wn(b)g)^s \int_{\mathbb{V}_{2,v}} r(g)\phi_{2,v}(x_2, u)\psi_v(b(uq(x_2) - a))d_ux_2db.$$

Here d_ux_2 is the self-dual measure for $(\mathbb{V}_{2,v}, uq)$. It follows that

$$W_{a,v}^\circ(s, 1, u) = \int_{F_v} \delta(wn(b))^s \int_{\mathbb{V}_{2,v}} \phi_{2,v}(x_2, u)\psi_v(b(uq(x_2) - a))d_ux_2db.$$

It suffices to verify the formulae for Whittaker functions under the condition that $u = 1$. The general case is obtained by replacing q by uq and $\phi_{2,v}(x_2)$ by

$\phi_{2,v}(x_2, u)$. We will drop the dependence on u to simplify the notation. Then we write

$$W_{a,v}^{\circ}(s,1) = \int_{F_v} \delta(wn(b))^s \int_{\mathbb{V}_{2,v}} \phi_{2,v}(x_2)\psi_v(b(q(x_2) - a))dx_2 db.$$

By

$$\delta(wn(b)) = \begin{cases} 1 & \text{if } b \in O_{F_v}, \\ |b|^{-1} & \text{otherwise}, \end{cases}$$

we will split the integral over F_v into the sum of an integral over O_{F_v} and an integral over $F_v - O_{F_v}$. Then

$$W_{a,v}^{\circ}(s,1) = \int_{O_{F_v}} \int_{\mathbb{V}_{2,v}} \phi_{2,v}(x_2)\psi_v(b(q(x_2) - a))dx_2 db$$

$$+ \int_{F_v - O_{F_v}} |b|^{-s} \int_{\mathbb{V}_{2,v}} \phi_{2,v}(x_2)\psi_v(b(q(x_2) - a))dx_2 db.$$

The second integral can be decomposed as

$$\sum_{n=1}^{\infty} \int_{p_v^{-n} - p_v^{-n+1}} N_v^{-ns} \int_{\mathbb{V}_{2,v}} \phi_{2,v}(x_2)\psi_v(b(q(x_2) - a))dx_2 db$$

$$= \sum_{n=1}^{\infty} \int_{p_v^{-n}} N_v^{-ns} \int_{\mathbb{V}_{2,v}} \phi_{2,v}(x_2)\psi_v(b(q(x_2) - a))dx_2 db$$

$$- \sum_{n=1}^{\infty} \int_{p_v^{-(n-1)}} N_v^{-ns} \int_{\mathbb{V}_{2,v}} \phi_{2,v}(x_2)\psi_v(b(q(x_2) - a))dx_2 db.$$

Combine with the first integral to obtain

$$W_{a,v}^{\circ}(s,1) = \sum_{n=0}^{\infty} \int_{p_v^{-n}} N_v^{-ns} \int_{\mathbb{V}_{2,v}} \phi_{2,v}(x_2)\psi_v(b(q(x_2) - a))dx_2 db$$

$$- \sum_{n=0}^{\infty} \int_{p_v^{-n}} N_v^{-(n+1)s} \int_{\mathbb{V}_{2,v}} \phi_{2,v}(x_2)\psi_v(b(q(x_2) - a))dx_2 db$$

$$= (1 - N_v^{-s}) \sum_{n=0}^{\infty} N_v^{-ns} \int_{p_v^{-n}} \int_{\mathbb{V}_{2,v}} \phi_{2,v}(x_2)\psi_v(b(q(x_2) - a))dx_2 db.$$

As for the last double integral, change the order of the integration. The integral on b is nonzero if and only if $q(x_2) - a \in p_v^n d_v^{-1}$. Here d_v is the local different of F over $\mathbb{Q}$, and also the conductor of ψ_v. Then we have

$$W_{a,v}^{\circ}(s,1) \quad = \quad (1 - N_v^{-s}) \sum_{n=0}^{\infty} N_v^{-ns} \text{vol}(p_v^{-n}) \int_{D_n(a)} \phi_{2,v}(x_2)dx_2$$

$$= \quad |d_v|^{\frac{1}{2}}(1 - N_v^{-s}) \sum_{n=0}^{\infty} N_v^{-ns+n} \int_{D_n(a)} \phi_{2,v}(x_2)dx_2.$$

It proves (1).

Now we show (2) using (1). The key is that only those $D_n(a)$ with $n \geq 0$ are involved in the formula. Recall that

$$D_n(a) = \{x_2 \in \mathbb{V}_{2,v} : uq(x_2) \in a + p_v^n d_v^{-1}\}.$$

If $v(a) < -v(d_v)$, then for every $x_2 \in D_n(a)$, we have $v(uq(x_2)) = v(a)$. Then $\phi_{2,v}(x_2, u) = 0$ if $v(a)$ is too small. It follows that $W_{a,v}^\circ(s, 1, u) = 0$ if $v(a)$ is too small. This is apparently true for all Schwartz function ϕ_v.

If $v(a) \geq -v(d_v)$, then for every $x_2 \in D_n(a)$, we have $v(uq(x_2)) \geq -v(d_v)$. By the assumption, $\phi_{2,v}(\cdot, u)$ is zero on $D_n(a)$. In that case, $W_{a,v}^\circ(s, 1, u) = 0$ identically. It proves the result. $\qquad\square$

PROOF OF PROPOSITION 6.8. Both results are obtained as a consequence of Proposition 6.10. We first look at (2). By linearity and Lemma 5.9, it suffices to consider the case that $\phi_v = \phi_{1,v} \otimes \phi_{2,v}$ with $\phi_{2,v}$ satisfies the condition of Proposition 6.10 (2). Set $k_{\phi_v}(1, y, u)$ to be zero if $y_2 = 0$. It is easy to see that it gives a Schwartz function by Proposition 6.10 (2).

Now we consider (1). It suffices to show that for any $a \in F_u(v)$,

$$W_{a,v}^{\circ\,\prime}(0, 1, u) = 1_{O_{F_v}}(a) 1_{O_{F_v}^\times}(u) \frac{v(a) + 1}{2}(1 + N_v^{-1}) \log N_v.$$

Use the formula in Proposition 6.10. We need to simplify

$$D_n(a) = \{x_2 \in \mathbb{V}_{2,v} : uq(x_2) - a \in p_v^n\}.$$

We first have $v(a) \neq v(q(x_2))$ because a is not represented by $uq(x_2)$. Actually $v(q(x_2))$ is always even and $v(a)$ must be odd. Then

$$v(q(x_2) - a) = \min\{v(a), v(q(x_2))\}, \quad \forall\, x_2 \in \mathbb{V}_{2,v}.$$

We see that $D_n(a)$ is empty if $v(a) < n$. Otherwise, it is equal to

$$D_n := \{x_2 \in \mathbb{V}_{2,v} : uq(x_2) \in p_v^n\}.$$

It follows that

$$W_{a,v}^\circ(s, 1, u) \;=\; (1 - N_v^{-s}) \sum_{n=0}^{v(a)} N_v^{-ns+n} \int_{D_n} \phi_{2,v}(x_2, u) d_u x_2.$$

It is a finite sum and we do not have any convergence problem. Then

$$W_{a,v}^{\circ\,\prime}(0, 1, u) \;=\; \log N_v \sum_{n=0}^{v(a)} N_v^n \int_{D_n} \phi_{2,v}(x_2, u) d_u x_2.$$

It is nonzero only if $u \in O_{F_v}^\times$ and $a \in O_{F_v}$. Identify $\mathbb{V}_{2,v}$ with E_v. Then

$$D_n = \{x_2 \in E_v : q(x_2) \in p_v^n\} = p_v^{\lceil \frac{n+1}{2} \rceil} O_{E_v},$$

and $\mathrm{vol}(D_n) = N_v^{-2[\frac{n+1}{2}]}$. Note that $v(a)$ is odd since it is not represented by q_2. Then it is easy to have

$$W_{a,v}^{\circ}{}'(0, 1, u) = \log N_v \sum_{n=0}^{v(a)} N_v^{n-2[\frac{n+1}{2}]} = \frac{v(a)+1}{2}(1 + N_v^{-1}).$$

□

6.3   ARCHIMEDEAN COMPONENTS

For an archimedean place v, the quaternion algebra $\mathbb{B}_v$ is isomorphic to the Hamiltonian quaternion. We will compute $k_{\phi_v}(g, y, u)$ for standard ϕ_v introduced in §4.1. The computation here is done by Proposition 2.11.

The result involves the exponential integral Ei defined by

$$\mathrm{Ei}(z) = \int_{-\infty}^{z} \frac{e^t}{t} dt, \quad z \in \mathbb{C}.$$

Another expression is

$$\mathrm{Ei}(z) = \gamma + \log(-z) + \int_0^z \frac{e^t - 1}{t} dt,$$

where γ is the Euler constant. It follows that it has a logarithmic singularity near 0. This fact is useful when we compare the result here with the archimedean local height, since we know that Green's functions have a logarithmic singularity.

PROPOSITION 6.11.

$$k_{\phi_v}(g, y, u) = \begin{cases} -\dfrac{1}{2}\mathrm{Ei}(4\pi uq(y_2)y_0)\, |y_0|e^{2\pi iuq(y)(x_0+iy_0)}e^{2i\theta} & \text{if } uy_0 > 0, \\ 0 & \text{if } uy_0 < 0, \end{cases}$$

for any

$$g = \begin{pmatrix} z_0 \\ & z_0 \end{pmatrix} \begin{pmatrix} y_0 & x_0 \\ & 1 \end{pmatrix} \begin{pmatrix} \cos\theta & \sin\theta \\ -\sin\theta & \cos\theta \end{pmatrix} \in \mathrm{GL}_2(F_v)$$

in the form of the Iwasawa decomposition.

PROOF. It suffices to show the formula in the case $g = 1$:

$$k_{\phi_v}(1, y, u) = \begin{cases} -\frac{1}{2}\mathrm{Ei}(4\pi uq(y_2))e^{-2\pi uq(y)} & \text{if } u > 0; \\ 0 & \text{if } u < 0. \end{cases}$$

The general case is obtained by Proposition 6.6 and the fact that $r(k_\theta)\phi_v = e^{2i\theta}\phi_v$.

It amounts to show that, for any $a \in F_u(v)$ (i.e., $ua < 0$),

$$W_{a,v}^{\circ}{}'(0,1,u) = \begin{cases} -\pi e^{-2\pi a} \mathrm{Ei}(4\pi a) & \text{if } u > 0; \\ 0 & \text{if } u < 0. \end{cases}$$

The case $u < 0$ is trivial. The essential case is $u > 0$ and $a < 0$. It is just the case of $d = 2$ in Proposition 2.11 (3). □

6.4 HOLOMORPHIC PROJECTION

In this section we consider the general theory of holomorphic projection which we will apply to the form $I'(0, g, \chi)$ in the next section. Denote by $\mathcal{A}(\mathrm{GL}_2(\mathbb{A}), \omega)$ the space of automorphic forms of central character ω, by $\mathcal{A}_0(\mathrm{GL}_2(\mathbb{A}), \omega)$ the subspace of cusp forms, and by $\mathcal{A}_0^{(2)}(\mathrm{GL}_2(\mathbb{A}), \omega)$ the subspace of holomorphic cusp forms of parallel weight two.

The usual Petersson inner product is just

$$(f_1, f_2)_{\mathrm{pet}} = \int_{Z(\mathbb{A})\mathrm{GL}_2(F)\backslash\mathrm{GL}_2(\mathbb{A})} f_1(g)\overline{f_2(g)}dg, \quad f_1, f_2 \in \mathcal{A}(\mathrm{GL}_2(\mathbb{A}), \omega).$$

Denote by $\mathcal{P}r : \mathcal{A}(\mathrm{GL}_2(\mathbb{A}), \omega) \to \mathcal{A}_0^{(2)}(\mathrm{GL}_2(\mathbb{A}), \omega)$ the orthogonal projection. Namely, for any $f \in \mathcal{A}(\mathrm{GL}_2(\mathbb{A}), \omega)$, the image $\mathcal{P}r(f)$ is the unique form in $\mathcal{A}_0^{(2)}(\mathrm{GL}_2(\mathbb{A}), \omega)$ such that

$$(\mathcal{P}r(f), \varphi)_{\mathrm{pet}} = (f, \varphi)_{\mathrm{pet}}, \quad \forall \varphi \in \mathcal{A}_0^{(2)}(\mathrm{GL}_2(\mathbb{A}), \omega).$$

We simply call $\mathcal{P}r(f)$ the holomorphic projection of f. Apparently $\mathcal{P}r(f) = 0$ if f is an Eisenstein series.

6.4.1 A general formula

For any automorphic form f for $\mathrm{GL}_2(\mathbb{A})$ we define a Whittaker function

$$f_{\psi,s}(g) = (4\pi)^{[F:\mathbb{Q}]} W^{(2)}(g_\infty) \int_{Z(F_\infty)N(F_\infty)\backslash\mathrm{GL}_2(F_\infty)} \delta(h)^s f_\psi(gh)\overline{W^{(2)}(h)}dh.$$

Here $W^{(2)}$ is the standard holomorphic Whittaker function of weight two at infinity, and f_ψ denotes the Whittaker function of f with respect to the character $\psi : F\backslash\mathbb{A} \to \mathbb{C}^\times$.

PROPOSITION 6.12. *Let* $f \in \mathcal{A}(\mathrm{GL}_2(\mathbb{A}), \omega)$ *be a form with asymptotic behavior*

$$f\left(\begin{pmatrix} a & 0 \\ 0 & 1 \end{pmatrix} g\right) = O_g(|a|_{\mathbb{A}}^{1-\epsilon})$$

as $a \in \mathbb{A}^\times$, $|a|_{\mathbb{A}} \to \infty$ *for some* $\epsilon > 0$. *Then the holomorphic projection* $\mathcal{P}r(f)$ *has the Whittaker function*

$$\mathcal{P}r(f)_\psi(g) = \lim_{s\to 0} f_{\psi,s}(g).$$

PROOF. For any Whittaker function W of $\mathrm{GL}_2(\mathbb{A})$ with decomposition $W(g) = W^{(2)}(g_\infty)W_f(g)$ such that $W^{(2)}(g_\infty)$ is standard holomorphic of weight 2 and that $W_f(g)$ is compactly supported modulo $Z(\mathbb{A}_f)N(\mathbb{A}_f)$, the Poincaré series is defined as

$$\varphi_W(g) := \lim_{s\to 0+} \sum_{\gamma \in Z(F)N(F)\backslash G(F)} W(\gamma g)\delta_\infty(\gamma g)^s,$$

where

$$\delta_\infty(g) = |a_\infty/d_\infty|^{\frac{1}{2}}, \qquad g_\infty = \begin{pmatrix} a_\infty & b_\infty \\ 0 & d_\infty \end{pmatrix} k_\infty$$

with k_∞ the maximal compact subgroup of $\mathrm{GL}_2(F_\infty)$.

Assume that W and f have the same central character. Since f has asymptotic behavior as in the proposition, their inner product can be computed as follows:

$$
\begin{aligned}
(f, \varphi_W)_{\mathrm{pet}} &= \int_{Z(\mathbb{A})\mathrm{GL}_2(F)\backslash \mathrm{GL}_2(\mathbb{A})} f(g)\overline{\varphi_W(g)}dg \\
&= \lim_{s\to 0} \int_{Z(\mathbb{A})N(F)\backslash \mathrm{GL}_2(\mathbb{A})} f(g)\overline{W(g)}\delta_\infty(g)^s dg \\
&= \lim_{s\to 0} \int_{Z(\mathbb{A})N(\mathbb{A})\backslash \mathrm{GL}_2(\mathbb{A})} f_\psi(g)\overline{W(g)}\delta_\infty(g)^s dg.
\end{aligned}
$$

We may apply this formula to $\mathcal{P}r(f)$ which has the same inner product with φ_W as f. Write

$$\mathcal{P}r(f)_\psi(g) = W^{(2)}(g_\infty)\mathcal{P}r(f)_\psi(g).$$

Then the above integral is a product of integrals over finite places and integrals at infinite places:

$$\int_{Z(\mathbb{R})N(\mathbb{R})\backslash \mathrm{GL}_2(\mathbb{R})} |W^{(2)}(g)|^2 dg = \int_0^\infty y^2 e^{-4\pi y} dy/y^2 = (4\pi)^{-1}.$$

In other words, we have

$$(f, \varphi_W)_{\mathrm{pet}} = (4\pi)^{-[F:\mathbb{Q}]} \int_{Z(\mathbb{A}_f)N(\mathbb{A}_f)\backslash \mathrm{GL}_2(\mathbb{A}_f)} \mathcal{P}r(f)_\psi(g)\overline{W_f(g)}dg.$$

As $\overline{W}_f$ can be any Whittaker function with compact support modulo $Z(\mathbb{A}_f)N(\mathbb{A}_f)$, the combinations of the above formulae give the proposition.   □

We introduce an operator $\mathcal{P}r'$ formally defined on the function space of $N(F)\backslash \mathrm{GL}_2(\mathbb{A})$. For any function $f : N(F)\backslash \mathrm{GL}_2(\mathbb{A}) \to \mathbb{C}$, denote as above

$$f_{\psi,s}(g) = (4\pi)^{[F:\mathbb{Q}]} W^{(2)}(g_\infty) \int_{Z(F_\infty)N(F_\infty)\backslash \mathrm{GL}_2(F_\infty)} \delta(h)^s f_\psi(gh)\overline{W^{(2)}(h)}dh$$

if it has meromorphic continuation around $s = 0$. Here f_ψ denotes the first Fourier coefficient of f. Denote

$$\mathcal{P}r'(f)_\psi(g) = \widetilde{\lim}_{s \to 0} f_{\psi,s}(g),$$

where the "quasi-limit" $\widetilde{\lim}_{s \to 0}$ denotes the constant term of the Laurent expansion at $s = 0$. Finally, we write

$$\mathcal{P}r'(f)(g) = \sum_{a \in F^\times} \mathcal{P}r'(f)_\psi(d^*(a)g).$$

The above result is just $\mathcal{P}r(f) = \mathcal{P}r'(f)$ under the growth condition. In general, $\mathcal{P}r'(f)$ is not automorphic when f is automorphic but fails the growth condition of Proposition 6.12.

6.4.2   Growth of the kernel function

Now we want to consider the growth of $I'(0, g, \chi, \phi)$ so that we can apply the formula. The growth condition in Proposition 6.12 is not satisfied for general $\phi \in \overline{\mathcal{S}}(\mathbb{B} \times \mathbb{A}^\times)$. However, we will prove that Assumption 5.4 miraculously fits the growth condition!

Recall from (5.1.2) that

$$I(s, g, \chi, \phi) = \int_{T(F)\backslash T(\mathbb{A})/Z(\mathbb{A})}^* I(s, g, r(t, 1)\phi) \, \chi(t) dt.$$

It has central character $\chi|_{\mathbb{A}_F^\times} = \omega_\sigma^{-1}$. The integral is essentially a finite sum. So the growth of $I'(0, g, \chi, \phi)$ is reduced to the growth of $I'(0, g, \phi)$.

If $\phi = \phi_1 \otimes \phi_2$, we have

$$I(s, g, \phi) = \sum_{u \in \mu_U^2 \backslash F^\times} \theta(g, u, \phi_1) E(s, g, u, \phi_2).$$

Denote the "absolute constant term"

$$I_{00}(s, g, \phi) := \sum_{u \in \mu_U^2 \backslash F^\times} I_{00}(s, g, u, \phi)$$

where

$$I_{00}(s, g, u, \phi) := \theta_0(g, u, \phi_1) E_0(s, g, u, \phi_2)$$

is the product of the constant terms of $\theta(g, u, \phi_1)$ and $E(s, g, u, \phi_2)$. Equivalently,

$$I_{00}(s, g, u, \phi) = \delta(g)^s r(g)\phi(0, u) + r_1(g)\phi_1(0, u)W_0(s, g, u, \phi_2).$$

Taking integration, we obtain

$$I_{00}(s, g, \chi, \phi) = \int_{T(F)\backslash T(\mathbb{A})/Z(\mathbb{A})}^* I_{00}(s, g, r(t, 1)\phi) \, \chi(t) dt.$$

Note that we use the notation I_{00} since it is only a part of the constant term I_0 of I.

For general $\phi \in \overline{\mathcal{S}}(\mathbb{B} \times \mathbb{A}^\times)$, extend the definition of $I_{00}(s, g, u, \phi)$ and $I_{00}(s, g, \phi)$ by linearity. Alternatively, we can use the formula

$$I_{00}(s, g, u, \phi) = \delta(g)^s r(g)\phi(0, u)$$
$$- \int_{\mathbb{A}} \delta(wn(b)g)^s \int_{V_2} r(g)\phi(x_2, u)\psi(buq(x_2))d_u x_2 db.$$

LEMMA 6.13. *For any $\phi \in \overline{\mathcal{S}}(\mathbb{B} \times \mathbb{A}^\times)$, the difference*

$$I'(0, g, \chi, \phi) - I'_{00}(0, g, \chi, \phi)$$

decays exponentially in the sense that there exists $\epsilon_g > 0$ depending on g such that

$$I'(0, d^*(a)g, \chi, \phi) - I'_{00}(0, d^*(a)g, \chi, \phi) = O(e^{-\epsilon_g |a|_\mathbb{A}}), \quad a \in \mathbb{A}^\times.$$

PROOF. It suffices to consider the case $\phi = \phi_1 \otimes \phi_2$. Fix an archimedean place τ of F.

First, we reduce the problem to the case that $a \in F_\tau^\times$. Fix g and view

$$\mathcal{I}_g(a) := I'(0, d^*(a)g, \chi, \phi) - I'_{00}(0, d^*(a)g, \chi, \phi)$$

as a function of $a \in \mathbb{A}^\times$. It is easy to verify that $\mathcal{I}_g(a)$ is invariant under the action of $F^\times$ on a. By smoothness, there exists a nontrivial open and compact subgroup Q of $\mathbb{A}_f^\times$ (depending on g) such that $\mathcal{I}_g(a)$ is invariant under the action of Q on a.

The quotient $F^\times \backslash \mathbb{A}^\times / F_\tau^\times Q$ is finite. Fix a set of representatives $a_i \in \mathbb{A}^\times$. It suffices to prove the result for a in each single coset $a_i F^\times F_\tau^\times Q$. By the invariance of $\mathcal{I}_g(a)$ and $|a|_\mathbb{A}$ under $F^\times Q$, we can assume that $a \in a_i F_\tau^\times$. The problem in this case becomes

$$\mathcal{I}_g(a_i b) = O(e^{-\epsilon_g |a_i b|_\mathbb{A}}), \quad b \in F_\tau^\times.$$

It is just

$$\mathcal{I}_{d^*(a_i)g}(b) = O(e^{-\epsilon_g |b|_\tau}), \quad b \in F_\tau^\times.$$

The problem is reduced to $F_\tau^\times$.

Second, reduce the problem to

$$\theta(d^*(a)g, u, \phi_1)E'(0, d^*(a)g, u, \phi_2)$$
$$- \theta_0(d^*(a)g, u, \phi_1)E'_0(0, d^*(a)g, u, \phi_2) = O(e^{-\epsilon_{g,u}|a|_\tau}), \quad a \in F_\tau^\times.$$

It follows from the fact that $I'(0, g, \chi, \phi)$ is essentially a finite sum of $I'(0, g, r(t, 1)\phi)$, and the fact that $I'(0, g, \phi)$ is essentially a finite sum of

$\theta(g, u, \phi_1)E'(0, g, u, \phi_2)$. The first finite sum is independent of g, and the second finite sum is independent of g_∞.

Finally, it is classical that $\theta(d^*(a)g, u, \phi_1) - \theta_0(d^*(a)g, u, \phi_1)$ and $E'(0, d^*(a)g, u, \phi_2) - E'_0(0, d^*(a)g, u, \phi_2)$ decay exponentially. It yields the result combining with

$$\theta_0(d^*(a)g, u, \phi_1) = O_{g,u}(|a|_\tau^{\frac{1}{2}}), \quad a \in F_\tau^\times,$$

$$E'_0(0, d^*(a)g, u, \phi_2) = O_{g,u}(|a|_\tau^{\frac{1}{2}} \log|a|_\tau), \quad a \in F_\tau^\times.$$

$\square$

PROPOSITION 6.14. *Under Assumption 5.4,*

$$I'_{00}(0, g, u, \phi) = 0, \quad \forall\, g \in \mathrm{GL}_2(\mathbb{A}), \; u \in F^\times.$$

Therefore, there exists $\epsilon_g > 0$ depending on g such that

$$I'(0, d^*(a)g, \chi, \phi) = O(e^{-\epsilon_g|a|_\mathbb{A}}), \quad a \in \mathbb{A}^\times.$$

PROOF. It suffices to prove the first statement. It amounts to get a more explicit expression of $I'_{00}(0, g, u, \phi)$.

For any $\phi \in \overline{\mathcal{S}}(\mathbb{B} \times \mathbb{A}^\times)$ of the form $\phi = \phi_1 \otimes \phi_2$,

$$I'_{00}(0, g, u, \phi) = \log\delta(g)\; r(g)\phi(0, u) + r_1(g)\phi_1(0, u)W'_0(0, g, u, \phi_2).$$

Recall that the computation in Proposition 6.7 gives

$$W'_0(0, g, u) = -c_0 r(g)\phi_2(0, u) - \sum_v W^\circ_{0,v}{}'(0, g_v, u)r_2(g^v)\phi_{2,v}(0, u)$$

with the constant

$$c_0 = \frac{d}{ds}\Big|_{s=0}\left(\log\frac{L(s, \eta)}{L(s+1, \eta)}\right).$$

It follows that

$$I'_{00}(0, g, u, \phi) = (\log\delta(g) - c_0)r(g)\phi(0, u)$$
$$- \sum_v r(g^v)\phi_v(0, u) \cdot r_1(g_v)\phi_{1,v}(0, u)W^\circ_{0,v}{}'(0, g_v, u, \phi_{2,v}).$$

By linearity,

$$\kappa^\circ_{\phi_v}(g_v, u) := r_1(g_v)\phi_{1,v}(0, u)W^\circ_{0,v}{}'(0, g_v, u, \phi_{2,v})$$

makes sense even if ϕ_v is not of the form $\phi_{1,v} \otimes \phi_{2,v}$. Thus we always have

$$I'_{00}(0, g, u, \phi) = (\log\delta(g) - c_0)r(g)\phi(0, u) - \sum_v r(g^v)\phi^v(0, u)\,\kappa^\circ_{\phi_v}(g, u).$$

Now $I'_{00}(0, g, u, \phi) = 0$ follows easily from Assumption 5.4.

$\square$

6.4.3   The work of Gross–Zagier and S. Zhang

Here we recall the treatment of the holomorphic projection of $I'(0, g, \chi, \phi)$ in [GZ] and [Zh1, Zh2, Zh3].

Without Assumption 5.4, the best bound is

$$I'_{00}(0, d^*(a)g, \chi, \phi) = O_g(|a|_{\mathbb{A}} \log |a|_{\mathbb{A}}).$$

Then Lemma 6.13 gives

$$I'(0, d^*(a)g, \chi, \phi) = O_g(|a|_{\mathbb{A}} \log |a|_{\mathbb{A}}).$$

Thus $I'(0, g, \chi, \phi)$ fails the growth condition of Proposition 6.12.

To apply the proposition, let $\mathcal{J}(s, g, u, \phi)$ be the Eisenstein series formed by $I_{00}(s, g, u, \phi)$, i.e.,

$$\mathcal{J}(s, g, u, \phi) := \sum_{\gamma \in P(F) \backslash \mathrm{GL}_2(F)} I_{00}(s, \gamma g, u, \phi).$$

Take integrations to get

$$\mathcal{J}(s, g, \phi) := \sum_{u \in \mu_U^2 \backslash F^\times} \mathcal{J}(s, g, u, \phi),$$

$$\mathcal{J}(s, g, \chi, \phi) := \int_{T(F) \backslash T(\mathbb{A})/Z(\mathbb{A})}^* \mathcal{J}(s, g, r(t, 1)\phi)\chi(t)dt.$$

Then $I'(0, g, \chi, \phi) - \mathcal{J}'(0, g, \chi, \phi)$ satisfies the growth condition, and we can apply the proposition to it.

Since $\mathcal{J}'(0, g, \chi)$ is orthogonal to all cusp forms, the definition of holomorphic projection yields

$$\mathcal{P}rI'(0, g, \chi) = \mathcal{P}r(I'(0, g, \chi) - \mathcal{J}'(0, g, \chi)).$$

The right-hand side can be computed as

$$\mathcal{P}r'(I'(0, g, \chi) - \mathcal{J}'(0, g, \chi)) = \mathcal{P}r'I'(0, g, \chi) - \mathcal{P}r'\mathcal{J}'(0, g, \chi).$$

Then one has to treat the "extra term" $\mathcal{P}r'\mathcal{J}'(0, g, \chi)$.

This extra term essentially "cancels" the height pairings involving the Hodge class. In [GZ], the cancellation is verified by explicit computation. In [Zh1, Zh2, Zh3], the cancellation is proved by a theory of derivation after detailed analysis of the structure of $\mathcal{P}r'\mathcal{J}'(0, g, \chi)$.

6.5   HOLOMORPHIC KERNEL FUNCTION

By Proposition 6.14, we can apply Proposition 6.12 to $I'(0, g, \chi)$. We get

$$\mathcal{P}rI'(0, g, \chi) = \int_{T(F) \backslash T(\mathbb{A})/Z(\mathbb{A})}^* \mathcal{P}r'I'(0, g, r(t, 1)\phi)\, \chi(t)dt.$$

It suffices to compute $\mathcal{P}r'I'(0, g, r(t, 1)\phi)$ or just $\mathcal{P}r'I'(0, g, \phi)$, where the operator $\mathcal{P}r'$ is defined after Proposition 6.12.

Recall that in Proposition 6.7, we have the simple decomposition

$$I'(0, g, \phi) = -\sum_{v \text{ nonsplit}} I'(0, g, \phi)(v), \quad \forall g \in P(F_{S_1})\mathrm{GL}_2(\mathbb{A}^{S_1}).$$

It is true under Assumption 5.3.

PROPOSITION 6.15. *Under Assumption 5.3,*

$$\mathcal{P}r'(I'(0, g, \phi)) = -\sum_{v|\infty} \overline{I}'(0, g, \phi)(v) - \sum_{v\nmid\infty \text{ nonsplit}} I'(0, g, \phi)(v),$$

$$\forall g \in P(F_{S_1})\mathrm{GL}_2(\mathbb{A}^{S_1}).$$

Here $I'(0, g, \phi)(v)$ is the same as in Proposition 6.5, and for any archimedean v,

$$\overline{I}'(0, g, \phi)(v) = 2\!\!\int_{Z(\mathbb{A})T(F)\backslash T(\mathbb{A})} \overline{\mathcal{K}}_\phi^{(v)}(g, (t, t))dt,$$

$$\overline{\mathcal{K}}_\phi^{(v)}(g, (t_1, t_2)) = \sum_{a\in F^\times} \widetilde{\lim}_{s\to 0} \sum_{y\in\mu_U\backslash(B(v)_+^\times - E^\times)} r(g, (t_1, t_2))\phi(y)_a \, k_{v,s}(y),$$

$$k_{v,s}(y) = \frac{\Gamma(s+1)}{2(4\pi)^s} \int_1^\infty \frac{1}{t(1-\lambda(y)t)^{s+1}}dt.$$

Here $\lambda(y) = q(y_2)/q(y)$ is viewed as an element of F_v, and $\phi(y)_a = \phi(y, aq(y)^{-1})$ is as before.

PROOF. It suffices to check that $\mathcal{P}r'(I'(0, g)(v)) = I'(0, g)(v)$ for finite v, and $\mathcal{P}r'(I'(0, g)(v)) = \overline{I}'(0, g)(v)$ for infinite v. They actually hold for all $g \in \mathrm{GL}_2(\mathbb{A})$.

By Proposition 6.5,

$$I'(0, g)(v) = 2\!\!\int_{Z(\mathbb{A})T(F)\backslash T(\mathbb{A})} \mathcal{K}_\phi^{(v)}(g, (t, t))dt$$

with

$$\mathcal{K}_\phi^{(v)}(g, (t_1, t_2)) = \sum_{u\in\mu_U^2\backslash F^\times} \sum_{y\in B(v)-E} k_{r(t_1,t_2)\phi_v}(g, y, u) r(g, (t_1, t_2))\phi^v(y, u).$$

Note that the integral above is just a finite sum.

We have a simple rule

$$r(n(b)g, (t_1, t_2))\phi^v(y, u) = \psi(uq(y)b) \, r(g, (t_1, t_2))\phi^v(y, u),$$

and its analogue

$$k_{r(t_1,t_2)\phi_v}(n(b)g,y,u) = \psi(uq(y)b)k_{r(t_1,t_2)\phi_v}(g,y,u)$$

shown in Proposition 6.6. By these rules it is easy to see that the first Fourier coefficient is given by

$$\mathcal{K}_\phi^{(v)}(g,(t_1,t_2))_\psi$$

$$= \sum_{(y,u)\in\mu_U\backslash((B(v)-E)\times F^\times)_1} k_{r(t_1,t_2)\phi_v}(g_v,y_v,u_v)r(g,(t_1,t_2))\phi^v(y,u).$$

If v is non-archimedean, all the infinite components are already holomorphic of weight two. So the operator $\mathcal{P}r'$ does not change $\mathcal{K}_\phi^{(v)}(g,(t_1,t_2))_\psi$ at all. Thus

$$\mathcal{P}r'(\mathcal{K}_\phi^{(v)}(g,(t_1,t_2))) = \sum_{a\in F^\times} \mathcal{K}_\phi^{(v)}(d^*(a)g,(t_1,t_2))_\psi.$$

It is easy to check that it is exactly equal to $\mathcal{K}_\phi^{(v)}(g,(t_1,t_2))$ by Proposition 6.6 that $k_{r(t_1,t_2)\phi_v}$ transforms according to the Weil representation under upper triangular matrices. We conclude that $\mathcal{P}r'$ does not change $\mathcal{K}_\phi^{(v)}(g,(t_1,t_2))$, and thus we have $\mathcal{P}r'(I'(0,g)(v)) = I'(0,g)(v)$.

Now we look at the case that v is archimedean. The only difference is that we need to replace $k_{\phi_v}(g,y,u)$ by some $\widetilde{k}_{\phi_v,s}(g,y,u)$, and then take a "quasi-limit" $\widetilde{\lim}$. It suffices to consider the case that $uq(y) = 1$. It is given by

$$\widetilde{k}_{\phi_v,s}(g,y,u) = 4\pi W^{(2)}(g_v)\int_{F_{v,+}} y_0^s e^{-2\pi y_0} k_{\phi_v}(d^*(y_0),y,u)\frac{dy_0}{y_0}.$$

Then $\widetilde{k}_{\phi_v,s}(g,y,u) \neq 0$ only if $u > 0$, since $k_{\phi_v}(d^*(y_0),y,u) \neq 0$ only if $u > 0$.

Assume that $u > 0$, which is equivalent to $q(y) > 0$ since we assume $uq(y) = 1$ for the moment. By Proposition 6.11,

$$\int_{F_{v,+}} y_0^s e^{-2\pi y_0} k_{\phi_v}(d^*(y_0),y,u)\frac{dy_0}{y_0}$$

$$= -\frac{1}{2}\int_{F_{v,+}} y_0^s e^{-2\pi y_0}\operatorname{Ei}(4\pi uq(y_2)y_0)\,y_0 e^{-2\pi y_0}\frac{dy_0}{y_0}$$

$$= -\frac{1}{2}\int_0^\infty y_0^{s+1} e^{-4\pi y_0}\operatorname{Ei}(-4\pi\alpha y_0)\frac{dy_0}{y_0} \qquad \left(\alpha = -uq(y_2) = -\frac{q(y_2)}{q(y)} > 0\right)$$

$$= \frac{1}{2}\int_0^\infty y_0^{s+1} e^{-4\pi y_0}\int_1^\infty t^{-1} e^{-4\pi\alpha y_0 t}dt\frac{dy_0}{y_0}$$

$$= \frac{1}{2}\int_1^\infty t^{-1}\int_0^\infty y_0^{s+1} e^{-4\pi(1+\alpha t)y_0}\frac{dy_0}{y_0}dt$$

$$= \frac{\Gamma(s+1)}{2(4\pi)^{s+1}}\int_1^\infty \frac{1}{t(1+\alpha t)^{s+1}}dt.$$

Hence,

$$\tilde{k}_{\phi_v,s}(g,y,u) = W^{(2)}(g_v)\frac{\Gamma(s+1)}{2(4\pi)^s}\int_1^\infty \frac{1}{t(1-\frac{q(y_2)}{q(y)}t)^{s+1}}dt = W^{(2)}(g_v)k_{v,s}(y).$$

This matches the result in the proposition. Since $k_{v,s}(y)$ is invariant under the multiplication action of $F^\times$ on y, it is easy to get

$$
\begin{aligned}
\mathcal{P}r'(\mathcal{K}_\phi^{(v)}(g,(t_1,t_2))) &= \overline{\mathcal{K}}_\phi^{(v)}(g,(t_1,t_2)), \\
\mathcal{P}r'(I'(0,g)(v)) &= \overline{I}'(0,g)(v).
\end{aligned}
$$

$\square$

Chapter Seven

Decomposition of the Geometric Kernel

Let $\phi = \phi_f \otimes \phi_\infty \in \overline{\mathcal{S}}(\mathbb{V} \times \mathbb{A}^\times)^{U \times U}$ be a Schwartz function with standard ϕ_∞. Assume that $-1 \notin U$ to simply notations. Recall that in §5.1 we have introduced the generating series

$$Z(g, \phi)_U = Z_0(g, \phi)_U + Z_*(g, \phi)_U, \quad g \in \mathrm{GL}_2(\mathbb{A}).$$

Here the non-constant part

$$Z_*(g, \phi)_U = \sum_{a \in F^\times} \sum_{x \in K \backslash \mathbb{B}_f^\times} r(g)\phi(x)_a \, Z(x)_U.$$

We further have the height series as follows:

$$Z(g, (h_1, h_2), \phi)_U = \langle Z(g, \phi)_U [h_1]_U^\circ, [h_2]_U^\circ \rangle_{\mathrm{NT}}, \quad h_1, h_2 \in \mathbb{B}^\times;$$

$$Z(g, \chi, \phi)_U = \int_{T(F) \backslash T(\mathbb{A})/Z(\mathbb{A})}^* Z(g, (t, 1), \phi)_U \, \chi(t) dt.$$

By Lemma 3.19, $Z(g, (h_1, h_2), \phi)_U$ is cuspidal in g. So we can replace $Z(g, \phi)_U$ by $Z_*(g, \phi)_U$ in the definition of $Z(g, (h_1, h_2), \phi)_U$. The constant term $Z_0(g, \phi)_U$ will be ignored in the rest of this book.

The goal of this chapter is to decompose the height series

$$Z(g, (t_1, t_2), \phi)_U = \langle Z_*(g, \phi)_U [t_1]_U^\circ, [t_2]_U^\circ \rangle_{\mathrm{NT}}, \quad t_1, t_2 \in \mathbb{B}^\times.$$

We presume the assumptions in §5.2. Then there is not horizontal self intersection in the above pairing, and the intersections with the Hodge bundles are zero. Then we have a decomposition to a sum of local heights by standard results in Arakelov theory.

In §7.1, we review the definition of the Néron–Tate height, and how to compute it by the arithmetic Hodge index theorem. When there is no horizontal self-intersection, the height pairing automatically decomposes to a summation of local pairings.

In §7.2, we decompose the height series $Z(g, (t_1, t_2), \phi)$ into local heights. To do the decomposition, we describe the arithmetic models we are using.

In §7.3, we prove that the contribution of the Hodge bundles in the height series is zero. It is true under the assumptions in §5.2.

In §7.4, we compare two kernel functions and state the computational result (Theorem 7.8) from the next chapter. We deduce the kernel identity from Theorem 7.8.

7.1   NÉRON–TATE HEIGHT

In this section we review some basic theory of the Néron–Tate heights, the arithmetic intersection theory on arithmetic surfaces by Arakelov [Ar] and Gillet-Soulé [GS], the arithmetic Hodge index theorem by Faltings [Fa2] and Hriljac [Hr], and the notion of admissible arithmetic extension by Zhang [Zh2].

The Hodge index theorem gives a way to "compute" the Néron–Tate height by flat arithmetic extensions. Flat arithmetic extensions are only for divisors of degree zero, so it is not enough in the computation. It is the reason to use admissible arithmetic extensions. If the divisors are disjoint on the generic fiber, we have a natural decomposition of the Néron–Tate height into local heights.

7.1.1   Néron–Tate height on abelian varieties

All the results here can be found in Serre's book [Se]. In the following, fix a number field F. Our normalization of heights depends on F.

The *standard height* on the projective space $\mathbb{P}^n_F$ is a map $h : \mathbb{P}^n(\overline{F}) \to \mathbb{R}$ defined by

$$h(x) := \frac{1}{[K : F]} \sum_w \log \max\{|x_0|_w, |x_1|_w, \cdots, |x_n|_w\}.$$

Here we represent $x = (x_0, \cdots, x_n)$ in terms of the homogeneous coordinate, the field K is any finite extension of F containing all $x_0, \cdots, x_n$, and the summation is over all places w of K with normalized absolute value $|\cdot|_w$. The choices of $x_0, \cdots, x_n$ and K are not unique, but the definition is independent of the choices.

Let Y be a projective variety over F, and L be a line bundle on Y. There is a Weil height $h_L : Y(\overline{F}) \to \mathbb{R}$ associated to L. It is uniquely determined by L up to a bounded function on $Y(\overline{F})$. To review the definition, we first assume that L is very ample. In that case, choose an embedding $i : Y \hookrightarrow \mathbb{P}^n$ such that $i^*O(1) \simeq L$, and set h_L to be the composition

$$Y(\overline{K}) \xrightarrow{i} \mathbb{P}^n(\overline{K}) \xrightarrow{h} \mathbb{R}.$$

The function h_L differs by a bounded function for a different choice of the embedding i. For a general line bundle L, write $L = L_1 \otimes L_2^{\otimes(-1)}$ for two very ample line bundles L_1 and L_2 on Y. Set $h_L = h_{L_1} - h_{L_2}$. Similarly, h_L differs by a bounded function for a different choice of (L_1, L_2) and (h_{L_1}, h_{L_2}).

Let A be an abelian variety over F. Let L be a line bundle on A which is symmetric in the sense that $[-1]^*L \simeq L$. Here $[-1]$ denotes the endomorphism of A given by the inverse in the group law. Define the *Néron–Tate height* $\widehat{h}_L : A(\overline{F}) \longrightarrow \mathbb{R}$ by

$$\hat{h}_L(x) := \lim_{N \to \infty} \frac{1}{N^2} h_L(Nx).$$

Here $h_L : A(\overline{F}) \to \mathbb{R}$ is any Weil height associated to L defined above. The limit exists and does not depend on the choice of h_L, so it is also called *the canonical height*.

THEOREM 7.1. *Assume that L is ample and symmetric on A. Then the following are true:*

(1) One has $\widehat{h}_L(x) \geq 0$ for any $x \in A(\overline{F})$. The equality holds if and only if x is a torsion point.

(2) The function $\widehat{h}_L : A(\overline{F}) \longrightarrow \mathbb{R}$ is quadratic in the sense that

$$\widehat{h}_L(x + y) + \widehat{h}_L(x - y) = 2\widehat{h}_L(x) + 2\widehat{h}_L(y), \quad \forall x, y \in A(\overline{F}).$$

By the theorem, we can define a bilinear pairing on $A(\overline{F})$ by

$$\langle x, y \rangle_L := \widehat{h}_L(x + y) - \widehat{h}_L(x) - \widehat{h}_L(y), \quad \forall x, y \in A(\overline{F}).$$

It is positive definite up to torsion points. It is called *the Néron–Tate height pairing*. By linearity, it extends to a unique positive definite Hermitian pairing on $A(\overline{F})_{\mathbb{C}}$.

There is a refinement of the Néron–Tate height pairing. Let A be as above, and let P be the Poincaré line bundle on $A \times A^{\vee}$. Then P is symmetric and its Néron–Tate height on $A \times A^{\vee}$ gives a map

$$\widehat{h}_P : A(\overline{F}) \times A^{\vee}(\overline{F}) \longrightarrow \mathbb{R}.$$

It is a bilinear pairing, and we also denote it by

$$\langle \cdot, \cdot \rangle_{\mathrm{NT}} : A(\overline{F}) \times A^{\vee}(\overline{F}) \longrightarrow \mathbb{R}.$$

It is called *the Néron–Tate height pairing between A and $A^{\vee}$.*

Go back to the situation of Theorem 7.1. The ample line bundle L gives a polarization

$$\phi_L : A \longrightarrow A^{\vee}, \quad x \longmapsto T_x^* L \otimes L^{\otimes(-1)}.$$

The following result asserts the compatibility of the two pairings we have just defined.

THEOREM 7.2. *Assume that L is ample and symmetric on A. Then*

$$\langle x, y \rangle_L = \langle x, \phi_L(y) \rangle_{\mathrm{NT}}, \quad \forall x, y \in A(\overline{F}).$$

Let $f : A \to B$ be a homomorphism of abelian varieties over F. The pullback $f^* : \mathrm{Pic}^0(B) \to \mathrm{Pic}^0(A)$ induces a homomorphism $f^{\vee} : B^{\vee} \to A^{\vee}$. The following projection formula asserts that $f^{\vee}$ is the adjoint of f under the height pairing.

PROPOSITION 7.3. *For any homomorphism $f : A \to B$ as above,*

$$\langle f(x), y \rangle_{\mathrm{NT}} = \langle x, f^{\vee}(y) \rangle_{\mathrm{NT}}, \quad \forall x \in A(\overline{F}),\ y \in B^{\vee}(\overline{F}).$$

PROOF. It follows from the interpretation of the pairing in the theorem on page 37 of [Se], and the functoriality in the first theorem on page 35 of [Se]. $\square$

7.1.2 Néron–Tate height on curves

Let X be a smooth algebraic curve over F of genus $g > 0$. Let $J = J(X)$ be the Jacobian variety of X. Let $j : X \hookrightarrow J$ be the usual embedding sending x to the divisor class of $x - a$. Here $a \in \mathrm{Div}(X_{\overline{F}})$ is a fixed divisor of degree 1. Recall that the theta divisor Θ on J is the image of the finite map

$$X^{g-1} \longrightarrow J, \quad (x_1, \cdots, x_{g-1}) \longmapsto x_1 + \cdots + x_{g-1} - (g-1)a.$$

The theta divisor Θ is ample. It depends on the choice of a.

The Néron–Tate height pairing

$$\langle \cdot, \cdot \rangle_{\mathrm{NT}} : J(\overline{F}) \times J(\overline{F}) \longrightarrow \mathbb{R}$$

is defined to be the composition of the isomorphism

$$(\mathrm{id}, \phi_\Theta) : J(\overline{F}) \times J(\overline{F}) \longrightarrow J(\overline{F}) \times J^\vee(\overline{F})$$

with the pairing

$$\langle \cdot, \cdot \rangle_{\mathrm{NT}} : J(\overline{F}) \times J^\vee(\overline{F}) \longrightarrow \mathbb{R}$$

introduced above. Here the polarization $\phi_\Theta : J \to J^\vee$ is the canonical isomorphism, which is independent of the choice of a.

Alternatively, define a new divisor $\widehat{\Theta} = \Theta + [-1]^*\Theta$ on J. It is symmetric by definition. One can check that it is independent of the choice of a. It gives the Néron–Tate height

$$\hat{h}_{\widehat{\Theta}} : J(\overline{F}) \longrightarrow \mathbb{R}.$$

Then the Néron–Tate height pairing

$$\langle \cdot, \cdot \rangle_{\mathrm{NT}} : J(\overline{F}) \times J(\overline{F}) \longrightarrow \mathbb{C}$$

is equal to the pairing

$$\frac{1}{2} \langle \cdot, \cdot \rangle_{\widehat{\Theta}} : J(\overline{F}) \times J(\overline{F}) \longrightarrow \mathbb{C}.$$

By the identification $J(\overline{F}) = \mathrm{Pic}^0(X_{\overline{F}})$, obtain

$$\langle \cdot, \cdot \rangle_{\mathrm{NT}} : \mathrm{Pic}^0(X_{\overline{F}}) \times \mathrm{Pic}^0(X_{\overline{F}}) \longrightarrow \mathbb{C}.$$

This is the pairing described by the arithmetic Hodge index theorem.

If X is a disjoint union of projective smooth curves $X_1, \cdots, X_r$ over F, the above theory can be extended to X. Recall that $\mathrm{Div}^0(X)$ (resp. $\mathrm{Pic}^0(X)$) are the subgroup of elements of $\mathrm{Div}(X)$ (resp. $\mathrm{Pic}(X)$) with degree zero on every connected component of X. Then we have canonical decompositions $\mathrm{Div}^0(X) = \oplus_i \mathrm{Div}^0(X_i)$, $\mathrm{Pic}^0(X) = \oplus_i \mathrm{Pic}^0(X_i)$ and $J(X) = \prod_i J(X_i)$. The Néron–Tate height pairings on $\mathrm{Pic}^0(X_i)$ extend uniquely to $\mathrm{Pic}^0(X)$ such that the direct sum $\mathrm{Pic}^0(X) = \oplus_i \mathrm{Pic}^0(X_i)$ gives an orthogonal decomposition.

7.1.3   Intersection theory on arithmetic surfaces

Let $\mathcal{X}$ be an arithmetic surface over O_F. Namely, $\mathcal{X}$ is a projective and flat scheme over $\mathrm{Spec}(O_F)$ with absolute dimension two. Assume that $\mathcal{X}$ is regular, but allow it to be disconnected.

Now we recall the arithmetic intersection theory defined by Gillet-Soulé [GS]. An *arithmetic divisor* on $\mathcal{X}$ is a pair $\widehat{D} = (D, g_D)$ where D is a divisor on $\mathcal{X}$, and $g_D : \mathcal{X}(\mathbb{C}) - |D(\mathbb{C})| \to \mathbb{R}$ is a *Green's function* of D. Namely, g_D is smooth with logarithmic singularity along $|D(\mathbb{C})|$ such that the current

$$\omega_D = \frac{\partial\bar{\partial}}{\pi i} g_D + \delta_{D(\mathbb{C})}$$

on $\mathcal{X}(\mathbb{C})$ is actually a smooth $(1,1)$-form on $\mathcal{X}(\mathbb{C})$. Note that $\mathcal{X}(\mathbb{C}) = \coprod_{\sigma:F \hookrightarrow \mathbb{C}} \mathcal{X}_\sigma(\mathbb{C})$. So g_D is just a collection of functions $g_{D,\sigma} : \mathcal{X}_\sigma(\mathbb{C}) - |D_\sigma(\mathbb{C})|$ satisfying similar properties. We call ω_D the *curvature form* of $\widehat{D}$.

An *arithmetic vertical divisor on* $\mathcal{X}$ is a divisor of the form (V, f) where f is a smooth function on $\mathcal{X}(\mathbb{C})$ and V is a vertical divisor on $\mathcal{X}$ in the sense that V is supported in finitely many fibers of $\mathcal{X}$ over $\mathrm{Spec}(O_F)$. In particular, $(V, 0)$ itself is an arithmetic vertical divisor. We sometimes write V for $(V, 0)$, and call it a *finite vertical divisor*. Similarly, we call $(0, f)$ an *infinite vertical divisor*.

Denote by $\widehat{\mathrm{Div}}(\mathcal{X})$ the group of arithmetic divisors on $\mathcal{X}$. For any rational function f on $\mathcal{X}$, the divisor

$$\widehat{\mathrm{div}}(f) := (\mathrm{div}(f), -\log|f|_\infty)$$

is an arithmetic divisor. Here $-\log|f|_\infty$ restricted to each $\mathcal{X}_\sigma(\mathbb{C})$ is just $-\log|f|_\sigma$. An arithmetic divisor of the form $\widehat{\mathrm{div}}(f)$ is called an *arithmetic principle divisor*. The group of principle divisors is denoted by $\widehat{\mathrm{Pr}}(\mathcal{X})$. The group of arithmetic divisor classes can be identified with the Picard group of isomorphism classes of Hermitian line bundles on $\mathcal{X}$ via the Chern class map

$$\widehat{\mathrm{Pic}}(\mathcal{X}) \longrightarrow \widehat{\mathrm{Div}}(\mathcal{X})/\widehat{\mathrm{Pr}}(\mathcal{X}), \qquad \overline{\mathcal{L}} \mapsto c_1(\overline{\mathcal{L}}).$$

More precisely, for any Hermitian line bundle $\overline{\mathcal{L}} = (\mathcal{L}, \|\cdot\|)$ on $\mathcal{X}$, its *arithmetic Chern class* $\hat{c}_1(\overline{\mathcal{L}})$ is defined by

$$\hat{c}_1(\overline{\mathcal{L}}) := (\mathrm{div}(s), -\log\|s\|) \pmod{\widehat{\mathrm{Pr}}(\mathcal{X})}.$$

Here s is any nonzero rational section of $\mathcal{L}$, and the class in $\widehat{\mathrm{Pic}}(\mathcal{X})$ does not depend on the choice of s. In this way, we can view any Hermitian line bundle as an arithmetic divisor class on $\mathcal{X}$.

The arithmetic intersection is a symmetric pairing

$$\widehat{\mathrm{Pic}}(\mathcal{X}) \times \widehat{\mathrm{Pic}}(\mathcal{X}) \longrightarrow \mathbb{R}.$$

In the case that two divisors $\widehat{D}_1 = (D_1, g_{D_1})$ and $\widehat{D}_2 = (D_2, g_{D_2})$ in $\widehat{\mathrm{Div}}(\mathcal{X})$ have no horizontal self-intersection in the sense that $|D_{1,F}| \cap |D_{2,F}|$ is empty, the intersection is defined by

$$\widehat{D}_1 \cdot \widehat{D}_2 = (D_1 \cdot D_2) + g_{D_1}(D_2(\mathbb{C})) + \int_{\mathcal{X}(\mathbb{C})} g_{D_2} \omega_{D_1}. \qquad (7.1.1)$$

Here g_{D_1} acts on $(D_2(\mathbb{C}))$ by linearity, and the finite part

$$(D_1 \cdot D_2) = \sum_{v \nmid \infty} m_v(D_1 \cdot D_2) \log N_v \qquad (7.1.2)$$

sums over all finite primes v of F. For archimedean v, we take the convention that $\log N_v$ equals 1 (resp. 2) if v is real (resp. complex).

For each v, the multiplicity

$$m_v(D_1 \cdot D_2) = \sum_{x \in \pi^{-1}(v)} m_x(D_1 \cdot D_2),$$

where $\pi : \mathcal{X} \to \mathrm{Spec}(\mathcal{O}_F)$ denotes the structure morphism, and $\pi^{-1}(v)$ is just the set of closed points of $\mathcal{X}$ lying above v. Here $m_x(D_1 \cdot D_2)$ is the usual intersection multiplicity in algebraic geometry.

In this case, we have a natural decomposition

$$\widehat{D}_1 \cdot \widehat{D}_2 = \sum_v (\widehat{D}_1 \cdot \widehat{D}_2)_v.$$

If v is non-archimedean,

$$(D_1 \cdot D_2)_v = m_v(D_1 \cdot D_2) \log N_v.$$

If v is archimedean,

$$(\widehat{D}_1 \cdot \widehat{D}_2)_v = g_{D_1,v}(D_{2,v}(\mathbb{C})) + \int_{\mathcal{X}_v(\mathbb{C})} g_{D_2,v} \omega_{D_1,v}.$$

If there are horizontal self-intersections, the above formula does not work and we can use principal divisors to "move" D_2 to convert to the above situation.

We further remark that the definitions still make sense if we allow the finite part D_1, D_2 to be divisors on $\mathcal{X}$ with coefficients in $\mathbb{Q}, \mathbb{R}$ or $\mathbb{C}$. In the case $\mathbb{C}$, we make the intersection to be Hermitian.

7.1.4   Arithmetic Hodge index theorem

Let F be any number field, and X be a complete smooth curve over F. Note that X may not be connected. Recall that $\mathrm{Div}^0(X)$ is the group of divisors on X that has degree zero on every connected component of X, and $\mathrm{Pic}^0(X)$ is the group of rational equivalence classes in $\mathrm{Div}^0(X)$. The Néron–Tate height gives a bilinear pairing $\langle \cdot, \cdot \rangle_{\mathrm{NT}}$ on $\mathrm{Pic}^0(X)$.

Let $\mathcal{X}$ be an integral model of X over O_F, i.e., $\mathcal{X}$ is an arithmetic surface over O_F with $\mathcal{X}_F = X$. Assume that $\mathcal{X}$ is regular with semistable reduction. An arithmetic divisor $\widehat{D}$ on $\mathcal{X}$ is called *flat* if its intersection with any vertical divisor on $\mathcal{X}$ is zero. Equivalently, its curvature form ω_D is zero and its intersection with any finite vertical divisor on $\mathcal{X}$ is zero.

It is easy to see that the flatness of $\widehat{D} = (D, g_D)$ forces the generic fiber D_F to lie in $\mathrm{Div}^0(X)$. Then it makes sense to talk about the Néron–Tate height of D_F. The following is the arithmetic Hodge index theorem of Faltings [Fa2] and Hriljac [Hr].

THEOREM 7.4. *Let* $\widehat{D}_1 = (D_1, g_{D_1})$ *and* $\widehat{D}_2 = (D_2, g_{D_2})$ *be two flat arithmetic divisors on* $\mathcal{X}$. *Then*

$$\widehat{D}_1 \cdot \widehat{D}_2 = -\langle D_{1,F}, D_{2,F} \rangle_{\mathrm{NT}}.$$

To recover the Néron–Tate height on $\mathrm{Pic}^0(X_{\overline{F}})$, it suffices to consider arithmetic intersections on integral models of X over all finite extensions L of F. The equality above will be modified by a factor $[L : F]$ due to our normalization.

7.1.5   Admissible arithmetic divisors

Let $\mathcal{X}$ be an integral model of X as above, and fix an arithmetic divisor class $\widehat{\xi} \in \widehat{\mathrm{Pic}}(\mathcal{X})$ whose generic fiber has degree one on any connected component of X.

Let $\widehat{D} = (D, g_D)$ be an arithmetic divisor on $\mathcal{X}$. We can always write $D = H + V$ where H is the horizontal part of D, and V is the vertical part of D. Namely, every irreducible component of V is contained in a fiber of X over $\mathrm{Spec}(O_F)$, but every irreducible component of H is mapped surjectively to $\mathrm{Spec}(O_F)$ via the structure morphism $X \to \mathrm{Spec}(O_F)$. The arithmetic divisor $\widehat{D} = (D, g_D)$ is called $\widehat{\xi}$-*admissible* if the following conditions hold on each connected component of $\mathcal{X}$:

- The difference $\widehat{D} - \deg D \cdot \widehat{\xi}$ is flat;

- The integral $\displaystyle \int_{\mathcal{X}_v(\mathbb{C})} g_D c_1(\widehat{\xi}) = 0$ at any archimedean place v of F;

- The intersection $(V \cdot \widehat{\xi})_v = 0$ at any non-archimedean place v of F.

Any divisor D in $\mathrm{Div}(X)$ has a unique $\widehat{\xi}$-admissible extension, which we denote by $\widehat{D}$. It depends on $\widehat{\xi}$. If D belongs to $\mathrm{Div}^0(X)$, then the extension $\widehat{D}$ is apparently flat. It is easy to check that for two $\widehat{\xi}$-admissible divisors $\widehat{D}_1 = (D_1, g_{D_1})$ and $\widehat{D}_2 = (D_2, g_{D_2})$, $\widehat{D}_1 = \widehat{D}_2$ if and only if $D_1 = D_2$.

Our admissibility on the Green's function is the same as what Arakelov [Ar] originally required, but we also put conditions on the finite part. By complex analysis, there is a unique symmetric function

$$g : X(\mathbb{C}) \times X(\mathbb{C}) - \Delta \longrightarrow \mathbb{R}$$

depending on the curvature form of $\widehat{\xi}$ which gives all the $\widehat{\xi}$-admissible Green's functions. Here Δ denotes the diagonal of $X(\mathbb{C}) \times X(\mathbb{C})$. Namely, for any $\widehat{\xi}$-admissible divisor $\widehat{D} = (D, g_D)$,

$$g_D(x) = g(D(\mathbb{C}), z), \quad \forall z \in X(\mathbb{C}) - |D(\mathbb{C})|.$$

Here $g(D(\mathbb{C}), z)$ makes sense by linearity of divisors. See [Ar] for more information.

For any two $\widehat{\xi}$-admissible divisors $\widehat{D}_1 = (D_1, g_{D_1})$ and $\widehat{D}_2 = (D_2, g_{D_2})$, the infinite part

$$(\widehat{D}_1 \cdot \widehat{D}_2)_\infty = g(D_1(\mathbb{C}), D_2(\mathbb{C})).$$

In fact, it suffices to check

$$\int_{\mathcal{X}(\mathbb{C})} g_{D_2} \omega_{D_1} = 0.$$

By definition, ω_{D_1} is a multiple of the curvature form $\omega_{\widehat{\xi}}$ of ξ on every connected component of $\mathcal{X}(\mathbb{C})$. Then the integration is zero by the admissibility of $\widehat{D}_2$.

7.1.6 Admissible extension

For the rest of this section, fix an arithmetic class $\widehat{\xi} \in \varinjlim \widehat{\mathrm{Pic}}(\mathcal{Y})$ whose generic fiber has degree one on any geometrically connected component. Here the inverse limit is taking over all integral models $\mathcal{Y}$ of X_L over O_L for all extensions L/F, and the arrows between different models are the pull-back maps on arithmetic divisors.

Let $D \in \mathrm{Div}(X_{\overline{F}})$. Assume that D is defined over a finite extension L of F and $\mathcal{Y}$ is an integral model of X_L over O_L. Assume further that $\widehat{\xi}$ is represented by an arithmetic divisor on $\mathcal{Y}$. We still denote it by $\widehat{\xi}$. Any arithmetic extension of D on $\mathcal{Y}$ is of the form $\widehat{D} = (\overline{D} + V, g_D)$, where $\overline{D}$ is the Zariski closure of D, V is some finite vertical divisor, and g_D is some Green's function of $\overline{D}$. There is a unique choice of V and g_D such that $\widehat{D}$ is an $\widehat{\xi}$-admissible arithmetic divisor, and we call it the $\widehat{\xi}$-admissible extension of D on $\mathcal{Y}$.

Assume that $D_1, D_2 \in \mathrm{Div}(X_{\overline{F}})$ are defined over L, and $\mathcal{Y}$ is an integral model over O_L as above. Define a pairing

$$\langle D_1, D_2 \rangle := -\frac{1}{[L : F]} \widehat{D}_1 \cdot \widehat{D}_2.$$

Here $\widehat{D}_1$ and $\widehat{D}_2$ are the $\widehat{\xi}$-admissible extensions on $\mathcal{Y}$. When $D_1, D_2 \in$ $\mathrm{Div}^0(X_{\bar{F}})$, the pairing is just the Néron–Tate height pairing by the arithmetic Hodge index theorem.

When varying L and $\mathcal{Y}$, the $\widehat{\xi}$-admissibility is preserved by pull-backs. Hence the pairing depends only on D_1, D_2 and the choice of $\widehat{\xi}$. The pairing does not factor through $\widehat{\mathrm{Pic}}(X_{\bar{F}})$, since rational equivalence does not keep $\widehat{\xi}$-admissibility.

7.1.7   Decomposition of the pairing

There is a non-canonical decomposition $\langle \cdot, \cdot \rangle = -i - j$ depending on the choice of integral models. Let $X, \widehat{\xi}$ be as above. Fix a regular and semistable integral model $\mathcal{Y}_0$ of X_{L_0} over O_{L_0} for some field extension L_0 of F with a fixed embedding $L_0 \hookrightarrow \bar{F}$. Assume that $\widehat{\xi}$ is realized as a divisor class on $\mathcal{Y}_0$.

Let $D_1, D_2 \in \mathrm{Div}(X_{\bar{F}})$ be two divisors with disjoint supports. Assume that D_2 is defined over L_0. Let L be any field extension of L_0 such that D_1 is defined over L. Then we can decompose $\langle D_1, D_2 \rangle$ according to the model $\mathcal{Y}_{0,O_L}$.

We first consider the case that $\mathcal{Y}_{0,O_L}$ is regular. Let $\widehat{D}_i = (\overline{D}_i + V_i, g_i)$ be the $\widehat{\xi}$-admissible extensions on the model. Note that $V_1 \cdot \widehat{D}_2 = 0$ since V_1 is orthogonal to both $\widehat{D}_2 - \deg D_2 \cdot \widehat{\xi}$ and $\widehat{\xi}$. It follows that

$$\langle D_1, D_2 \rangle = -\frac{1}{[L:F]} \left(\overline{D}_1 \cdot \overline{D}_2 + \overline{D}_1 \cdot V_2 + g(D_1(\mathbb{C}), D_2(\mathbb{C})) \right).$$

Here g is introduced above. Define

$$i(D_1, D_2) := \frac{1}{[L:F]} (\overline{D}_1 \cdot \overline{D}_2 + g(D_1(\mathbb{C}), D_2(\mathbb{C}))),$$

$$j(D_1, D_2) := \frac{1}{[L:F]} \overline{D}_1 \cdot V_2.$$

Then we have a decomposition

$$\langle D_1, D_2 \rangle = -i(D_1, D_2) - j(D_1, D_2).$$

The decomposition is still valid even if $\mathcal{Y}_{0,O_L}$ is not regular. We have the $\widehat{\xi}$-admissible extension of D_2 on the regular model $\mathcal{Y}_0$. Pull it back to $\mathcal{Y}_{0,O_L}$. We get the extension $\widehat{D}_2 = (\overline{D}_2 + V_2, g_2)$ on $\mathcal{Y}_{0,O_L}$. All divisors on $\mathcal{Y}_0$ are Cartier divisors since it is regular. It follows that $\overline{D}_2$ and V_2 are Cartier divisors on $\mathcal{Y}_{0,O_L}$ since they are pull-backs of Cartier divisors. Hence the intersections $i(D_1, D_2)$ and $j(D_1, D_2)$ are well-defined. To verify the equality, we first decompose the pairing on any desingularization of $\mathcal{Y}_{0,O_L}$, and then use the projection formula.

By the definition of arithmetic intersection theory above, we have further decompositions to local heights:

$$i(D_1, D_2) = \sum_{w \in S_L} i_w(D_1, D_2) \log N_w,$$

$$j(D_1, D_2) = \sum_{w \in S_L} j_w(D_1, D_2) \log N_w.$$

Here S_L denote the set of all places of F. The local multiplicities are defined according to (7.1.2). More precisely,

$$i_w(D_1, D_2) = \begin{cases} \dfrac{1}{[L:F]}(\overline{D}_1 \cdot \overline{D}_2)_w, & w \nmid \infty, \\ \dfrac{1}{[L:F]}g(D_{1,w}(\mathbb{C}), D_{2,w}(\mathbb{C})), & w \mid \infty, \end{cases}$$

and

$$j_w(D_1, D_2) = \begin{cases} \dfrac{1}{[L:F]}(\overline{D}_1 \cdot V_2)_w, & w \nmid \infty, \\ 0, & w \mid \infty. \end{cases}$$

Note that i_w and j_w are local intersection multiplicities of i and j over the model $\mathcal{Y}_{0,O_L}$.

It is convenient to rearrange the decomposition in terms of places of F. Namely,

$$i(D_1, D_2) = \sum_{v \in S_F} i_v(D_1, D_2) \log N_v, \quad j(D_1, D_2) = \sum_{v \in S_F} j_v(D_1, D_2) \log N_v$$

with

$$i_v(D_1, D_2) = \frac{1}{\#S_{L_v}} \sum_{w \in S_{L_v}} i_w(D_1, D_2),$$

$$j_v(D_1, D_2) = \frac{1}{\#S_{L_v}} \sum_{w \in S_{L_v}} j_w(D_1, D_2).$$

Here S_F denotes the set of all places of F, and S_{L_v} denotes the set of places of L lying over v.

Fix an embedding $\overline{F} \hookrightarrow \overline{F}_v$, or equivalently fix an extension $\overline{v}$ of the valuation v to $\overline{F}$. By varying the fields as in the global case, we have well-defined pairings $i_{\overline{v}}$ and $j_{\overline{v}}$ on $\mathrm{Div}(X_{\overline{F}_v})$ for proper intersections. It is the same as considering intersections on the model $\mathcal{Y}_{0,O_{\overline{F}_v}}$. The formulae above have the following equivalent forms:

$$i_v(D_1, D_2) = \int_{\mathrm{Gal}(\overline{F}/F)} i_{\overline{v}}(D_1^\sigma, D_2^\sigma) d\sigma,$$

$$j_v(D_1, D_2) = \int_{\mathrm{Gal}(\overline{F}/F)} j_{\overline{v}}(D_1^\sigma, D_2^\sigma) d\sigma.$$

Here the integral on the Galois group takes the Haar measure with total volume one.

If D_1, D_2 have common irreducible components, we can define $i(D_1, D_2)$ and $j(D_1, D_2)$ by a slightly different method. The pairing $i(D_1, D_2)$ cannot be decomposed to a sum of local heights any more while the decomposition for $j(D_1, D_2)$ is still valid. This case does not happen in the computation of this book. In any case, j_v is identically zero if v is archimedean or the model $\mathcal{Y}_0$ is smooth over all primes of L_0 dividing v.

7.2   DECOMPOSITION OF THE HEIGHT SERIES

Go back to the setting of Shimura curve X_U. Our goal in the geometric side is to compute

$$Z(g, (t_1, t_2)) = \langle Z_*(g)(t_1 - \xi_{t_1}), t_2 - \xi_{t_2} \rangle_{\mathrm{NT}}, \qquad t_1, t_2 \in C_U.$$

Here we write $\xi_t = \xi_{U,q(t)}$ as above. In this section, we decompose the above pairing into a sum of local heights and some global pairings with ξ.

7.2.1   Arithmetic model of the Shimura curve

Recall that we have a Shimura curve X_U over F. For each archimedean place τ of F, the set of complex points under $\tau : F \to \mathbb{C}$ forms a Riemann surface uniformized by

$$X_{U,\tau}^{\mathrm{an}} = B(\tau)^\times \backslash \mathcal{H}^\pm \times \mathbb{B}_f^\times / U \cup \{\mathrm{cusps}\}.$$

Here $B(\tau)$ is the nearby quaternion algebra over F.

We assume that U satisfies the conditions in Assumption 5.6. In particular, the last condition assures that the Hodge bundle

$$L_U = \omega_{X_U/F} + \{\mathrm{cusps}\}.$$

Here $\omega_{X_U/F}$ is the canonical bundle, and each cusp has exactly multiplicity one.

Recall that X_U has a canonical regular and semistable integral model $\mathcal{X}_U$ over O_F. At each finite place v of F, the base change $\mathcal{X}_{U,O_{F_v}}$ to O_{F_v} parametrizes p-divisible groups with level structures. Locally at a geometric point, $\mathcal{X}_{U,O_{F_v}}$ is the universal deformation space of certain p-divisible groups. See [Car, Zh1] for example.

The Hodge bundle L_U can be canonically extended to a Hermitian line bundle on $\mathcal{X}_U$. Take advantage of the fact that L_U is isomorphic to the sum of the canonical bundle $\omega_{X_U/F}$ with the cusps. Define $\mathcal{L}_U$ to be the sum of the relative dualizing sheaf $\omega_{\mathcal{X}_U/O_F}$ with the Zariski closure of each cusp of X_U in $\mathcal{X}_U$. It is an integral model of L_U. Define

$$\widehat{L}_U := (\mathcal{L}_U, \|\cdot\|_{\mathrm{Pet}}).$$

Here the Petersson metric $\|\cdot\|_{\mathrm{Pet}}$ on $\mathcal{L}_U$ is defined such that its restriction to every connected component of $\mathcal{X}_U(\mathbb{C}) = \coprod_{\tau : F \to \mathbb{C}} X_{U,\tau}(\mathbb{C})$ takes the form

$$\|f(z)dz\|_{\mathrm{Pet}} = 4\pi \cdot \mathrm{Im}\, z \cdot |f(z)|.$$

See Lemma 3.1 for more details. Then we define

$$\widehat{\xi}_U = \frac{1}{\kappa_U^\circ} \widehat{L}_U.$$

Here κ_U° is the degree of L_U on any geometrically connected component of X_U as before.

We will denote by $\langle \cdot, \cdot \rangle$ the $\hat{\xi}_U$-admissible pairing explained in the last section. To consider the decomposition $i + j$ and its corresponding local components, we need an integral model over a field where the CM points are rational. Let $H = H_U$ be the minimal field extension of E which contains the fields of definition of all $t \in C_U$. Then H is an abelian extension over E given by the reciprocity law. We will use the regular integral model $\mathcal{Y}_U$ of X_U over O_H introduced in the following to get the decomposition $i + j$.

Recall that we have assumed in Assumption 5.6 that U_v is of the form $(1 + \varpi_v^r O_{\mathbb{B}_v})^\times$ for every finite place v, where $O_{\mathbb{B}_v}$ is a maximal order of $\mathbb{B}_v$. To describe $\mathcal{Y}_U$, it suffices to describe the corresponding local model $\mathcal{Y}_{U,w} = \mathcal{Y}_U \times O_{H_w}$ for any finite place w of H. Let v be the place of F induced from w. Then $\mathcal{Y}_{U,w}$ is defined by the following process:

- Let $U^0 = U^v U_v^0$ with $U_v^0 = O_{\mathbb{B}_v}^\times$ the maximal compact subgroup. Then X_{U^0} has a canonical regular model $\mathcal{X}_{U^0,v}$ over O_{F_v}. It is smooth if $\mathbb{B}_v$ is the matrix algebra. Let $\mathcal{X}'_U$ be the normalization of $\mathcal{X}_{U^0,v}$ in the function field of X_{U,H_w}.

- Make a minimal desingularization of $\mathcal{X}'_U$ to get $\mathcal{Y}_{U,w}$.

With the model $\mathcal{Y}_U$, we can always write

$$\langle \beta, t \rangle = -i(\beta, t) - j(\beta, t), \quad \beta \in \mathrm{CM}_U, t \in C_U.$$

We can further write i, j in terms of their corresponding local components if $\beta \neq t$.

7.2.2 Decomposition of the kernel function

Go back to

$$Z(g, (t_1, t_2)) = \langle Z_*(g)(t_1 - \xi_{t_1}), t_2 - \xi_{t_2} \rangle_{\mathrm{NT}}, \quad t_1, t_2 \in C_U.$$

We first write

$$Z(g, (t_1, t_2)) = \langle Z_*(g)t_1, t_2 \rangle - \langle Z_*(g)t_1, \xi_{t_2} \rangle - \langle Z_*(g)\xi_{t_1}, t_2 \rangle + \langle Z_*(g)\xi_{t_1}, \xi_{t_2} \rangle.$$

We will prove that the last three terms are zero under Assumption 5.4. But we first look at the first term $\langle Z_*(g)t_1, t_2 \rangle$.

We claim that under Assumption 5.3, the pairing $\langle Z_*(g)t_1, t_2 \rangle$ does not involve any self-intersection for any $g \in 1_{S_1} \mathrm{GL}_2(\mathbb{A}^{S_1})$. In other words, the multiplicity of $[t_2]$ in $Z_*(g)t_1$ is zero. Recall that

$$Z_*(g)t_1 \quad = \quad \sum_{a \in F^\times} \sum_{x \in \mathbb{B}_f^\times/U} r(g)\phi(x)_a [t_1 x].$$

Note that the factor w_U disappears since we assume $-1 \notin U$ in Assumption 5.3 to simplify the notations.

Apparently $[t_1 x] = [t_2]$ as CM points on X_U if and only if $x \in t_1^{-1} t_2 E^\times U$. It follows that the coefficient of $[t_2]$ in $Z_*(g)t_1$ is equal to

$$\sum_{a \in F^\times} \sum_{x \in t_1^{-1} t_2 E^\times U / U} r(g)\phi(x)_a \;=\; \sum_{a \in F^\times} \sum_{y \in E^\times / (E^\times \cap U)} r(g)\phi(t_1^{-1} t_2 y)_a.$$

As in the analytic side, under Assumption 5.3, $r(g)\phi(t_1^{-1} t_2 y)_a = 0$ for all $y \in E(\mathbb{A})$ and $g \in 1_{S_1} \mathrm{GL}_2(\mathbb{A}^{S_1})$. So there is no self-intersection under the assumption.

Now the decomposition of $\langle Z_*(g)t_1, t_2 \rangle$ is routine. By the model $\mathcal{Y}$ over O_H, we can decompose

$$\langle Z_*(g)t_1, t_2 \rangle = -i(Z_*(g)t_1, t_2) - j(Z_*(g)t_1, t_2).$$

The j-part is always a sum of local pairings over places and Galois orbits, and so is the i-part if there are no self-intersections occurring.

In the following we list decomposition of $i(Z_*(g)t_1, t_2)$ as a sum of local heights under Assumption 5.3 and for $g \in 1_{S_1} \mathrm{GL}_2(\mathbb{A}^{S_1})$. All the notations and decompositions apply to $j(Z_*(g)t_1, t_2)$ even without these assumptions.

Note that Galois conjugates of points in C_U over E are described easily by multiplication by elements of $T(F)\backslash T(\mathbb{A}_f)$ via the reciprocity law. It is convenient to group local intersections according to places of E. We first write

$$i(Z_*(g)t_1, t_2) \;=\; \frac{1}{2} \sum_{v \in S_E} i_v(Z_*(g)t_1, t_2) \log N_v,$$

$$i_v(Z_*(g)t_1, t_2) \;=\; \fint_{\mathrm{Gal}(E^{\mathrm{ab}}/E)} i_{\bar{v}}(Z_*(g)t_1^\sigma, t_2^\sigma) d\sigma.$$

Here S_E denotes the set of all places of E, and E^{ab} denotes the maximal abelian extension of E. In the integration, we have replaced $\overline{E}$ by E^{ab} since all CM points and their conjugates are defined over E^{ab}. The definition of $i_{\bar{v}}$ depends on fixed embeddings $H \hookrightarrow \overline{E}$ and $\overline{E} \hookrightarrow \overline{E}_v$, and can be viewed as intersections on $\mathcal{Y}_U \times_{O_H} O_{\overline{E}_v}$.

By class field theory, we have a continuous surjective homomorphism

$$T(F)\backslash T(\mathbb{A}) \longrightarrow \mathrm{Gal}(E^{\mathrm{ab}}/E)$$

whose kernel contains $T(F_\infty)$. It follows that we can further write

$$i_v(Z_*(g)t_1, t_2) \;=\; \fint_{T(F)\backslash T(\mathbb{A})/Z(\mathbb{A})} i_{\bar{v}}(Z_*(g)tt_1, tt_2) dt.$$

The regularized averaging integral on $T(F)\backslash T(\mathbb{A})$ is introduced in §1.6.

To compare with the analytic kernel, we also need to group the pairing in terms of places of F. We have:

$$i(Z_*(g)t_1, t_2) = \sum_{v \in S_F} i_v(Z_*(g)t_1, t_2) \log N_v,$$

$$i_v(Z_*(g)t_1, t_2) = \oint_{T(F) \backslash T(\mathbb{A})/Z(\mathbb{A})} i_{\overline{v}}(Z_*(g)tt_1, tt_2) dt,$$

$$i_{\overline{v}}(Z_*(g)t_1, t_2) = \frac{1}{\#S_{E_v}} \sum_{v \in S_{E_v}} i_{\overline{v}}(Z_*(g)t_1, t_2).$$

Here S_{E_v} denotes the set of places of E lying over v. It has one or two elements. The local pairing

$$i_{\overline{v}}(Z_*(g)t_1, t_2) = \sum_{a \in F^\times} \sum_{x \in \mathbb{B}_f^\times/U} r(g)\phi(x)_a \sum_v i_{\overline{v}}(t_1 x, t_2)$$

is our main goal for the next chapter. We will divide it into a few cases and discuss them in different sections. We will have explicit expressions for $i_{\overline{v}}$ in the case that v is archimedean or the Shimura curve has good reduction at v.

7.3 VANISHING OF THE CONTRIBUTION OF THE HODGE CLASSES

Recall that

$$Z(g, (t_1, t_2)) = \langle Z_*(g)t_1, t_2 \rangle - \langle Z_*(g)\xi_{t_1}, t_2 \rangle + \langle Z_*(g)\xi_{t_1}, \xi_{t_2} \rangle - \langle Z_*(g)t_1, \xi_{t_2} \rangle.$$

Here the pairings on the right-hand side are $\hat{\xi}$-admissible pairings. We have considered a decomposition of the first term on the right-hand side in the last section. The main result here asserts the vanishing of the other terms.

PROPOSITION 7.5. *Assuming Assumption 5.4, then*

$$\langle Z_*(g)\xi_{t_1}, t_2 \rangle = \langle Z_*(g)\xi_{t_1}, \xi_{t_2} \rangle = \langle Z_*(g)t_1, \xi_{t_2} \rangle = 0, \quad \forall \, t_1, t_2 \in T(\mathbb{A}_f).$$

The vanishing follows from the vanishing of the degree of the components of $Z_*(g)$.

7.3.1 Vanishing of the degree of the generating series

For any $\alpha \in F_+^\times \backslash \mathbb{A}_f^\times / q(U)$, the α-component of the generating series is given by $Z(g, \phi)_{U,\alpha} = Z(g, \phi)|_{M_{K,\alpha}}$. Recall that in Proposition 4.2 we have

$$\deg Z(g, \phi)_{U,\alpha} = -\frac{1}{2} \kappa_U^\circ E(0, g, r(h)\phi)_U.$$

Here h is any element of $\mathbb{B}_f^\times$ such that $q(h) \in \alpha^{-1} F_+^\times q(U)$. The weight two Eisenstein series are defined as follows:

$$E(s, g, u, \phi) = \sum_{\gamma \in P^1(F) \backslash \mathrm{SL}_2(F)} \delta(\gamma g)^s r(\gamma g) \phi(0, u),$$

$$E(s, g, \phi)_U = \sum_{u \in \mu_U^2 \backslash F^\times} E(s, g, u, \phi).$$

These series vanish immediately by Assumption 5.4. Thus we have the following lemma.

LEMMA 7.6. *Under Assumption 5.4, the constant term* $Z_0(g, \phi)_{U,\alpha} = 0$ *and the degree* $\deg Z(g, \phi)_{U,\alpha} = 0$ *for any* $\alpha \in F_+^\times \backslash \mathbb{A}_f^\times / q(U)$.

By the lemma, it is easy to get the vanishing of the first two terms in Proposition 7.5. They actually follow from the vanishing of $Z_*(g)\xi_{t_1}$. The correspondences $Z(x)$ are étale on the generic fiber; it keeps the canonical bundle up to a multiple under pull-back and push-forward. Then the Hodge bundles are eigenvectors of all Hecke operators up to translation of components. More precisely, one has

$$Z(x)\xi_{t_1} = (\deg Z(x))\xi_{t_1 x}, \quad \forall x \in \mathbb{B}_f^\times.$$

It follows that for any $\alpha \in F_+^\times \backslash \mathbb{A}_f^\times / q(U)$,

$$Z_*(g)_{U,\alpha} \, \xi_{t_1} = \deg Z_*(g)_{U,\alpha} \, \xi_{\alpha t_1} = 0.$$

It remains to treat $\langle Z_*(g) t_1, \xi_{t_2} \rangle$. Our idea is to "move" the action of $Z_*(g)$ to ξ_{t_2} so that we can apply the above argument. It will be achieved by the projection formula on the integral models. So it depends on extensions of Hecke operators to the integral model.

7.3.2   Integral models of Hecke operators

For convenience, we denote $Z(x) := Z(1^P x)$ for any $x \in \mathbb{B}_P^\times$. Here P can be any subset of S_F such that either P or $S_F - P$ is a finite set. Then it is easy to verify that $Z(x) = Z(x_P)Z(x^P) = Z(x^P)Z(x_P)$ as correspondences for any $x \in \mathbb{B}$.

Let S be a finite subset of non-archimedean places of F containing all v such that U_v is not maximal or that E is ramified at v. Then the models $\mathcal{X}_U$ and $\mathcal{Y}_U$ are smooth away from S. We will have "good extension" of $Z(x)$ away from S.

For any $x \in \mathbb{B}_P^\times$, let $\mathcal{Z}(x)$ be the Zariski closure of $Z(x)$ in $\mathcal{X}_U \times_{O_F} \mathcal{X}_U$. If $x \in (\mathbb{B}_f^S)^\times$, then many good properties of $\mathcal{Z}(x)$ are obtained in [Zh1]. For example, it has a canonical moduli interpretation, and satisfies

(1) $\mathcal{Z}(x_1)$ commutes with $\mathcal{Z}(x_2)$ for any $x_1, x_2 \in (\mathbb{B}_f^S)^\times$;

(2) $\mathcal{Z}(x) = \prod_{v \notin S} \mathcal{Z}(x_v)$ for any $x \in (\mathbb{B}_f^S)^\times$;

(3) for any $x \in (\mathbb{B}_f^S)^\times$, both structure projections from $\mathcal{Z}(x)$ to $\mathcal{X}_U$ are finite everywhere, and étale above the set of places v with $x_v \in U_v$.

Fix $x \in (\mathbb{B}_f^S)^\times$. Define an arithmetic class $D(x)$ on $\mathcal{X}_U$ by

$$D(x) := \mathcal{Z}(x)\hat{\xi} - \widehat{\mathcal{Z}(x)\xi} = \mathcal{Z}(x)\hat{\xi} - \deg Z(x)\ \hat{\xi}.$$

Then $D(x)$ is a vertical divisor since it is zero on the generic fiber. We claim that $D(x)$ is a *constant divisor*, i.e., the pull-back of an arithmetic divisor D on $\mathrm{Spec}(O_F)$. Then it only depends on $\widehat{\deg}(D)$, and sometimes we identify it with this number.

Now we explain why $D(x)$ is a constant divisor. First, $D(x)$ is constant at archimedean places because the Petersson metric on the upper half plane is invariant under the action of $\mathrm{GL}_2(\mathbb{R})_+$. Now we look at non-archimedean places. Both structure morphisms of $\mathcal{Z}(x)$ are étale above S, so the pull-back and push-forward keep $\mathcal{L}_U$ above S since they keep both the relative dualizing sheaf and the Zariski closure of the cusps. Then the finite part of $D(x)$ is lying above primes not in S. Note that $\mathcal{X}_U$ is smooth outside S; its special fibers outside S are irreducible. Hence the finite part of $D(x)$ is a linear combination of these special fibers which are constant.

LEMMA 7.7. *Let v be a non-archimedean place outside S. For any $x \in \mathbb{B}_v^\times$ and any $D \in \mathrm{Div}(X_{U,\overline{F}})$,*

$$\langle Z(x)D, \xi \rangle = \deg Z(x)\ \langle D, \xi \rangle - \deg(D)\ D(x).$$

Here $D(x)$ is viewed as a constant, and $\deg(D)$ is the sum of the degrees of D on all geometrically connected components.

PROOF. We first reduce the problem to the case that D is defined over F. Indeed, since ξ and $Z(x)$ are defined over F, both sides of the equality are invariant under Galois actions on D. So it suffices to show the result for the sum of all Galois conjugates of D.

Assume that D is defined over F. By §7.1,

$$\langle Z(x)D, \xi \rangle = \langle \overline{Z(x)D},\ \hat{\xi} \rangle.$$

Here $\overline{Z(x)D}$ is the Zariski closure in $\mathcal{X}_U$, and we denote the normalized intersection

$$\langle D_1, D_2 \rangle := -D_1 \cdot D_2$$

for any arithmetic divisors D_1, D_2 on $\mathcal{X}_U$.

The correspondence $\mathcal{Z}(x)$ keeps Zariski closure by finiteness of its structure morphisms. It follows that $\overline{Z(x)D} = \mathcal{Z}(x)\overline{D}$. By the projection formula, we get

$$\langle Z(x)D, \xi \rangle = \langle \mathcal{Z}(x)\overline{D},\ \hat{\xi} \rangle = \langle \overline{D},\ \mathcal{Z}(x)^{\mathrm{t}}\ \hat{\xi} \rangle = \langle \overline{D},\ \mathcal{Z}(x^{-1})\ \hat{\xi} \rangle.$$

Here $\mathcal{Z}(x)^{\mathrm{t}}$ denotes the transpose of $\mathcal{Z}(x)$ as correspondences.

We claim that $\mathcal{Z}(x)^t \, \hat{\xi} = \mathcal{Z}(x)\hat{\xi}$ as elements in $\widehat{\mathrm{Pic}}(\mathcal{X}_U)_{\mathbb{Q}}$. We first have $\mathcal{Z}(x)^t = \mathcal{Z}(x^{-1})$ by $Z(x)^t = Z(x^{-1})$. Since U_v is maximal, we know that

$$U_v x^{-1} U_v = q(x)^{-1} U_v \overline{x} U_v = q(x)^{-1} U_v x U_v.$$

Then $Z(x)$ is the composition of $Z(x)^t$ with $Z(q(x)^{-1})$. Note that $Z(q(x)^{-1})$ acts by right multiplication by $q(x)$, which gives a Galois automorphism on $\mathcal{X}_U$ by the reciprocity law. It suffices to show $\mathcal{Z}(q(x)^{-1})$ acts trivially on $\hat{\xi}$, or equivalently the constant $D(q(x)^{-1}) = 0$. It is true since the automorphism has a finite order.

Go back to $\langle Z(x)D, \xi \rangle = \langle \overline{D}, \, \mathcal{Z}(x) \, \hat{\xi} \rangle$. By the definition of $D(x)$,

$$\langle Z(x)D, \xi \rangle = \deg Z(x) \langle \overline{D}, \, \hat{\xi} \rangle + \langle \overline{D}, \, D(x) \rangle.$$

The second pairing
$$\langle \overline{D}, D(x) \rangle = -\deg(D) \; D(x),$$

because $D(x)$ is a constant divisor. $\square$

There is an explicit way to compute $D(x)$ for $x \in (\mathbb{B}_f^S)^\times$ by [Zh1]. We briefly mention it here, since we do not need it in this book. First, $D(x)$ satisfies a "product rule":

$$D(x) = \sum_{v \notin S} \deg Z(x^v) \; D(x_v), \quad x \in (\mathbb{B}_f^S)^\times.$$

It follows from the definition, and is called a "derivation" in the paper. Then the computation of $D(x)$ amounts to each single $D(x_v)$. It can be computed by the deformation method in Proposition 4.3.2 in [Zh1].

7.3.3   Vanishing of the third term

Now we are ready to prove $\langle Z_*(g)t_1, \xi_{t_2} \rangle = 0$. By Galois action of t_1 via the reciprocity law, we have $\langle Z_*(g)t_1, \xi_{t_2} \rangle = \langle Z_*(g)[t_1 t_2^{-1}], \xi_1 \rangle$. So it suffices to consider $\langle Z_*(g)t, \, \xi_1 \rangle$ for any $t \in T(\mathbb{A}_f)$. After separating components, we get

$$\langle Z_*(g)t, \, \xi_1 \rangle = \langle Z_*(g)_{q(1/t)}t, \, \xi_1 \rangle = \langle Z_*(g)_{q(1/t)}t, \, \xi \rangle.$$

It is also equal to $\langle Z_*(g)_{q(1/t)}[1], \, \xi \rangle$ by the Galois action. Hence it only depends on the geometrically connected component of t in X_U.

By (4.2.4),

$$Z_*(g)_{q(1/t)} = \sum_{u \in \mu'_U \backslash F^\times} \sum_{a \in F_+^\times} W_a^{(2)}(g_\infty) \sum_{y \in K^t \backslash \mathbb{B}_f(a)} r(g, (t, 1)) \phi_f(y, u) Z(t^{-1}y).$$

For fixed (a, u), denote the correspondence

$$A := \sum_{y \in K^t \backslash \mathbb{B}_f(a)} r(g, (t, 1)) \phi(y, u) Z(t^{-1}y).$$

We will prove that each $\langle A[1],\ \xi \rangle = 0$.

For each non-archimedean place v, we have the correspondence

$$A_v := \sum_{y \in K_v^t \backslash \mathbb{B}_v(a)} r(g, (t,1)) \phi_v(y, u) Z(t_v^{-1} y).$$

Then A_v equals the identity at almost all v. Furthermore, $A = \prod_{v \nmid \infty} A_v$ as a composition of correspondences.

Use the identity $U_v t_v^{-1} y U_v / U_v = t_v^{-1}(K_v^t y / U_v^1)$ as before. We obtain

$$\deg A_v = \sum_{y \in \mathbb{B}_v(a)/U_v^1} r(g, (t,1)) \phi_v(y, u).$$

It follows that $\deg A_v = 0$ for any $v \in S_2$ by Assumption 5.4 and Lemma 5.10.

Fix a place $v \in S_2$. Then $\deg A_v = 0$ and $\deg A^v = 0$ since S_2 has two elements. We will see that these are the reason for $\langle A[1],\ \xi \rangle = 0$. Write

$$\langle A[1],\ \xi \rangle = \langle A_v A^v[1],\ \xi \rangle = \sum_{y_v \in K_v^t \backslash \mathbb{B}_v(a)} r(g, (t,1)) \phi_v(y_v, u) \langle Z(t_v^{-1} y_v) A^v[1],\ \xi \rangle.$$

Apply Lemma 7.7 to $Z(t_v^{-1} y_v)$. We have

$$\begin{aligned}
\langle A[1],\ \xi \rangle \ =\ & \sum_{y_v \in K_v^t \backslash \mathbb{B}_v(a)} r(g, (t,1)) \phi_v(y_v, u) \deg Z(t_v^{-1} y_v)\ \langle A^v[1],\ \xi \rangle \\
& - \sum_{y_v \in K_v^t \backslash \mathbb{B}_v(a)} r(g, (t,1)) \phi_v(y_v, u)\ \deg(A^v[1])\ D(t_v^{-1} y_v).
\end{aligned}$$

In a manner similar to Lemma 7.7, it is just

$$\langle A[1],\ \xi \rangle \ =\ (\deg A_v)\ \langle A^v[1],\ \xi \rangle - (\deg A^v)\ D(A_v),$$

with

$$D(A_v) = \sum_{y_v \in K_v^t \backslash \mathbb{B}_v(a)} r(g, (t,1)) \phi_v(y_v, u)\ D(t_v^{-1} y_v).$$

The result follows since $\deg A_v = \deg A^v = 0$.

7.4   THE GOAL OF THE NEXT CHAPTER

7.4.1   Difference of the kernel functions

Now let us try to prove Theorem 5.1. We need to prove

$$(I'(0, g, \chi, \phi)_U,\ \varphi(g))_{\mathrm{Pet}} = 2\,(Z(g, \chi, \phi)_U,\ \varphi(g))_{\mathrm{Pet}}\,, \quad \forall \varphi \in \sigma.$$

By Theorem 5.7, we can assume that (ϕ, U) satisfies all the assumptions in §5.2.

Recall that in (5.1.1) and (5.1.2) we have introduced

$$I(s,g,\phi)_U = \sum_{u\in\mu_U^2\backslash F^\times} \sum_{\gamma\in P^1(F)\backslash SL_2(F)} \delta(\gamma g)^s \sum_{x_1\in E} r(\gamma g)\phi(x_1,u),$$

$$I(s,g,\chi,\phi)_U = \int_{T(F)\backslash T(\mathbb{A})/Z(\mathbb{A})}^* I(s,g,r(t,1)\phi)_U\ \chi(t)dt.$$

On the other hand, we have introduced a generating series

$$Z(g,\phi)_U = Z_0(g,\phi)_U + \sum_{a\in F^\times} \sum_{x\in U\backslash \mathbb{B}_f^\times/U} r(g)\phi(x)_a Z(x)_U.$$

It is an automorphic form of $GL_2(\mathbb{A})$ holomorphic of parallel weight two. The coefficients of $Z(g,\phi)_U$ are Hecke operators on the Shimura curve X_U. Then we have the height series

$$Z(g,(t_1,t_2),\phi)_U = \langle Z(g,\phi)_U\ t_1^\circ,\ t_2^\circ\rangle_{NT},$$

$$Z(g,\chi,\phi)_U = \int_{T(F)\backslash T(\mathbb{A})/Z(\mathbb{A})}^* Z(g,(t,1),\phi)_U\ \chi(t)dt.$$

For simplicity, we will omit the dependence of the series on U as before.

By holomorphic projection, it suffices to prove

$$(\mathcal{P}rI'(0,g,\chi,\phi) - 2Z(g,\chi,\phi),\ \varphi(g))_{\text{Pet}} = 0. \tag{7.4.1}$$

The holomorphic projection functors $\mathcal{P}r$ and $\mathcal{P}r'$ are introduced in §6.4. Here $\mathcal{P}r$ denotes the holomorphic projection defined by functoriality, and $\mathcal{P}r'$ is just an algorithm to compute $\mathcal{P}r$. Proposition 6.12 asserts $\mathcal{P}r = \mathcal{P}r'$ under certain growth conditions. In particular, $I'(0,g,\chi,\phi)$ satisfies the growth condition by Proposition 6.14. It follows that

$$\mathcal{P}rI'(0,g,\chi,\phi) = \mathcal{P}r'I'(0,g,\chi,\phi)$$

$$= \int_{T(F)\backslash T(\mathbb{A})/Z(\mathbb{A})}^* \mathcal{P}r'I'(0,g,r(t,1)\phi)\ \chi(t)dt.$$

Note that the integration is actually a finite summation. As for $\mathcal{P}r'I'(0,g,r(t,1)\phi)$, Proposition 6.15 gives

$$\mathcal{P}r'I'(0,g,\phi) = -\sum_{v|\infty}\overline{I'}(0,g,\phi)(v) - \sum_{v\nmid\infty \text{ nonsplit}} I'(0,g,\phi)(v),$$

$$\forall g\in P(F_{S_1})GL_2(\mathbb{A}^{S_1}).$$

Here

$$\overline{I'}(0,g,\phi)(v) = 2\!\!\fint_{Z(\mathbb{A})T(F)\backslash T(\mathbb{A})} \overline{\mathcal{K}}_\phi^{(v)}(g,(t,t))dt, \quad v\mid\infty;$$

$$I'(0,g,\phi)(v) = 2\!\!\fint_{Z(\mathbb{A})T(F)\backslash T(\mathbb{A})} \mathcal{K}_\phi^{(v)}(g,(t,t))dt, \quad v\nmid\infty.$$

The series $\mathcal{K}_\phi^{(v)}(g, (t_1, t_2))$ is a pseudo-theta series on the nearby quaternion algebra $B(v)$. Note that the integrals are essentially finite sums.

On the other hand, Proposition 7.5 asserts

$$Z(g, (t_1, t_2), \phi) = \langle Z_*(g, \phi)t_1, t_2 \rangle.$$

The right-hand side is the $\hat{\xi}$-admissible pairing on the integral model $\mathcal{Y}_U$. By §7.2, it has a decomposition in terms of local arithmetic intersections:

$$
\begin{aligned}
&\langle Z_*(g, \phi)t_1, t_2 \rangle \\
&= -\sum_v i_v(Z_*(g, \phi)t_1, t_2) \log N_v - \sum_v j_v(Z_*(g, \phi)t_1, t_2) \log N_v.
\end{aligned}
$$

Furthermore,

$$
\begin{aligned}
i_v(Z_*(g, \phi)t_1, t_2) &= \fint_{T(F) \backslash T(\mathbb{A})/Z(\mathbb{A})} i_{\bar{v}}(Z_*(g, \phi)tt_1, tt_2) dt, \\
j_v(Z_*(g, \phi)t_1, t_2) &= \fint_{T(F) \backslash T(\mathbb{A})/Z(\mathbb{A})} j_{\bar{v}}(Z_*(g, \phi)tt_1, tt_2) dt.
\end{aligned}
$$

All these identities are valid for $g \in 1_{S_1} \mathrm{GL}_2(\mathbb{A}^{S_1})$.

Putting these two parts together, we have

$$\mathcal{P}r I'(0, g, \chi, \phi) - 2Z(g, \chi, \phi) = 2 \int_{[T]}^* \fint_{[T]} D(g, (tt_1, t), \phi) \chi(t_1) dt_1 dt, \quad (7.4.2)$$

where

$$
\begin{aligned}
&D(g, (t_1, t_2), \phi) \\
&= \sum_{v|\infty} \overline{\mathcal{K}}_\phi^{(v)}(g, (t_1, t_2)) + \sum_{v \nmid \infty \text{ nonsplit}} \overline{\mathcal{K}}_\phi^{(v)}(g, (t_1, t_2)) \\
&\quad - \sum_v i_{\bar{v}}(Z_*(g, \phi)t_1, t_2) \log N_v - \sum_v j_{\bar{v}}(Z_*(g, \phi)t_1, t_2) \log N_v.
\end{aligned}
$$

Here we write $[T]$ for $T(F) \backslash T(\mathbb{A})/Z(\mathbb{A})$ for simplicity.

7.4.2 Main result on local computations of the next chapter

We have already seen that the computation of

$$\mathcal{P}r'(I'(0, g, r(t_1, t_2)\phi)) - 2\langle Z(g, \phi) \, t_1^\circ, t_2^\circ \rangle_{\mathrm{NT}}$$

amounts to the computation of the difference

$$\mathcal{K}_\phi^{(v)}(g, (t_1, t_2)) - i_{\bar{v}}(Z_*(g, \phi)t_1, t_2) \log N_v - j_{\bar{v}}(Z_*(g, \phi)t_1, t_2) \log N_v.$$

In the archimedean case, $\mathcal{K}_\phi^{(v)}$ should be replaced by $\overline{\mathcal{K}}_\phi^{(v)}$.

The following is the main result of the next chapter.

THEOREM 7.8. *Assume all the assumptions in §5.2. Then the following are true for all $t_1, t_2 \in T(\mathbb{A}_f)$ and all $g \in 1_{S_1} \mathrm{GL}_2(\mathbb{A}^{S_1})$:*

(1) If $v \in S_{\mathrm{split}}$, then

$$i_{\overline{v}}(Z_*(g, \phi)t_1, t_2) = j_{\overline{v}}(Z_*(g, \phi)t_1, t_2) = 0.$$

(2) If $v \in S_\infty$, then

$$\overline{\mathcal{K}}_\phi^{(v)}(g, (t_1, t_2)) = i_{\overline{v}}(Z_*(g, \phi)t_1, t_2), \quad j_{\overline{v}}(Z_*(g, \phi)t_1, t_2) = 0.$$

(3) If $v \in S_{\mathrm{nonsplit}} - S_1$, then

$$\overline{\mathcal{K}}_\phi^{(v)}(g, (t_1, t_2)) = i_{\overline{v}}(Z_*(g, \phi)t_1, t_2) \log N_v, \quad j_{\overline{v}}(Z_*(g, \phi)t_1, t_2) = 0.$$

(4) If $v \in S_1$, then there exist Schwartz functions $k_{\phi_v}, m_{\phi_v}, l_{\phi_v} \in \overline{\mathcal{S}}(B(v)_v \times F_v^\times)$ depending on ϕ_v and U_v such that:

$$\mathcal{K}_\phi^{(v)}(g, (t_1, t_2)) = \theta(g, (t_1, t_2), k_{\phi_v} \otimes \phi^v),$$
$$i_{\overline{v}}(Z_*(g, \phi)t_1, t_2) = \theta(g, (t_1, t_2), m_{\phi_v} \otimes \phi^v),$$
$$j_{\overline{v}}(Z_*(g, \phi)t_1, t_2) = \theta(g, (t_1, t_2), l_{\phi_v} \otimes \phi^v).$$

We recall that $B(v)$ is the nearby quaternion algebra of $\mathbb{B}$. The theta series

$$\theta(g, (t_1, t_2), \phi') = \sum_{u \in \mu_U^2 \backslash F^\times} \sum_{y \in B(v)} r(g, (t_1, t_2))\phi'(y, u)$$

for any $\phi' \in \overline{\mathcal{S}}(B(v)_\mathbb{A} \times \mathbb{A}^\times)$. It is automorphic for $g \in \mathrm{GL}_2(\mathbb{A})$.

Some results of the theorem are known. By definition, $j_{\overline{v}} = 0$ in (2) and (3). In (4), the result for $\mathcal{K}_\phi^{(v)}$ has been proved in Corollary 6.9. We know explicit expressions of $\overline{\mathcal{K}}_\phi^{(v)}$ in (2) and $\mathcal{K}_\phi^{(v)}$ in (3) before, so the result in these cases are proved by explicit computation of $i_{\overline{v}}(Z_*(g, \phi)t_1, t_2)$.

7.4.3 Completion of the proof

Recall that we only need to prove Theorem 5.1 under the assumption that the ramification set of $\mathbb{B}$ is equal to Σ. It is a consequence of Theorem 1.3. See §3.6.5 for more details.

Go back to (7.4.2) for the proof of Theorem 5.1. By Theorem 7.8, almost all terms in the summation defining $D(g, (t_1, t_2), \phi)$ are canceled. The remaining terms are just some theta series. More precisely,

$$D(g, (t_1, t_2), \phi) = \sum_{v \in S_1} \theta(g, (t_1, t_2), \phi^v \otimes d_{\phi_v}), \quad \forall g \in 1_{S_1} \mathrm{GL}_2(\mathbb{A}^{S_1}).$$

Here
$$d_{\phi_v} = k_{\phi_v} - (\log N_v) m_{\phi_v} - (\log N_v) l_{\phi_v}$$
is a Schwartz function in $\overline{\mathcal{S}}(B(v)_v \times F_v^\times)$.

In summary, we have proved that

$$\mathcal{P}rI'(0, g, \chi, \phi) - 2Z(g, \chi, \phi) = 2 \int_{[T]}^* \!\!\! \oint_{[T]} \sum_{v \in S_1} \theta(g, (tt_1, t), \phi^v \otimes d_{\phi_v}) \chi(t_1) dt_1 dt$$

is true for $g \in 1_{S_1} \mathrm{GL}_2(\mathbb{A}^{S_1})$. The simple argument mentioned in the introduction asserts that it is true for all $g \in \mathrm{GL}_2(\mathbb{A})$. In fact, the equality is true for all $g \in \mathrm{GL}_2(F)\mathrm{GL}_2(\mathbb{A}^S)$ since both sides are automorphic. Then it is true for all $g \in \mathrm{GL}_2(\mathbb{A})$ because $\mathrm{GL}_2(F)\mathrm{GL}_2(\mathbb{A}^S)$ is dense in $\mathrm{GL}_2(\mathbb{A})$.

To prove (7.4.1), it suffices to prove for all $v \in S_1$,

$$(I(0, g, \chi, \phi^v \otimes d_{\phi_v}), \varphi(g))_{\mathrm{Pet}} = 0.$$

Here

$$I(0, g, \chi, \phi^v \otimes d_{\phi_v}) = \int_{[T]}^* \!\!\! \oint_{[T]} \theta(g, (tt_1, t), \phi^v \otimes d_{\phi_v}) \chi(t_1) dt_1 dt.$$

To make use of the Shimizu lifting, we change the space $\overline{\mathcal{S}}(B(v)_\mathbb{A} \times \mathbb{A}^\times)$ to $\mathcal{S}(B(v)_\mathbb{A} \times \mathbb{A}^\times)$. Let Φ_∞ be a Schwartz function in $\mathcal{S}(\mathbb{B}_\infty \times F_\infty^\times)$ such that $\overline{\Phi}_\infty$ is standard. Then $\Phi(v) := \Phi_\infty \otimes \phi_f^v \otimes d_{\phi_v}$ lies in $\mathcal{S}(B(v)_\mathbb{A} \times \mathbb{A}^\times)$. Introduce

$$I(0, g, \chi, \Phi(v)) = \int_{T(F)\backslash T(\mathbb{A})} \int_{T(F)\backslash T(\mathbb{A})/Z(\mathbb{A})} \theta(g, (tt_1, t), \Phi(v)) \chi(t_1) dt_1 dt,$$

where

$$\theta(g, (t_1, t_2), \Phi(v)) = \sum_{u \in F^\times} \sum_{y \in B(v)} r(g, (t_1, t_2)) \Phi(v)(y, u).$$

Similar to (4.4.4), we can show that there is an explicit positive constant c such that

$$I(0, g, \chi, \Phi(v)) = c \, I(0, g, \chi, \phi^v \otimes d_{\phi_v}).$$

Thus it is reduced to prove

$$(I(0, g, \chi, \Phi(v)), \varphi(g))_{\mathrm{Pet}} = 0.$$

Now the result follows from Theorem 1.3, the result of Tunnell [Tu] and Saito [Sa]. The key is that the theta series $\theta(g, (t_1, t_2), \Phi(v))$ is defined on the nearby quaternion algebra $B(v)$. In fact,

$$(I(0, g, \chi, \Phi(v)), \varphi(g))_{\mathrm{Pet}}$$
$$= \int_{T(F)\backslash T(\mathbb{A})/Z(\mathbb{A})} \int_{T(F)\backslash T(\mathbb{A})/Z(\mathbb{A})} \theta'(\Phi(v) \otimes \varphi)(t_1, t_2) \chi(t_1) \chi^{-1}(t_2) dt_1 dt_2$$

where for $h_1, h_2 \in B(v)_{\mathbb{A}}^{\times}$, the Shimizu lifting

$$\theta'(\Phi(v) \otimes \varphi)(h_1, h_2) = \int_{\mathrm{GL}_2(F) \backslash \mathrm{GL}_2(\mathbb{A})} \theta(g, (h_1, h_2), \Phi(v)) \varphi(g) dg.$$

Note here we use θ' because our normalization is different from (2.2.1). Then we must have

$$\theta'(\Phi(v) \otimes \varphi) = \sum_{i=1}^{r} f_i \otimes f_i', \quad f_i \in \pi(v), \ f_i' \in \tilde{\pi}(v).$$

Here $\pi(v)$ denotes the automorphic representation of $B(v)_{\mathbb{A}}^{\times}$ obtained from σ by the Jacquet-Langland correspondence. It follows that

$$(I(0, g, \chi, \Phi(v)), \ \varphi(g))_{\mathrm{Pet}} = \sum_{i=1}^{r} P_{\chi}(f_i) P_{\chi^{-1}}(f_i').$$

Here

$$P_{\chi}(f_i) = \int_{T(F) \backslash T(\mathbb{A})/Z(\mathbb{A})} f_i(t) dt$$

is the period integral considered in Theorem 1.4. Note that the ramification set of $B(v)$ is not equal to Σ, since we have assumed that $\mathbb{B}$ matches Σ. As a consequence, Theorem 1.3 implies

$$\mathcal{P}(\pi(v), \chi) = \mathrm{Hom}_{E\mathbb{A}^{\times}}(\pi(v) \otimes \chi, \mathbb{C}) = 0.$$

It follows that $P_{\chi}(f_i) = 0$ for all $f \in \pi(v)$. The result is proved.

7.4.4   Rough idea of our proof of Theorem 7.8

We will follow the work of Gross–Zagier [GZ] and its extension in [Zh2] to compute the local intersection numbers. The first step of the proof is to write $i_{\bar{v}}(Z_*(g, \phi)t_1, t_2)$ and $j_{\bar{v}}(Z_*(g, \phi)t_1, t_2)$ in the form of a pseudo-theta series. It is achieved by the v-adic uniformization of the integral model or reduction of the Shimura curve.

To illustrate the idea, consider $i_{\bar{v}}(Z_*(g, \phi)t_1, t_2)$ for a place v split in $\mathbb{B}$ but nonsplit in E. Denote by $B = B(v)$ the nearby quaternion algebra. The height pairing is based on the natural isomorphism

$$\bar{v} : \mathrm{CM}_U = E^{\times} \backslash \mathbb{B}_f^{\times}/U \longrightarrow B^{\times} \backslash (B^{\times} \times_{E^{\times}} \mathbb{B}_v^{\times}/U_v) \times \mathbb{B}_f^{v \times}/U^v.$$

There is a local multiplicity function m on $(B_v^{\times} - E_v^{\times}) \times \mathbb{B}_v^{\times}/U_v$ such that

$$i_{\bar{v}}(\beta_1, t_2) = \sum_{\gamma \in \mu_U \backslash B^{\times}} m(\gamma t_{2v}, \beta_{1v}^{-1}) 1_{U^v}((\beta_1^v)^{-1} \gamma t_2^v)$$

for any two distinct CM points $\beta_1 \in \mathrm{CM}_U$ and $t_2 \in C_U$. By this formula, a simple computation gives

$$i_{\bar{v}}(Z_*(g)t_1, t_2) = \sum_{u \in \mu_U^2 \backslash F^\times} \sum_{y \in B-E} r(g, (t_1, t_2))\phi^v(y, u) \, m_{r(g,(t_1,t_2))\phi_v}(y, u)$$

where the desired m_{ϕ_v} is given by

$$m_{\phi_v}(y, u) = \int_{\mathbb{B}_v^\times} m(y, x^{-1})\phi_v(x, uq(y)/q(x))dx, \quad y \in B_v - E_v, \ u \in F_v^\times.$$

To prove Theorem 7.8 (3) for the i-part, we compute $m_{\phi_v}(y, u)$ and verify that it is the same as the result for $k_{\phi_v}(1, y, u)$ given in Corollary 6.8. The key in this unramified case is the explicit expression of the multiplicity function m of Zhang [Zh2], which was obtained by the technique of Gross [Gro1].

To prove Theorem 7.8 (4) for the i-part, it suffices to prove that m_{ϕ_v} extends to a Schwartz function on $B_v \times F_v^\times$. A priori, m_{ϕ_v} is locally constant in $(B_v - E_v) \times F_v^\times$ with logarithmic singularity along $E_v \times F_v^\times$. The singularity comes from the singularity of m, which corresponds to the case of self-intersection. However, the singularity of m_{ϕ_v} is killed by Assumption 5.3.

Chapter Eight

Local Heights of CM Points

The goal of this chapter is to prove Theorem 7.8, namely, to compute the local heights and compare them with the derivatives computed before. We check the theorem place by place. We assume all the assumptions in §5.2 throughout this chapter.

According to the reduction of the Shimura curve, we divide the situation to the following four cases:

- archimedean case: v is archimedean;

- supersingular case: v isnonsplit in E but split in $\mathbb{B}$;

- superspecial case: v isnonsplit in both E and $\mathbb{B}$;

- ordinary case: v is split in both E and $\mathbb{B}$.

The treatments in different cases are similar in spirit, except that the fourth case is slightly different. Of course, the supersingular case is divided into two subcases: unramified case ($v \notin S_1$) and ramified case ($v \in S_1$).

8.1 ARCHIMEDEAN CASE

In this section we want to describe local heights of CM points at any archimedean place v. Denote $B = B(v)$ and fix an identification $B(\mathbb{A}_f) = \mathbb{B}_f$. We will use the uniformization

$$X_{U,v}(\mathbb{C}) = B_+^\times \backslash \mathcal{H} \times \mathbb{B}_f^\times / U.$$

We follow the treatment of Gross–Zagier [GZ]. See also [Zh2].

8.1.1 Multiplicity function

For any two points $z_1, z_2 \in \mathcal{H}$, the hyperbolic cosine of the hyperbolic distance between them is given by

$$d(z_1, z_2) = 1 + \frac{|z_1 - z_2|^2}{2\mathrm{Im}(z_1)\mathrm{Im}(z_2)}.$$

It is invariant under the action of $\mathrm{GL}_2(\mathbb{R})$. For any $s \in \mathbb{C}$ with $\mathrm{Re}(s) > 0$, denote

$$m_s(z_1, z_2) = Q_s(d(z_1, z_2)),$$

where

$$Q_s(t) = \int_0^\infty \left(t + \sqrt{t^2 - 1} \cosh u \right)^{-1-s} du$$

is the Legendre function of the second kind. Note that

$$Q_0(1 + 2\lambda) = \frac{1}{2} \log(1 + \frac{1}{\lambda}), \quad \lambda > 0.$$

We see that $m_0(z_1, z_2)$ has the right logarithmic singularity.

For any two distinct points of

$$X_{U,v}(\mathbb{C}) = B_+^\times \backslash \mathcal{H} \times B_{\mathbb{A}_f}^\times / U$$

represented by $(z_1, \beta_1), (z_2, \beta_2) \in \mathcal{H} \times B_{\mathbb{A}_f}^\times$, we denote

$$g_s((z_1, \beta_1), (z_2, \beta_2)) = \sum_{\gamma \in \mu_U \backslash B_+^\times} m_s(z_1, \gamma z_2) \, 1_U(\beta_1^{-1} \gamma \beta_2).$$

It is easy to see that the sum is independent of the choice of the representatives $(z_1, \beta_1), (z_2, \beta_2)$, and hence defines a pairing on $X_{U,v}(\mathbb{C})$. Then the local height is given by

$$i_{\bar{v}}((z_1, \beta_1), (z_2, \beta_2)) = \widetilde{\lim}_{s \to 0} \, g_s((z_1, \beta_1), (z_2, \beta_2)).$$

Here $\widetilde{\lim}_{s \to 0}$ denotes the constant term at $s = 0$ of $g_s((z_1, \beta_1), (z_2, \beta_2))$, which converges for $\mathrm{Re}(s) > 0$ and has meromorphic continuation to $s = 0$ with a simple pole.

The definition above uses adelic language, but it is not hard to convert it to the classical language. Observe that $g_s((z_1, \beta_1), (z_2, \beta_2)) \neq 0$ only if there is a $\gamma_0 \in B_+^\times$ such that $\beta_1 \in \gamma_0 \beta_2 U$, which just means that $(z_1, \beta_1), (z_2, \beta_2)$ are in the same connected component. Assuming this, then $(z_2, \beta_2) = (z_2', \beta_1)$ where $z_2' = \gamma_0 z_2$. We have

$$g_s((z_1, \beta_1), (z_2, \beta_2)) = g_s((z_1, \beta_1), (z_2', \beta_1))$$
$$= \sum_{\gamma \in \mu_U \backslash B_+^\times} m_s(z_1, \gamma z_2') \, 1_U(\beta_1^{-1} \gamma \beta_1) = \sum_{\gamma \in \mu_U \backslash \Gamma} m_s(z_1, \gamma z_2').$$

Here we denote $\Gamma = B_+^\times \cap \beta_1 U \beta_1^{-1}$. The connected component of these two points is exactly

$$B_+^\times \backslash \mathcal{H} \times B_+^\times \beta_1 U / U \approx \Gamma \backslash \mathcal{H}, \qquad (z, b\beta_1 U) \mapsto b^{-1} z.$$

The stabilizer of $\mathcal{H}$ in Γ is exactly $\Gamma \cap F^\times = \mu_U$. Now we see that the formula is the same as those in [GZ] and [Zh2].

Next, we consider the special case of CM points. For any $\gamma \in B_{v,+}^\times - E_v^\times$, we have

$$1 + \frac{|z_0 - \gamma z_0|^2}{2\mathrm{Im}(z_0)\mathrm{Im}(\gamma z_0)} = 1 - 2\lambda(\gamma).$$

Here $\lambda(\gamma) = q(\gamma_2)/q(\gamma)$ is introduced at the end of the introduction.

Thus it is convenient to denote

$$m_s(\gamma) = Q_s(1 - 2\lambda(\gamma)), \qquad \gamma \in B_v^\times - E_v^\times.$$

For any two distinct CM points $\beta_1, \beta_2 \in \mathrm{CM}_U$, we obtain

$$g_s(\beta_1, \beta_2) = \sum_{\gamma \in \mu_U \backslash B_+^\times} m_s(\gamma)\, 1_U(\beta_1^{-1}\gamma\beta_2),$$

and

$$i_{\overline{v}}(\beta_1, \beta_2) = \widetilde{\lim}_{s \to 0} g_s(\beta_1, \beta_2).$$

Note that $m_s(\gamma)$ is not well-defined for $\gamma \in E^\times$. The above summation is understood to be

$$g_s(\beta_1, \beta_2) = \sum_{\gamma \in \mu_U \backslash B_+^\times,\ \beta_1^{-1}\gamma\beta_2 \in U} m_s(\gamma).$$

Then it still makes sense because $\beta_1 \neq \beta_2$ implies that $\beta_1^{-1}\gamma\beta_2 \notin U$ for all $\gamma \in E^\times$. Anyway, it is safer to write

$$g_s(\beta_1, \beta_2) = \sum_{\gamma \in \mu_U \backslash (B_+^\times - E^\times)} m_s(\gamma)\, 1_U(\beta_1^{-1}\gamma\beta_2).$$

Now the right-hand side is well-defined even for $\beta_1 = \beta_2$. In this case it can be interpreted as a contribution of some local height by an arithmetic adjunction formula, but we do not need this fact here since there is no self-intersection under our assumptions in §5.2.

8.1.2 Kernel function

We are going to compute the local height

$$i_{\overline{v}}(Z_*(g)t_1, t_2) \quad = \quad \sum_{a \in F^\times} \sum_{x \in \mathbb{B}_f^\times / U} r(g)\phi(x)_a i_{\overline{v}}(t_1 x, t_2).$$

It is well-defined under Assumption 5.3 which kills the self-intersections. The goal is to show that it is equal to $\overline{\mathcal{K}}_\phi^{(v)}(g, (t_1, t_2))$ obtained in Proposition 6.15. We still assume the assumption.

PROPOSITION 8.1. *For any* $t_1, t_2 \in C_U$,

$$i_{\overline{v}}(Z_*(g)t_1, t_2) := \sum_{a \in F^\times} \widetilde{\lim}_{s \to 0} \sum_{y \in \mu_U \backslash (B_+^\times - E^\times)} r(g, (t_1, t_2))\phi(y)_a m_s(y).$$

In particular, $i_{\overline{v}}(Z_*(g)t_1, t_2) = \overline{\mathcal{K}}_\phi^{(v)}(g, (t_1, t_2))$.

PROOF. By the above formula,

$$i_{\overline{v}}(Z_*(g)t_1, t_2)$$

$$= \sum_{a \in F^\times} \sum_{x \in \mathbb{B}_f^\times / U} r(g)\phi(x)_a \widetilde{\lim}_{s \to 0} \sum_{\gamma \in \mu_U \backslash (B_+^\times - E^\times)} m_s(\gamma) \, 1_U(x^{-1}t_1^{-1}\gamma t_2)$$

$$= \sum_{a \in F^\times} \widetilde{\lim}_{s \to 0} \sum_{\gamma \in \mu_U \backslash (B_+^\times - E^\times)} r(g)\phi(t_1^{-1}\gamma t_2)_a m_s(\gamma)$$

$$= \sum_{a \in F^\times} \widetilde{\lim}_{s \to 0} \sum_{\gamma \in \mu_U \backslash (B_+^\times - E^\times)} r(g, (t_1, t_2))\phi(\gamma)_a m_s(\gamma).$$

Here the second equality is obtained by replacing x by $t_1^{-1}\gamma t_2$. $\qquad\square$

We want to compare the above result with the holomorphic projection

$$\overline{\mathcal{K}}_\phi^{(v)}(g, (t_1, t_2)) = \sum_{a \in F^\times} \widetilde{\lim}_{s \to 0} \sum_{y \in \mu_U \backslash (B_+^\times - E^\times)} r(g, (t_1, t_2))\phi(y)_a \, k_{v,s}(y)$$

computed in Proposition 6.15.

It amounts to compare

$$m_s(y) = Q_s(1 - 2\lambda(y))$$

with

$$k_{v,s}(y) = \frac{\Gamma(s+1)}{2(4\pi)^s} \int_1^\infty \frac{1}{t(1 - \lambda_v(y)t)^{s+1}} dt.$$

By the result of Gross–Zagier,

$$\int_1^\infty \frac{1}{t(1 - \lambda t)^{s+1}} dt = 2Q_s(1 - 2\lambda) + O(|\lambda|^{-s-2}), \quad \lambda \to -\infty,$$

and the error term vanishes at $s = 0$. We conclude that

$$\overline{\mathcal{K}}_\phi^{(v)}(g, (t_1, t_2)) = i_{\overline{v}}(Z_*(g)t_1, t_2).$$

8.2   SUPERSINGULAR CASE

Let v be a finite prime of F nonsplit in E but split in $\mathbb{B}$. We consider the local pairing $i_{\overline{v}}$, which depends on the fixed embeddings $H \subset \overline{E} \subset \overline{E}_v$ and the model $\mathcal{Y}_U$ over O_H. It actually depends only on the local integral model $\mathcal{Y}_{U,w} = \mathcal{Y}_U \times_{O_H} O_{H_w}$ where w is the place of H induced by the embeddings. We will use the local multiplicity functions treated in Zhang [Zh2]. For more details, we refer to that paper.

8.2.1   Multiplicity function

Let $B = B(v)$ be the nearby quaternion algebra over F. Make an identification $B(\mathbb{A}^v) = \mathbb{B}^v$. Then the set of supersingular points on X_K over v is parametrized by

$$\mathbb{S}_U = B^\times \backslash (F_v^\times / \det(U_v)) \times (\mathbb{B}_f^{v\times}/U^v).$$

We have a natural isomorphism

$$\bar{v} : \mathrm{CM}_U = E^\times \backslash \mathbb{B}_f^\times / U \longrightarrow B^\times \backslash (B^\times \times_{E^\times} \mathbb{B}_v^\times / U_v) \times \mathbb{B}_f^{v\times}/U^v$$

sending β to $(1, \beta_v, \beta^v)$. The reduction map $\mathrm{CM}_U \to \mathbb{S}_U$ is given by taking norm on the first factor:

$$q : B^\times \times_{E^\times} \mathbb{B}_v^\times \longrightarrow F_v^\times, \quad (b, \beta) \longmapsto q(b)q(\beta).$$

The intersection pairing is given by a multiplicity function m on

$$\mathcal{H}_{U_v} := B_v^\times \times_{E_v^\times} \mathbb{B}_v^\times / U_v.$$

More precisely, the intersection of two points $(b_1, \beta_1), (b_2, \beta_2) \in \mathcal{H}_{U_v}$ is given by

$$g_v((b_1, \beta_1), (b_2, \beta_2)) = m(b_1^{-1}b_2, \beta_1^{-1}\beta_2).$$

The multiplicity function m is defined everywhere in $\mathcal{H}_{U_v}$ except at the image of $(1, 1)$. It satisfies the property

$$m(b, \beta) = m(b^{-1}, \beta^{-1}).$$

LEMMA 8.2. *For any two distinct CM points $\beta_1 \in \mathrm{CM}_U$ and $t_2 \in C_U$, their local height is given by*

$$i_{\bar{v}}(\beta_1, t_2) = \sum_{\gamma \in \mu_U \backslash B^\times} m(\gamma t_{2v}, \beta_{1v}^{-1}) 1_{U^v}((\beta_1^v)^{-1}\gamma t_2^v).$$

PROOF. Like the archimedean case, we compute the height by pulling back to $\mathcal{H}_{U_v} \times \mathbb{B}_f^{v\times}$. The height is the sum over $\gamma \in \mu_U \backslash B^\times$ of the intersection of $(1, \beta_{1v}, \beta_1^v)$ with $\gamma(1, t_{2v}, t_2^v) = (\gamma, t_{2v}, \gamma t_2^v) = (\gamma t_{2v}, 1, \gamma t_2^v)$ on $\mathcal{H}_{U_v} \times \mathbb{B}_f^{v\times}$. □

Analogous to the archimedean case, the summation is well-defined for all $\beta_1 \neq t_2$. Indeed, assume that there is a $\gamma \in E^\times$ such that $(\beta_1^v)^{-1}\gamma t_2^v \in U^v$ and $m(\gamma t_{2v}, \beta_{1v}^{-1})$ is not well-defined. Then we must have $\gamma t_{2v} \in E_v^\times$ and $\beta_{1v}^{-1}\gamma t_{2v} \in U^v$. It forces $\gamma \in E^\times$ and $\gamma t_2 \in \beta_1 U$, which implies that $\beta_1 = t_2 \in \mathrm{CM}_U$.

8.2.2   The kernel function

Now we compute

$$i_{\overline{v}}(Z_*(g)t_1, t_2) = \sum_{a \in F^\times} \sum_{x \in \mathbb{B}_f^\times/U} r(g)\phi(x)_a i_{\overline{v}}(t_1 x, t_2).$$

As in the archimedean case, we assume that ϕ is degenerate at two different finite places v_1, v_2 of F which arenonsplit in E, and only consider g in $P(F_{v_1,v_2})\mathrm{GL}_2(\mathbb{A}^{v_1,v_2})$.

By the above formula, we have

$$i_{\overline{v}}(Z_*(g)t_1, t_2)$$
$$= \sum_{a \in F^\times} \sum_{x \in \mathbb{B}_f^\times/U} r(g)\phi(x)_a \sum_{\gamma \in \mu_U \backslash B^\times} m(\gamma t_2, x^{-1}t_1^{-1}) 1_{U^v}(x^{-1}t_1^{-1}\gamma t_2).$$

Replace x^v by $t_1^{-1}\gamma t_2$ and then get

$$\sum_{a \in F^\times} \sum_{\gamma \in \mu_U \backslash B^\times} r(g)\phi^v(t_1^{-1}\gamma t_2)_a \sum_{x_v \in \mathbb{B}_v^\times/U_v} r(g)\phi_v(x_v)_a m(t_1^{-1}\gamma t_2, x^{-1})$$
$$= \sum_{u \in \mu_U^2 \backslash F^\times} \sum_{\gamma \in B^\times} r(g,(t_1,t_2))\phi^v(\gamma, u)$$
$$\cdot \sum_{x_v \in \mathbb{B}_v^\times/U_v} r(g)\phi_v(x_v, uq(\gamma)/q(x_v)) m(t_1^{-1}\gamma t_2, x^{-1}).$$

For convenience, we introduce the following notation.

NOTATION 8.3.

$$m_{\phi_v}(y, u) = \sum_{x \in \mathbb{B}_v^\times/U_v} m(y, x^{-1})\phi_v(x, uq(y)/q(x))$$
$$= \sum_{x \in \mathbb{B}_v^\times/U_v} m(y^{-1}, x)\phi_v(x, uq(y)/q(x)), \quad (y, u) \in (B_v - E_v) \times F_v^\times.$$

Notice that $m_{\phi_v}(y, u)$ is well-defined for $y \notin E_v$ since $m(y, x^{-1})$ has no singularity for such y. By this notation, we obtain the following result.

PROPOSITION 8.4.   For $g \in P(F_{S_1})\mathrm{GL}_2(\mathbb{A}^{S_1})$,

$$i_{\overline{v}}(Z_*(g)t_1, t_2) = \sum_{u \in \mu_U^2 \backslash F^\times} \sum_{y \in B-E} r(g,(t_1,t_2))\phi^v(y, u)\, m_{r(g,(t_1,t_2))\phi_v}(y, u).$$

Here we can change the summation to $y \in B - E$ since the contribution of $y \in E$ is zero by Assumption 5.3. We should compare the following result with Lemma 6.6 for $k_{\phi_v}(g, y, u)$.

LEMMA 8.5. *The function $m_{\phi_v}(y, u)$ behaves like Weil representation under the action of $P(F_v) \times (E_v^\times \times E_v^\times)$ on (y, u). Namely,*

$$m_{r(g,(t_1,t_2))\phi_v}(y, u) = r(g, (t_1, t_2))m_{\phi_v}(y, u), \quad (g, (t_1, t_2)) \in P(F_v) \times (E_v^\times \times E_v^\times).$$

More precisely,

$$
\begin{aligned}
m_{r(m(a))\phi_v}(y, u) &= |a|^2 m_{\phi_v}(ay, u), \quad a \in F_v^\times, \\
m_{r(n(b))\phi_v}(y, u) &= \psi(buq(y))m_{\phi_v}(y, u), \quad b \in F_v, \\
m_{r(d(c))\phi_v}(y, u) &= |c|^{-1} m_{\phi_v}(y, c^{-1}u), \quad c \in F_v^\times, \\
m_{r(t_1,t_2)\phi_v}(g, y, u) &= m_{\phi_v}(t_1^{-1} y t_2, q(t_1 t_2^{-1})u), \quad (t_1, t_2) \in E_v^\times \times E_v^\times.
\end{aligned}
$$

PROOF. They follow from some basic properties of the multiplicity function $m(x, y)$. We only verify the first identity:

$$
\begin{aligned}
&m_{r(m(a))\phi_v}(y, u) \\
&= \sum_{x \in \mathbb{B}_v^\times / U_v} m(y^{-1}, x) r(g_v) \phi_v(ax, uq(y)/q(x))|a|^2 \\
&= |a|^2 \sum_{x \in \mathbb{B}_v^\times / U_v} m(y^{-1}, a^{-1}x) r(g_v) \phi_v(x, uq(y)/q(a^{-1}x)) \\
&= |a|^2 \sum_{x \in \mathbb{B}_v^\times / U_v} m((ay)^{-1}, x) r(g_v) \phi_v(x, uq(ay)/q(x)) \\
&= |a|^2 m_{\phi_v}(ay, u).
\end{aligned}
$$

$\square$

8.2.3   Unramified case

Fix an isomorphism $\mathbb{B}_v = M_2(F_v)$. In this subsection we compute $m_{\phi_v}(y, u)$ in the following unramified case:

(1) ϕ_v is the characteristic function of $M_2(O_{F_v}) \times O_{F_v}^\times$;

(2) U_v is the maximal compact subgroup $GL_2(O_{F_v})$.

By [Zh2], there is a decomposition

$$GL_2(F_v) = \coprod_{c=0}^{\infty} E_v^\times h_c GL_2(O_{F_v}), \quad h_c = \begin{pmatrix} 1 & 0 \\ 0 & \varpi^c \end{pmatrix}. \tag{8.2.1}$$

The following result is Lemma 5.5.2 in [Zh2]. There is a small mistake in the original statement. Here is the corrected one.

LEMMA 8.6. *The multiplicity function $m(b, \beta) \neq 0$ only if $q(b)q(\beta) \in O_{F_v}^\times$.*
In this case, assume that $\beta \in E_v^\times h_c \mathrm{GL}_2(O_{F_v})$. Then

$$m(b, \beta) = \begin{cases} \frac{1}{2}(\mathrm{ord}_v \lambda(b) + 1) & \text{if } c = 0; \\ N_v^{1-c}(N_v + 1)^{-1} & \text{if } c > 0, \, E_v/F_v \text{ is unramified}; \\ \frac{1}{2} N_v^{-c} & \text{if } c > 0, \, E_v/F_v \text{ is ramified}. \end{cases}$$

Here $\lambda(b) = q(b_2)/q(b)$ and b_2 is given by the orthogonal decomposition $b = b_1 + b_2$ with respect to $B_v = E_v + E_v j_v$.

PROPOSITION 8.7. *The function $m_{\phi_v}(y, u) \neq 0$ only if $(y, u) \in O_{B_v} \times O_{F_v}^\times$.*
In this case,

$$m_{\phi_v}(y, u) = \frac{1}{2}(\mathrm{ord}_v q(y_2) + 1).$$

PROOF. We will use Lemma 8.6. Recall that

$$m_{\phi_v}(y, u) = \sum_{x \in \mathrm{GL}_2(F_v)/U_v} m(y^{-1}, x)\phi_v(x, uq(y)/q(x)).$$

Note that $m(y^{-1}, x) \neq 0$ only if $\mathrm{ord}_v(q(x)/q(y)) = 0$. Under this condition, $\phi_v(x, uq(y)/q(x)) \neq 0$ if and only if $u \in O_{F_v}^\times$ and $x \in M_2(O_{F_v})$. It follows that $m_{\phi_v}(y, u) \neq 0$ only if $u \in O_{F_v}^\times$ and $n = \mathrm{ord}(q(y)) \geq 0$. Assuming these two conditions, we have

$$m_{\phi_v}(y, u) = \sum_{x \in M_2(O_{F_v})_n/U_v} m(y^{-1}, x),$$

where $M_2(O_{F_v})_n$ denotes the set of integral matrices whose determinants have valuation n. Using decomposition (8.2.1), we obtain

$$m_{\phi_v}(y, u) = \sum_{c=0}^\infty m(y^{-1}, h_c)\mathrm{vol}(E_v^\times h_c \mathrm{GL}_2(O_{F_v}) \cap M_2(O_{F_v})_n).$$

We first consider the case that E_v/F_v is unramified. The set in the right-hand side is non-empty only if $n - c$ is even and non-negative. In this case it is given by

$$\varpi^{(n-c)/2} O_{E_v}^\times h_c U_v.$$

The volume of this set is 1 if $c = 0$ and $N_v^{c-1}(N_v+1)$ if $c > 0$ by the computation of [Zh2, p. 101]. It follows that, for $c > 0$ with $2 \mid (n - c)$,

$$m(y^{-1}, h_c)\mathrm{vol}(E_v^\times h_c \mathrm{GL}_2(O_{F_v}) \cap M_2(O_{F_v})_n) = 1.$$

If n is even,

$$m_{\phi_v}(y, u) = \frac{1}{2}(\mathrm{ord}_v \lambda(y) + 1) + \frac{n}{2} = \frac{1}{2}(\mathrm{ord}_v q(y_2) + 1).$$

If n is odd, $m_{\phi_v}(y, u) = \frac{1}{2}(n + 1)$. It is easy to see that $\mathrm{ord}_v q(y_1)$ is even and $\mathrm{ord}_v q(y_2)$ is odd. Then $n = \mathrm{ord}_v q(y_2)$, since

$$n = \mathrm{ord}_v q(y) = \min\{\mathrm{ord}_v q(y_1), \mathrm{ord}_v q(y_2)\}$$

is odd. We still get $m_{\phi_v}(y, u) = \frac{1}{2}(\mathrm{ord}_v q(y_2) + 1)$ in this case.

Now assume that E_v/F_v is ramified. Then the condition that $2 \mid (n - c)$ is unnecessary, and $\mathrm{vol}(E_v^\times h_c \mathrm{GL}_2(O_{F_v}) \cap M_2(O_{F_v})_n) = N_v^c$. Thus

$$m_{\phi_v}(y, u) = \frac{1}{2}(\mathrm{ord}_v \lambda(y) + 1) + n \cdot \frac{1}{2} = \frac{1}{2}(\mathrm{ord}_v q(y_2) + 1).$$

$\square$

We immediately see that in the unramified case, m_{ϕ_v} matches the analytic kernel k_{ϕ_v} computed in Proposition 6.8.

PROPOSITION 8.8. *Let* $v \in S_{\mathrm{nonsplit}} - S_1$ *be as in Assumption 5.5. Then*

$$k_{r(t_1, t_2)\phi_v}(g, y, u) = m_{r(g, (t_1, t_2))\phi_v}(y, u) \log N_v,$$

and thus

$$\mathcal{K}_\phi^{\cdot(v)}(g, (t_1, t_2)) = i_{\overline{v}}(Z_*(g)t_1, t_2) \log N_v.$$

PROOF. The case $(g, t_1, t_2) = (1, 1, 1)$ follows from the above result and Corollary 6.8. It is also true for $g \in \mathrm{GL}_2(O_{F_v})$ since it is easy to see that such g has the same kernel functions as 1 for standard ϕ_v. For the general case, apply the action of $P(F_v)$ and $E_v^\times \times E_v^\times$. The equality follows from Proposition 6.6 and Lemma 8.5. $\square$

8.2.4   Ramified case

Now we consider general U_v. By Proposition 8.7 for the unramified case, we know that m_{ϕ_v} may have logarithmic singularity around the boundary $E_v \times F_v^\times$. The singularity is caused by self-intersections in the computation of local multiplicity. However, we will see that there is no singularity if $\phi_v \in \overline{\mathcal{S}}^1(\mathbb{B}_v \times F_v^\times)$ is degenerate.

PROPOSITION 8.9. *Assume that* $\phi_v \in \overline{\mathcal{S}}^1(\mathbb{B}_v \times F_v^\times)$ *and it is invariant under the right action of* U_v. *Then* $m_{\phi_v}(y, u)$ *can be extended to a Schwartz function for* $(y, u) \in B_v \times F_v^\times$.

PROOF. By the choice of ϕ_v, there is a constant $c > 0$ such that $\phi_v(x, u) \neq 0$ only if $-c < v(q(x)) < c$ and $-c < v(u) < c$. Recall that

$$m_{\phi_v}(y, u) = \frac{1}{\mathrm{vol}(U_v)} \int_{\mathbb{B}_v^\times} m(y^{-1}, x)\phi_v(x, uq(y)/q(x))dx.$$

In order that $m(y^{-1}, x) \neq 0$, we have to make $q(y)/q(x) \in q(U_v)$ and thus $v(q(y)) = v(q(x))$. It follows that $\phi_v(x, uq(y)/q(x)) \neq 0$ only if $-c < v(u) < c$. The same is true for $m_{\phi_v}(y, u)$ by looking at the integral above. Then it is easy to see that $m_{\phi_v}(y, u)$ is Schwartz for $u \in F_v^\times$.

On the other hand, $m_{\phi_v}(y, u) \neq 0$ only if $-c < v(q(y)) < c$, which follows from the fact that $\phi_v(x, uq(y)/q(x)) \neq 0$ only if $-c < v(q(x)) < c$. Extend m_{ϕ_v} to $B_v \times F^\times$ by taking zero outside $B_v^\times \times F^\times$. We only need to show that it is locally constant in $B_v^\times \times F^\times$.

We have $\phi_v(E_v U_v, F_v^\times) = 0$ by the Assumption 5.3 and the invariance of ϕ_v under U_v. Thus

$$m_{\phi_v}(y, u) = \frac{1}{\mathrm{vol}(U_v)} \int_{\mathbb{B}_v^\times} m(y^{-1}, x)(1 - 1_{E_v^\times U_v}(x))\phi_v(x, uq(y)/q(x))dx.$$

It is locally constant in $B_v^\times \times F_v^\times$, since $m(y^{-1}, x)(1 - 1_{E_v^\times U_v}(x))$ is locally constant as a function on $B_v^\times \times \mathbb{B}_v^\times$. This completes the proof. $\qquad\square$

As in the analytic case, we want to approximate the above pseudo-theta series for $i_{\bar{v}}(Z_*(g)t_1, t_2)$ by the usual theta series

$$\theta(g, (t_1, t_2), m_{\phi_v} \otimes \phi^v)$$
$$= \sum_{u \in \mu_U^2 \backslash F^\times} \sum_{y \in V} r(g, (t_1, t_2))m_{\phi_v}(y, u)\, r(g, (t_1, t_2))\phi^v(y, u).$$

The following result is parallel to Corollary 6.9.

COROLLARY 8.10. *Let $v \in S_1 - \Sigma$ satisfy Assumption 5.3. Then*

$$i_{\bar{v}}(Z_*(g, \phi)t_1, t_2) = \theta(g, (t_1, t_2), m_{\phi_v} \otimes \phi^v), \quad \forall g \in 1_{S_1} \mathrm{GL}_2(\mathbb{A}^{S_1}).$$

8.3   SUPERSPECIAL CASE

Let v be a finite prime of F nonsplit in both $\mathbb{B}$ and E, and we consider the local height $i_{\bar{v}}(Z_*(g)t_1, t_2)$. The Shimura curve always has a bad reduction at v due to the ramification of the quaternion algebra. We only control the singularities as in Proposition 8.9. It is enough for approximation since there are only finitely many places nonsplit in $\mathbb{B}$.

In fact, all such places lie in S_1 and we can use Assumption 5.3 as above. Most of the definitions and computations are similar to the supersingular case and will be mentioned briefly. Meanwhile, we will pay special attention to the parts that are different to the supersingular case.

8.3.1   Kernel function

Denote by $B = B(v)$ the nearby quaternion algebra. We fix identifications $B_v \simeq M_2(F_v)$ and $B(\mathbb{A}_f^{\,v}) \simeq \mathbb{B}_f^v$. The intersection pairing is given by a multiplicity function m on

$$\mathcal{H}_{U_v} := B_v^\times \times_{E_v^\times} \mathbb{B}_v^\times / U_v.$$

More precisely, the intersection of two points $(b_1, \beta_1), (b_2, \beta_2) \in \mathcal{H}_{U_v}$ is given by

$$g_v((b_1, \beta_1), (b_2, \beta_2)) = m(b_1^{-1}b_2, \beta_1^{-1}\beta_2).$$

The multiplicity function m is defined everywhere on $\mathcal{H}_{U_v}$ except at the image of $(1, 1)$. It satisfies the property

$$m(b, \beta) = m(b^{-1}, \beta^{-1}).$$

For any two distinct CM points $\beta_1 \in \mathrm{CM}_U$ and $t_2 \in C_U$, their local height is given by

$$i_{\overline{v}}(\beta_1, t_2) = \sum_{\gamma \in \mu_U \backslash B^\times} m(\gamma t_{2v}, \beta_{1v}^{-1}) 1_{U^v}((\beta_1^v)^{-1}\gamma t_2^v).$$

Analogous to Proposition 8.4, we have the following result.

PROPOSITION 8.11. *For* $g \in P(F_{S_1})\mathrm{GL}_2(\mathbb{A}^{S_1})$,

$$i_{\overline{v}}(Z_*(g, \phi)t_1, t_2) = \sum_{u \in \mu_U^2 \backslash F^\times} \sum_{y \in B-E} r(g, (t_1, t_2))\phi^v(y, u) \, m_{r(g,(t_1,t_2))\phi_v}(y, u).$$

Here we use the same notation:

$$m_{\phi_v}(y, u) = \sum_{x \in \mathbb{B}_v^\times / U_v} m(y^{-1}, x)\phi_v(x, uq(y)/q(x)), \quad (y, u) \in (B_v - E_v) \times F_v^\times.$$

Lemma 8.5 is still true. It says that the action of $P(F_v) \times (E_v^\times \times E_v^\times)$ on $m_{r(g,(t_1,t_2))\phi_v}(y, u)$ behaves like Weil representation.

The following is a basic result used to control the singularity of the series. Its proof is given in the next two subsections.

LEMMA 8.12.   *(1) If* v *is unramified in* E, *then* $m(b, \beta) \neq 0$ *only if*

$$\mathrm{ord}_v(q(b)q(\beta)) = 0, \quad b \in F_v^\times \mathrm{GL}_2(O_{F_v}).$$

(2) If v *is ramified in* E, *then* $m(b, \beta) \neq 0$ *only if*

$$\mathrm{ord}_v(q(b)q(\beta)) = 0, \quad b \in F_v^\times \mathrm{GL}_2(O_{F_v}) \bigcup \begin{pmatrix} & 1 \\ \varpi_v & \end{pmatrix} F_v^\times \mathrm{GL}_2(O_{F_v}).$$

The main result below is parallel to Proposition 8.9.

PROPOSITION 8.13. *Assume* $\phi_v \in \overline{\mathcal{S}}^1(\mathbb{B}_v \times F_v^\times)$ *is invariant under the right action of* U_v. *Then* $m_{\phi_v}(y, u)$ *can be extended to a Schwartz function for* $(y, u) \in B_v \times F_v^\times$.

PROOF. The proof is very similar to Proposition 8.9. By the argument of Proposition 8.9, there is a constant $C > 0$ such that $m_{\phi_v}(y, u) \neq 0$ only if $-C < v(q(y)) < C$ and $-C < v(u) < C$. Extend m_{ϕ_v} to $B_v \times F^\times$ by taking zero outside $B_v^\times \times F^\times$. The same method shows that it is locally constant on $B_v^\times \times F^\times$. It is compactly supported in y by Lemma 8.12 since $v(q(y))$ is bounded.   □

As in the analytic case and the supersingular case, denote

$$\theta(g,(t_1,t_2),m_{\phi_v}\otimes\phi^v) = \sum_{u\in\mu_U^2\backslash F^\times}\sum_{y\in V} r(g,(t_1,t_2))m_{\phi_v}(y,u)\, r(g,(t_1,t_2))\phi^v(y,u).$$

Then it approximates the original series as in Corollary 6.9 and Corollary 8.10.

COROLLARY 8.14. *Let* $v\in S_1\cap\Sigma$ *satisfy Assumption 5.3. Then*

$$i_{\overline{v}}(Z_*(g,\phi)t_1,t_2) = \theta(g,(t_1,t_2),m_{\phi_v}\otimes\phi^v),\quad \forall g\in 1_{S_1}\mathrm{GL}_2(\mathbb{A}^{S_1}).$$

8.3.2 Support of the multiplicity function: unramified quadratic extension

Here we verify Lemma 8.12 assuming that v is unramified in E. The case that v is ramified in E is slightly different and considered in the next subsection. The idea is very simple: two points in $\mathcal{H}_{U_v} = B_v^\times \times_{E^\times}\mathbb{B}_v^\times/U_v$ have a nonzero intersection only if they specialize to the same irreducible component of the special fiber in the related formal neighborhood.

We first look at the case of full level, for which the integral model is very clear by the Cherednik–Drinfeld uniformization. We can easily have an explicit expression for the multiplicity function, but we do not need it in this book.

Assume that $U_v = O_{\mathbb{B}_v}^\times$ is maximal and U^v is sufficiently small. By the reciprocity law, all points in CM_U are defined over $\widehat{E}_v^{\mathrm{ur}}$, the completion of the maximal unramified extension of E_v. We have $\widehat{E}_v^{\mathrm{ur}} = \widehat{F}_v^{\mathrm{ur}}$ since E_v is unramified over F_v. In particular, the field H_w is unramified over F_v, and the model $\mathcal{Y}_{U,w} = \mathcal{X}_U\times_{O_F}O_{H_w}$ since the latter is still regular. It suffices to compute the intersections over $\mathcal{X}_U\times_{O_F}O_{\widehat{F}_v^{\mathrm{ur}}}$.

The rigid analytic space X_U^{an} has the uniformization

$$X_U^{\mathrm{an}}\widehat{\otimes}\widehat{F}_v^{\mathrm{ur}} = B^\times\backslash(\Omega\widehat{\otimes}\widehat{F}_v^{\mathrm{ur}})\times\mathbb{Z}\times\mathbb{B}_f^{v\times}/U^v.$$

Here Ω is the rigid analytic upper half plane over F_v. More importantly, it has the integral version

$$\widehat{\mathcal{X}}_U\widehat{\otimes}O_{\widehat{F}_v^{\mathrm{ur}}} = B^\times\backslash(\widehat{\Omega}\widehat{\otimes}O_{\widehat{F}_v^{\mathrm{ur}}})\times\mathbb{Z}\times\mathbb{B}_f^{v\times}/U^v.$$

Here $\widehat{\mathcal{X}}_U$ denotes the formal completion of the integral model $\mathcal{X}_U$ along the special fiber over v, and $\widehat{\Omega}$ is the formal model of Ω over O_{F_v} obtained by successive blowing-ups of rational points on the special fibers of the scheme $\mathbb{P}^1_{O_{F_v}}$.

The formal model $\widehat{\Omega}$ is regular and semistable. Its special fiber is a union of $\mathbb{P}^1$'s indexed by scalar equivalence class of O_{F_v}-lattices of F_v^2, and acted transitively by $\mathrm{GL}_2(F_v)$. Hence these irreducible components are parametrized by

$$\mathrm{GL}_2(F_v)/F_v^\times\mathrm{GL}_2(O_{F_v}).$$

It follows that the set $\mathcal{V}_U$ of irreducible components of $\mathcal{X}_U \times O_{\widehat{F}_v^{\mathrm{ur}}}$ can be indexed as

$$\mathcal{V}_U = B^\times \backslash (\mathrm{GL}_2(F_v)/F_v^\times \mathrm{GL}_2(O_{F_v})) \times \mathbb{Z} \times \mathbb{B}_f^{v\times}/U^v.$$

Consider the set of CM points

$$\mathrm{CM}_U = E^\times \backslash B_{\mathbb{A}_f}^\times / U = B^\times \backslash (B^\times \times_{E^\times} \mathbb{B}_v^\times / U_v) \times \mathbb{B}_f^{v\times}/U^v.$$

The embedding $\mathrm{CM}_U \to X_U^{\mathrm{an}}$ is given by

$$\mathcal{H}_{U_v} \longrightarrow (\Omega \widehat{\otimes} \widehat{F}_v^{\mathrm{ur}}) \times \mathbb{Z}, \quad (b, \beta) \longmapsto (bz_0, \mathrm{ord}_v(q(b)q(\beta))).$$

Here $\mathcal{H}_{U_v} = B_v^\times \times_{E^\times} \mathbb{B}_v^\times / U_v$ is the space where the multiplicity function is defined, and z_0 is a point in $\Omega(E_v)$ fixed by $E_v^\times$. Since all CM points are defined over $\widehat{F}_v^{\mathrm{ur}}$, their reductions are smooth points. The reduction map $\mathrm{CM}_U \to \mathcal{V}_U$ is given by the $B_v^\times$-equivariant map

$$\mathcal{H}_{U_v} \longrightarrow (\mathrm{GL}_2(F_v)/F_v^\times \mathrm{GL}_2(O_{F_v})) \times \mathbb{Z}, \quad (b, \beta) \longmapsto (bb_0, \mathrm{ord}_v(q(b)q(\beta))).$$

Here b_0 represents the irreducible component of the reduction of z_0.

Consider the multiplicity function $m(b, \beta)$. It is equal to the intersection of Zariski closures of the points (b, β) and $(1, 1)$ in $(\widehat{\Omega} \widehat{\otimes} O_{\widehat{F}_v^{\mathrm{ur}}}) \times \mathbb{Z}$. If their intersection is nonzero, they have to lie in the same irreducible component on the special fiber. It is true if and only if $b \in F_v^\times \mathrm{GL}_2(O_{F_v})$ and $\mathrm{ord}_v(q(b)q(\beta)) = 0$. It gives the lemma.

Next, we assume that $U_v = 1 + p_v^r O_{\mathbb{B}_v}$ is general. Denote $U^0 = O_{\mathbb{B}_v}^\times \times U^v$. By the construction in §7.2, there is a morphism $\mathcal{Y}_{U,w} \to \mathcal{X}_{U^0,v}$. It induces a map $\mathcal{V}_U \to \mathcal{V}_{U^0}$ on the sets of irreducible components on the special fibers. Composing with the reduction map $\mathrm{CM}_U \to \mathcal{V}_U$, we obtain a map $\mathrm{CM}_U \to \mathcal{V}_{U^0}$ which is also the composition of $\mathrm{CM}_U \to \mathrm{CM}_{U^0}$ and $\mathrm{CM}_{U^0} \to \mathcal{V}_{U^0}$. Hence it is induced by the $B_v^\times$-equivariant map

$$\begin{aligned}
\mathcal{H}_{U_v} &\longrightarrow (\mathrm{GL}_2(F_v)/F_v^\times \mathrm{GL}_2(O_{F_v})) \times \mathbb{Z}, \\
(b, \beta) &\longmapsto (bb_0, \mathrm{ord}_v(q(b)q(\beta))).
\end{aligned}$$

It has the same form as above. Then $m(b, \beta)$ is nonzero only if (b, β) and $(1, 1)$ have the same image in the map, which still implies $\mathrm{ord}_v(q(b)q(\beta)) = 0$ and $b \in F_v^\times \mathrm{GL}_2(O_{F_v})$.

8.3.3 Support of the multiplicity function: ramified quadratic extension

Assume that v is ramified in E. We consider Lemma 8.12. Similar to the unramified case, the general case essentially follows from the case of full level.

We first assume that $U_v = O_{\mathbb{B}_v}^\times$ is maximal. Then all points in CM_U are still defined over $\widehat{E}_v^{\mathrm{ur}}$, and $H_w \subset \widehat{E}_v^{\mathrm{ur}}$. But $\widehat{E}_v^{\mathrm{ur}}$ is a quadratic extension of $\widehat{F}_v^{\mathrm{ur}}$

this time. The model $\mathcal{Y}_{U,w}$ is obtained from $\mathcal{X} \times_{O_F} O_{H_w}$ by blowing-up all the ordinary double points on the special fiber.

We consider uniformizations over $\widehat{E}_v^{\mathrm{ur}}$. The uniformization on the generic fiber does not change:

$$X_U^{\mathrm{an}} \widehat{\otimes} \widehat{E}_v^{\mathrm{ur}} = B^{\times} \backslash (\Omega' \widehat{\otimes} \widehat{E}_v^{\mathrm{ur}}) \times \mathbb{Z} \times \mathbb{B}_f^{v\times}/U^v.$$

Here $\Omega' = \Omega \widehat{\otimes} E_v$. Let $\widehat{\Omega}'$ be the blowing-up of all double points on the special fiber of $\widehat{\Omega} \otimes O_{E_v}$. It is regular and semistable. Then the formal completion of $\widehat{\mathcal{Y}}_{U,w}$ along its special fiber is uniformized by

$$\widehat{\mathcal{Y}}_{U,w} \widehat{\otimes} O_{\widehat{E}_v^{\mathrm{ur}}} = B^{\times} \backslash (\widehat{\Omega}' \widehat{\otimes} O_{\widehat{E}_v^{\mathrm{ur}}}) \times \mathbb{Z} \times \mathbb{B}_f^{v\times}/U^v.$$

The special fiber of $\widehat{\Omega}'$ consists of the strict transforms of the irreducible components on the special fiber of $\widehat{\Omega}$ and exceptional components coming from the blowing-up. The reduction map sends CM_U to the set $\mathcal{V}_U'$ of exceptional components. The exceptional components are indexed by double points in $\widehat{\Omega}$, and each double point corresponds to a pair of adjacent lattices in F_v^2. The action of $\mathrm{GL}_2(F_v)$ on the double points is transitive. Then $\mathcal{V}_U' \cong \mathrm{GL}_2(F_v)/S_v$ where S_v is the stabilizer of any double point.

Similar to the unramified case, the reduction map $\mathrm{CM}_U \to \mathcal{V}_U'$ is given by the $\mathrm{GL}_2(F_v)$-equivariant map

$$\mathcal{H}_{U_v} \longrightarrow (\mathrm{GL}_2(F_v)/S_v) \times \mathbb{Z}, \quad (b, \beta) \longmapsto (bb_0, \mathrm{ord}_v(q(b)q(\beta))).$$

Use the same argument that two points intersect on the special fiber if and only if they reduce to the same irreducible component. We see that $m(b, \beta) \neq 0$ only if $b \in S_v$ and $\mathrm{ord}_v(q(b)q(\beta)) = 0$. It suffices to bound S_v.

Take S_v to be the stabilizer of the double point corresponding to the edge between the lattices $O_{F_v} \oplus O_{F_v}$ and $O_{F_v} \oplus \varpi_v O_{F_v}$. The action of $h_v = \begin{pmatrix} 0 & 1 \\ \varpi_v & 0 \end{pmatrix}$ switch these two lattices. Then it is easy to see that

$$S_v \subset F_v^{\times} \mathrm{GL}_2(O_{F_v}) \cup h_v F_v^{\times} \mathrm{GL}_2(O_{F_v}).$$

The result is verified in this case.

We remark that the group S_v is generated by the center $F_v^{\times}$, the element h_v, and the subgroup

$$\Gamma_0(p_v) = \left\{ \begin{pmatrix} a & b \\ c & d \end{pmatrix} \in \mathrm{GL}_2(O_{F_v}) : c \in p_v \right\}.$$

Now we consider the general case $U_v = 1 + p_v^r O_{\mathbb{B}_v}$. It is similar to the unramified case. Still compare it with $U^0 = O_{\mathbb{B}_v}^{\times} \times U^v$. The reduction map $\mathrm{CM}_U \to \mathcal{V}_{U^0}'$ is given by the $B_v^{\times}$-equivariant map

$$\mathcal{H}_{U_v} \longrightarrow (\mathrm{GL}_2(F_v)/S_v) \times \mathbb{Z}, \quad (b, \beta) \longmapsto (bb_0, \mathrm{ord}_v(q(b)q(\beta))).$$

The multiplicity function $m(b, \beta)$ is nonzero only if $\mathrm{ord}_v(q(b)q(\beta)) = 0$ and $b \in F_v^{\times} \mathrm{GL}_2(O_{F_v})$.

8.4   ORDINARY CASE

In this section we consider the case that v is a finite prime of F split in E. The local height is expected to vanish because there is no corresponding v-part in the analytic kernel in this case.

PROPOSITION 8.15. *Under Assumption 5.3,*

$$i_{\overline{v}}(Z_*(g)t_1, t_2) = 0, \quad \forall\, g \in 1_{S_1} \mathrm{GL}_2(\mathbb{A}^{S_1}).$$

Let ν_1 and ν_2 be the two primes of E lying over v. They correspond to two places w_1 and w_2 of H via our fixed embedding $\overline{E} \hookrightarrow \overline{F}_v$. For $\ell = 1, 2$, the intersection multiplicity $i_{\overline{\nu}_\ell}$ is computed on the model $\mathcal{Y}_U \times_{O_H} O_{\overline{E}_{\nu_\ell}}$ where the fiber product is taken according to the fixed inclusions $H \subset \overline{E} \subset \overline{E}_{\nu_\ell}$. It is actually a base change of the local integral model $\mathcal{Y}_{U,w} = \mathcal{Y}_U \times_{O_H} O_{H_w}$ for w a place of H induced by ν_ℓ.

By the embedding $E_v \to \mathbb{B}_v$ we see that $\mathbb{B}_v$ is split. Fix an identification $\mathbb{B}_v \cong \mathrm{M}_2(F_v)$ under which $E_v = \begin{pmatrix} F_v & \\ & F_v \end{pmatrix}$. Assume that ν_1 corresponds to the ideal $\begin{pmatrix} F_v & \\ 0 & \end{pmatrix}$ and ν_2 corresponds to $\begin{pmatrix} 0 & \\ & F_v \end{pmatrix}$ of E_v. It suffices to show that $i_{\overline{\nu}_1}(Z_*(g)t_1, t_2) = 0$.

We still make use of the results of [Zh2]. The reduction map of CM points to ordinary points at ν_1 is given by

$$E^\times \backslash \mathbb{B}_f^\times / U \longrightarrow E^\times \backslash (N(F_v) \backslash \mathrm{GL}_2(F_v)) \times \mathbb{B}_f^{v\times}/U.$$

The intersection multiplicity is a function $\mathfrak{g}_{\nu_1} : (N(F_v)U_v/U_v)^2 \to \mathbb{R}$. An explicit expression of $\mathfrak{g}_{\nu_1}$ for general U_v are proved in [Zh2, Lemma 6.3.2]. But we do not need it here. The local height pairing of two distinct CM points $\beta_1, \beta_2 \in E^\times \backslash \mathbb{B}_f^\times / U$ is given by

$$i_{\overline{\nu}_1}(\beta_1, \beta_2) = \sum_{\gamma \in \mu_U \backslash E^\times} \mathfrak{g}_{\nu_1}(\beta_1, \gamma\beta_2) 1_{U^v}(\beta_1^{-1}\gamma\beta_2).$$

Unlike other cases, the above summation has nothing to do with the nearby quaternion algebra. It is only a "small" sum for $\gamma \in E^\times$. This is the key for the vanishing of the local height series.

Now we can look at

$$i_{\overline{\nu}_1}(Z_*(g)t_1, t_2) = \sum_{a \in F^\times} \sum_{x \in \mathbb{B}_f^\times / U} r(g)\phi(x)_a \sum_{\gamma \in \mu_U \backslash E^\times} \mathfrak{g}_{\nu_1}(t_1 x, \gamma t_2) 1_{U^v}(x^{-1} t_1^{-1} \gamma t_2).$$

A nonzero term has to satisfy $x^v \in t_1^{-1} \gamma t_2 U^v$. For such x^v, Assumption 5.3 gives

$$\phi_{S_1}(x)_a = \phi_{S_1}(t_1^{-1}\gamma t_2)_a = 0.$$

It follows that $i_{\overline{\nu}_1}(Z_*(g)t_1, t_2) = 0$.

8.5   THE *J*-PART

Now we consider $j_{\bar{v}}(Z_*(g)t_1, t_2)$ for a non-archimedean place v of F. It is nonzero only if X_U has a bad reduction at v. Similar to the i-part, we will show that it can be approximated by theta series under Assumption 5.3.

The pairing $j_{\bar{v}}(Z_*(g)t_1, t_2)$ is the average of $j_{\bar{\nu}}(Z_*(g)t_1, t_2)$ for each non-archimedean place ν of E lying over v. Similar to the computation of the i-part, the intersection is computed on the model $\mathcal{Y}_{U,w}$ introduced in §7.2. Here w is the place of H induced by $\bar{\nu}$. By the definition in §7.1,

$$j_{\bar{\nu}}(Z_*(g)t_1, t_2) = \frac{1}{[H:F]} \overline{Z_*(g)t_1} \cdot V_{t_2}.$$

Here $\overline{Z_*(g)t_1}$ is the Zariski closure in $\mathcal{Y}_{U,w}$ and V_{t_2} is a linear combination of irreducible components in the special fibers of $\mathcal{Y}_{U,w}$ which gives the $\hat{\xi}$-admissible arithmetic extension of t_2.

It suffices to treat $(Z_*(g)t, C)$ for any $t \in C_U$ and any irreducible component C in special fiber of $\mathcal{Y}_{U,w}$. Here we use the notation $(D, C) = \frac{1}{[H:F]} \overline{D} \cdot C$ for convenience. It is essentially a question about the reduction of CM points in CM_U to irreducible components on the special fiber on the model $\mathcal{Y}_{U,w} \times O_{\overline{E}_\nu}$. Many cases have been described explicitly in [Zh2].

By the construction in §7.2, there is a morphism $\mathcal{Y}_{U,w} \to \mathcal{X}_{U^0, v}$, where $U^0 = O_{\mathbb{B}_v}^\times \times U^v$. By this map, we classify the irreducible component C on the special fiber into the following three categories:

- C is *ordinary* if the image of C in $\mathcal{X}_{U^0, v}$ is not a point.

- C is *supersingular* if v is split $\mathbb{B}$ and C maps to a point in $\mathcal{X}_{U^0, v}$. Notice this point must be supersingular.

- C is *superspecial* if v isnonsplit in $\mathbb{B}$.

Let $\mathcal{V}_U^{\mathrm{ord}}$, $\mathcal{V}_U^{\mathrm{sing}}$, and $\mathcal{V}_U^{\mathrm{spe}}$ denote the set of these components.

8.5.1   Ordinary Components

We first consider $(Z_*(g)t, C)$ in the case that C is an ordinary component. It is nonzero only if points in CM_U reduce to ordinary components, which happens exactly when E is split at v. Let ν_1, ν_2 be the two places of E above v, and we use the convention of the splitting $E_v = F_v \oplus F_v$ at the beginning of §8.4. It suffices to consider $j_{\bar{\nu}_1}$. The treatment is very similar to Proposition 4.2 by separating geometrically connected components.

By Lemma 5.4.2 in [Zh2], the ordinary components are parametrized by

$$\mathcal{V}_U^{\mathrm{ord}} = F_+^\times \backslash \mathbb{A}_f^\times / q(U) \times P(F_v) \backslash \mathrm{GL}_2(F_v) / U_v.$$

Note that the first double coset is exactly the set of geometrically connected components. The reduction $\mathrm{CM}_U \longrightarrow \mathcal{V}_U^{\mathrm{ord}}$ is given by the natural map:

$$E^\times \backslash \mathbb{B}_f^\times / U \longrightarrow F_+^\times \backslash \mathbb{A}_f^\times / q(U) \times P(F_v) \backslash \mathrm{GL}_2(F_v) / U_v,$$
$$g \longmapsto (\det g, g_v).$$

For any $\beta \in \mathrm{CM}_U$, the intersection $(\beta, C) \neq 0$ only if β and C are in the same geometrically connected component. Once this is true, it is given by a locally constant function l_C for $\beta_v \in \mathbb{B}_v^\times$. Moreover, the function l_C factors through $P(F_v) \backslash \mathrm{GL}_2(F_v) / U_v$. In summary, we have

$$(\beta, C) = l_C(\beta_v) 1_{F_+^\times q(\beta_C) q(U)}(q(\beta)).$$

Here $\beta_C \in \mathbb{B}_f^\times$ is any element such that $q(\beta_C)$ gives the geometrically connected component containing C.

Therefore, we have

$$(Z_*(g)t, C) = (Z_*(g)_\alpha t, C)$$

where $\alpha = q(t)^{-1} q(\beta_C)$. By (4.3.1),

$$Z_*(g)_\alpha t = \sum_{u \in \mu_U' \backslash F^\times} \sum_{a \in F_+^\times} \sum_{y \in \mathbb{B}_f(a)/U^1} r(g, (\beta_C^{-1}t, 1)) \phi(y, u) \, [\beta_C y].$$

Thus

$$(Z_*(g)t, C) = \sum_{u \in \mu_U' \backslash F^\times} \sum_{a \in F_+^\times} \sum_{y \in \mathbb{B}_f(a)/U^1} r(g, (\beta_C^{-1}t, 1)) \phi(y, u) l_C(\beta_C y).$$

Similar to Lemma 7.6, it is zero by Assumption 5.4. In fact, fixing a place $v' \in S_2 - \{v\}$, the inside sum has a factor

$$\sum_{y \in \mathbb{B}_{v'}(a)/U_{v'}^1} r(g, (t, 1)) \phi_{v'}(y, u) = \frac{1}{\mathrm{vol}(U_{v'}^1)} \int_{\mathbb{B}_{v'}(a)} r(g, (t, 1)) \phi_{v'}(y, u) dy = 0.$$

8.5.2 Supersingular components

Now we consider $(Z_*(g)t, C)$ in the case that C is a supersingular component. It is nonzero only if points in CM_U reduce to supersingular components, which happens exactly when both $\mathbb{B}$ and E arenonsplit at v. The treatment is similar to §8.2. Denote $B = B(v)$ and fix an isomorphism $B(\mathbb{A}_f^v) \simeq \mathbb{B}_f^v$ as usual.

The key is to characterize the reduction map

$$\mathrm{CM}_U \longrightarrow \mathcal{V}_U^{\mathrm{sing}} \longrightarrow \mathbb{S}_U.$$

Here $\mathbb{S}_U$ is the set of supersingular points in $\mathcal{X}_{U,v}$. Recall that

$$\mathrm{CM}_U = E^\times \backslash \mathbb{B}_f^\times / U = B^\times \backslash (B^\times \times_{E^\times} \mathbb{B}_v^\times) \times (\mathbb{B}_f^v)^\times / U \hookrightarrow B^\times \backslash \mathcal{H}_v \times (\mathbb{B}_f^v)^\times / U$$

where $\mathcal{H}_v = B_v^\times \times_{E_v^\times} \mathbb{B}_v^\times$. We also have the parametrization

$$\mathbb{S}_U = B^\times \backslash \mathbb{B}_f^\times / U = B^\times \backslash F_v^\times \times (\mathbb{B}_f^v)^\times / U.$$

Then the reduction map $\mathrm{CM}_U \to \mathbb{S}_U$ is given by the product of determinants:

$$\mathcal{H}_v \longrightarrow F_v^\times, \qquad (b, \beta) \longmapsto q(b)q(\beta).$$

Fix one supersingular point z on the special fiber of $\mathcal{X}_{U,v}$. Denote by $\mathcal{N}_{U_v}$ the formal completion of $\mathcal{X}_{U,\overline{O}_{F_v}}$ along z. It depends only on U_v, and represents the deformation of the formal O_{F_v}-module of height two with Drinfeld U_v-level structures. The formal completion $\widehat{\mathcal{X}}_{U,\overline{O}_{F_v}}$ of $\mathcal{X}_{U,v}$ along its supersingular locus $\mathbb{S}_U$ is given by

$$\widehat{\mathcal{X}}_{U,\overline{O}_{F_v}} = B^\times \backslash (\mathcal{N}_{U_v} \times \mathbb{Z}) \times (\mathbb{B}_f^v)^\times / U^v.$$

To check this identity, note that both sides are formal schemes supported at finite sets, then it essentially follows from the coset expression of $\mathbb{S}_U$.

Let $\widetilde{\mathcal{N}}_{U_v}$ be the minimal desingularization of $\mathcal{N}_{U_v} \widehat{\otimes} O_{H_w}$. Then the formal completion of $\mathcal{Y}_{U,w}$ along the union of fibers in $\mathcal{V}_U^{\mathrm{sing}}$ can be described as

$$\widehat{\mathcal{Y}}_U = B^\times \backslash (\widetilde{\mathcal{N}}_{U_v} \times \mathbb{Z}) \times (\mathbb{B}_f^v)^\times / U.$$

Let $\widetilde{\mathcal{V}}$ be the set of irreducible components on the special fiber of $\widetilde{\mathcal{N}}_{U_v} \times \mathbb{Z}$. It also admits an action by $B_v^\times \times U_v$. Then we have a description

$$\mathcal{V}_U^{\mathrm{sing}} = B^\times \backslash \widetilde{\mathcal{V}} \times (\mathbb{B}_f^v)^\times / U.$$

Our conclusion is that the map

$$\mathrm{CM}_U \longrightarrow \mathcal{V}_U^{\mathrm{sing}} \longrightarrow \mathbb{S}_U$$

is given by $(B_v^\times \times U_v)$-equivariant maps

$$\mathcal{H}_v \longrightarrow \widetilde{\mathcal{V}} \longrightarrow \mathbb{Z}.$$

We are now applying the above result to compute the intersection pairing $(Z_*(g)t, C)$. It is very similar to our treatment of the i-part. Let $(C_0, \beta_C) \in \widetilde{\mathcal{V}} \times (\mathbb{B}_f^v)^\times$ be a couple representing $C \in \mathcal{V}_U^{\mathrm{sing}}$. The intersection with C_0 defines a locally constant function l_{C_0} on $\mathcal{H}_{U_v} = B_v^\times \times_{E_v^\times} \mathbb{B}_v^\times / U_v$. Unlike the multiplicity function m, l_{C_0} has no singularity on $\mathcal{H}_{U_v}$. For any CM point $\beta \in \mathrm{CM}_U$, the intersection pairing is given by

$$(\beta, C) = \sum_{\gamma \in \mu_U \backslash B^\times} l_{C_0}(\gamma, \beta_v) 1_{U^v}(\beta_C^{-1} \gamma \beta^v).$$

Hence, we obtain

$$(Z_*(g)t, C) = \sum_{a \in F^\times} \sum_{x \in \mathbb{B}_f^\times / U} r(g)\phi(x)_a \sum_{\gamma \in \mu_U \backslash B^\times} l_{C_0}(\gamma, t_v x_v) 1_{U^v}(\beta_C^{-1} \gamma t^v x^v).$$

Now we convert the above to a pseudo-theta series. The process is the same as the i-part. We sketch it here. Change the order of the summations. Note that $1_{U^v}(\beta_C^{-1}\gamma t^v x^v) \neq 0$ if and only if $x^v \in t^{-1}\gamma^{-1}\beta_C U^v$. Put it into the sum. We have

$$(Z_*(g)t, C) = \sum_{a \in F^\times} \sum_{\gamma \in \mu_U \backslash B^\times} r(g)\phi^v(t^{-1}\gamma^{-1}\beta_C)_a \sum_{x \in \mathbb{B}_v^\times / U_v} r(g)\phi_v(x)_a l_{C_0}(\gamma, t_v x_v).$$

Denote

$$l_{\phi_v}(y, u) = l_{C_0, \phi_v}(y, u) := \int_{\mathbb{B}_v^\times} \phi_v\left(x, \frac{uq(y)}{q(x)}\right) l_{C_0}(y^{-1}, x) dx.$$

Then

$$(Z_*(g)t, C) = \sum_{a \in F^\times} \sum_{y \in \mu_U \backslash B^\times} r(g, (t, \beta_C))\phi^v(y)_a l_{r(g,(t,1))\phi_v}(y)_a$$

$$= \sum_{u \in \mu_U^2 \backslash F^\times} \sum_{y \in B^\times} r(g, (t, \beta_C))\phi^v(y, u) l_{r(g,(t,1))\phi_v}(y, u).$$

It is a pseudo-theta series.

We claim that if $\phi_v \in \overline{\mathcal{S}}^1(\mathbb{B}_v \times F_v^\times)$, then l_{ϕ_v} extends to a Schwartz function for $(y, u) \in B_v \times F_v^\times$. The proof of Proposition 8.9 applies here. We only explain that $v(q(y)^{-1}q(x))$ is a constant on the support of $l_{C_0}(y^{-1}, x)$, which is needed for l_{ϕ_v} to be compactly supported. In fact, $l_{C_0}(y^{-1}, x) \neq 0$ only if the point $(y^{-1}, x) \in \mathcal{H}_v$ and $C_0 \in \mathcal{V}_H$ have the same image in $F_v^\times$. It determines the coset $q(y)^{-1}q(x)q(U_v)$ in $F_v^\times$ uniquely.

Similar to all the pseudo-theta series we treated before, our conclusion is

$$(Z_*(g, \phi)t, C) = \theta(g, (t, \beta_C), r(\beta_c^{-1}, 1)l_{\phi_v} \otimes \phi^v), \quad \forall g \in P(F_{S_1})\mathrm{GL}_2(\mathbb{A}^{S_1}).$$

In our particular case, C is in the geometrically connected components of t_2, and thus we can take $\beta_C = t_2$.

8.5.3 Superspecial components

Now we consider $(Z_*(g)t, C)$ in the case that C is a superspecial component. It happens when v isnonsplit in $\mathbb{B}$. Resume the notations in the treatment of the i-part. It is similar to the supersingular case.

The curve $\mathcal{X}_{U^0, v}$ has the explicit uniformization as formal schemes:

$$\widehat{\mathcal{X}}_{U^0, v} \widehat{\otimes} O_{\widehat{F}_v^{\mathrm{ur}}} = B^\times \backslash (\widehat{\Omega} \widehat{\otimes} O_{\widehat{F}_v^{\mathrm{ur}}}) \times \mathbb{Z} \times (\mathbb{B}_f^v)^\times / U^v.$$

For general levels, the uniformization is easily done in the level of rigid spaces:

$$X_U^{\mathrm{an}} \widehat{\otimes} \widehat{F}_v^{\mathrm{ur}} = B^\times \backslash \Sigma_r \times \mathbb{B}_f^{v\times} / U^v.$$

Here Σ_r is some etale rigid-analytic cover of $(\Omega \widehat{\otimes} \widehat{F}_v^{\mathrm{ur}}) \times \mathbb{Z}$ depending on r. Take the normalization of the formal scheme $(\widehat{\Omega} \widehat{\otimes} \widehat{O}_{F_v^{\mathrm{ur}}}) \times \mathbb{Z}$ in the rigid space $\Sigma_r \widehat{\otimes}_{\widehat{F}_v^{\mathrm{ur}}} \widehat{H}_w^{\mathrm{ur}}$, and make a minimal resolution of singularities. We obtain a regular formal scheme $\widehat{\Sigma}_r$ over $O_{H_w^{\mathrm{ur}}}$. The construction is compatible with the algebraic construction of $\mathcal{Y}_{U,w}$, i.e.,

$$\widehat{\mathcal{Y}}_{U,w} \widehat{\otimes} O_{\widehat{H}_w^{\mathrm{ur}}} = B^\times \backslash \widehat{\Sigma}_r \times \mathbb{B}_f^{v\times} / U^v.$$

Here $\widehat{\mathcal{Y}}_{U,w}$ is the formal completion of the $\mathcal{Y}_{U,w}$ along its special fiber. The uniformizations here are not explicit at all, but we only need some group-theoretical properties.

Let $\widetilde{\mathcal{V}}$ be the set of irreducible components of $\widehat{\Sigma}_r$. Then the reduction $\mathrm{CM}_U \to \mathcal{V}^{\mathrm{spe}}$ is given by a $B_v^\times$-equivariant map $\mathcal{H}_{U_v} \to \widetilde{\mathcal{V}}$. Under Assumption 5.3, the same calculation as in supersingular case will show that $(Z_*(g)t, C)$ can be approximated by a theta series on the quadratic space B.

Bibliography

[Ar]   S. J. Arakelov: *Intersection theory of divisors on an arithmetic surface.*
          Math. USSR Izvest. 8 (1974), 1167–1180.

[BC]   J.-F. Boutot; H. Carayol: *Uniformisation p-adique des courbes de
          Shimura: les théorèmes de Čerednik et de Drinfel'd.* Courbes modu-
          laires et courbes de Shimura (Orsay, 1987/1988). Astérisque No. 196-
          197 (1991), 7, 45–158 (1992).

[BCDT]   C. Breuil; B. Conrad; F. Diamond; R. Taylor: *On the modularity of
          elliptic curves over* $\mathbb{Q}$*: wild 3-adic exercises.* J. Amer. Math. Soc. 14
          (2001), 843-939.

[Bor]   R. Borcherds: *The Gross-Kohnen-Zagier theorem in higher dimensions.*
          Duke Math. J. 97 (1999), no. 2, 219–233.

[Car]   H. Carayol: *Sur la mauvaise réduction des courbes de Shimura.* Com-
          positio Math. 59 (1986), no. 2, 151–230.

[Cas]   W. Casselman: *On some results of Atkin and Lehner.* Math. Ann. 201
          (1973), 301–314.

[Dr]   V. G. Drinfel'd: *Two theorems on modular curves.* Funkcional. Anal. i
          Priložen. 7 (1973), no. 2, 83–84.

[Fa1]   G. Faltings: *Endlichkeitssätze für abelsche Varietäten über Zahlkörpern.*
          Invent. Math. 73 (1983), no. 3, 349–366.

[Fa2]   G. Faltings: *Calculus on arithmetic surfaces.* Ann. of Math. (2) 119
          (1984), 387-424.

[Gra]   L. Grafakos: *Classical Fourier analysis.* Second edition. Graduate Texts
          in Mathematics, 249. Springer, New York, 2008.

[GKZ]   B. Gross; W. Kohnen; D. Zagier: *Heegner points and derivatives of
          L-series. II.* Math. Ann. 278 (1987), no. 1-4, 497–562.

[Gro1]   B. Gross: *On canonical and quasi-canonical liftings.* Invent. Math. 84
          (1986), no. 2, 321–326.

[Gro2]   B. Gross: *Heegner points and representation theory.* Heegner points
         and Rankin *L*-series, 37–65, Math. Sci. Res. Inst. Publ., 49. Cambridge
         Univ. Press, Cambridge, 2004.

[GS]   H. Gillet; C. Soulé: *Arithmetic intersection theory.* Publ. Math. IHES,
         72 (1990), 93-174.

[GZ]   B. Gross; D. Zagier: *Heegner points and derivatives of L-series.* Invent.
         Math. 84 (1986), no. 2, 225–320.

[Hr]   P. Hriljac: *Heights and Arakelov's intersection theory.* Amer. J. Math.
         107 (1985), no. 1, 23–38.

[HZ]   F. Hirzebruch; D. Zagier: *Intersection numbers of curves on Hilbert
         modular surfaces and modular forms of Nebentypus.* Invent. Math. 36
         (1976), 57–113.

[JL]   H. Jacquet; R. P. Langlands: *Automorphic forms on GL(2).* Lecture
         Notes in Mathematics, Vol. 114. Springer-Verlag, Berlin-New York,
         1970.

[KM1]   S. Kudla; J. Millson: *The theta correspondence and harmonic forms. I.*
         Math. Ann. 274 (1986), no. 3, 353–378.

[KM2]   S. Kudla; J. Millson: *The theta correspondence and harmonic forms.
         II.* Math. Ann. 277 (1987), no. 2, 267–314.

[KM3]   S. Kudla; J. Millson: *Intersection numbers of cycles on locally sym-
         metric spaces and Fourier coefficients of holomorphic modular forms in
         several complex variables.* Inst. Hautes Études Sci. Publ. Math. No. 71
         (1990), 121–172.

[KR1]   S. Kudla; S. Rallis: *On the Weil-Siegel formula.* J. Reine Angew. Math.
         387 (1988), 1-68.

[KR2]   S. Kudla; S. Rallis: *On the Weil-Siegel formula II: the isotropic con-
         vergent case.* J. Reine Angew. Math. 391 (1988), 65-84.

[KR3]   S. Kudla; S. Rallis: *A regularized Siegel-Weil formula: the first term
         identity.* Ann. of Math. (2) 140 (1994), no. 1, 1–80.

[KRY1]   S. Kudla; M. Rapoport; T. Yang: *On the derivative of an Eisenstein
         series of weight one.* Internat. Math. Res. Notices (1999), no. 7, 347–
         385.

[KRY2]   S. Kudla; M. Rapoport; T. Yang: *Modular forms and special cycles on
         Shimura curves.* Annals of Mathematics Studies, 161. Princeton Uni-
         versity Press, Princeton, NJ, 2006.

[KRY3] S. Kudla; M. Rapoport; T. Yang: *Derivatives of Eisenstein series and Faltings heights*. Compos. Math. 140 (2004), no. 4, 887–951.

[Ku1] S. Kudla: *Algebraic cycles on Shimura varieties of orthogonal type*. Duke Math. J. 86 (1997), no. 1, 39-78.

[Ku2] S. Kudla: *Central derivatives of Eisenstein series and height pairings*. Ann. of Math. (2) 146 (1997), no. 3, 545–646.

[Ku3] S. Kudla: *Special cycles and derivatives of Eisenstein series*. Heegner points and Rankin *L*-series, 243–270, Math. Sci. Res. Inst. Publ., 49. Cambridge Univ. Press, Cambridge, 2004.

[Ku4] S. Kudla: *Some extensions of the Siegel-Weil formula*. Eisenstein series and applications, 205–237, Progr. Math., 258. Birkhäuser, Boston, MA, 2008.

[KY] S. Kudla; M. Rapoport; T. Yang: *Eisenstein series for SL(2)*. Science China Math., vol. 53, no. 9, 2275–2316.

[La] Y. T. Lam: *A first course in noncommutative rings*. Graduate Texts in Mathematics, 131. Springer-Verlag, New York, 1991.

[Le] N. N. Lebedev: *Special functions and their applications*. Dover, New York, 1972.

[Li] D. I. Lieberman: *Higher Picard varieties*. Amer. J. Math. 90 (1968), 1165–1199.

[Ma] J. I. Manin: *Parabolic points and zeta functions of modular curves*. Izv. Akad. Nauk SSSR Ser. Mat. 36 (1972), 19-66.

[Mu] D. Mumford: *Abelian Varieties*. Tata Institute of Fundamental Research Studies in Mathematics, No. 5. Published for the Tata Institute of Fundamental Research, Bombay; Oxford University Press, London, 1970.

[Ra] S. Rallis: *L-functions and the Oscillator Representation*. Lecture Notes in Math., vol. 1245. Springer-Verlag, New York, 1987.

[Sa] H. Saito: *On Tunnell's formula for characters of* GL(2). Compositio Math. 85 (1993), no. 1, 99–108.

[Sch] B. Schoeneberg: *Elliptic modular functions: an introduction*. Translated from the German by J. R. Smart and E. A. Schwandt. Die Grundlehren der mathematischen Wissenschaften, Band 203. Springer-Verlag, New York-Heidelberg, 1974.

[Se] J. P. Serre: *Lectures on the Mordell-Weil theorem*. Friedr. Vieweg & Sohn, Braunschweig, 1989.

[Shi]   G. Shimura: *Confluent hypergeometric functions on tube domains.*
         Math. Ann. 260 (1982), 269–302.

[Si]   C. L. Siegel: *Gesammelte Abhandlungen I,II,III.* Springer-Verlag,
         Berlin, Heidelberg, New York, 1966.

[Tu]   J. Tunnell: *Local ε-factors and characters of* GL(2). Amer. J. Math.
         105 (1983), no. 6, 1277–1307.

[TW]   R. Taylor; A. Wiles: *Ring-theoretic properties of certain Hecke algebras.*
         Ann. Math. 141 (1995), 553-572.

[Vi]   M. Vignéras: *Arithmétique des algèbres de quaternions.* Lecture Notes
         in Mathematics, 800. Springer, Berlin, 1980.

[Wa]   J. Waldspurger: *Sur les valeurs de certaines fonctions L automorphes
         en leur centre de symétrie.* Compositio Math. 54 (1985), no. 2, 173–242.

[We1]   A. Weil: *Sur certains groupes d'opérateurs unitaires.* Acta Math. 111
         (1964), 143–211.

[We2]   A. Weil: *Sur la formule de Siegel dans la théorie des groupes classiques.*
         Acta Math. 113 (1965), 1–87.

[Wi]   A. Wiles: *Modular elliptic curves and Fermat's last theorem.* Ann.
         Math. 142 (1995), 443-551.

[YZZ]   X. Yuan; S. Zhang; W. Zhang: *The Gross-Kohnen-Zagier theorem over
         totally real fields.* Compositio Math. 145 (2009), 1147-1162.

[Zh1]   S. Zhang: *Heights of Heegner points on Shimura curves.* Ann. of Math.
         (2) 153 (2001), no. 1, 27–147.

[Zh2]   S. Zhang: *Gross–Zagier formula for* GL_2. Asian J. Math. 5 (2001), no.
         2, 183–290.

[Zh3]   S. Zhang: *Gross–Zagier formula for* GL(2). II. Heegner points and
         Rankin *L*-series, 191–214, Math. Sci. Res. Inst. Publ., 49. Cambridge
         Univ. Press, Cambridge, 2004.

[Zha]   W. Zhang: *Modularity of Generating Functions of Special Cycles on
         Shimura Varieties.* Preprint.

Index